System Engineering: Probabilistic Models and Applications

System Engineering: Probabilistic Models and Applications

Editor: Reynolds Stewart

STATES
ACADEMIC PRESS
www.statesacademicpress.com

Published by States Academic Press
109 South 5th Street,
Brooklyn, NY 11249, USA
www.statesacademicpress.com

System Engineering: Probabilistic Models and Applications
Edited by Reynolds Stewart

International Standard Book Number: 978-1-63989-506-9 (Hardback)

Cataloging-in-Publication Data

System engineering : probabilistic models and applications / edited by Reynolds Stewart.
p. cm.
Includes bibliographical references and index.
ISBN 978-1-63989-506-9
1. Systems engineering. 2. System analysis. 3. Industrial engineering.
4. Engineering. I. Stewart, Reynolds.
TA168 .S97 2022
620.001 171--dc23

Contents

Preface

System engineering is an interdisciplinary field of engineering and engineering management, which focuses on designing, integrating and managing complex systems over their life cycles. Fundamentally, it utilizes the principles of systems theory to organize this body of knowledge. An engineered system is the outcome of such efforts, a combination of components that collaborate to collectively perform a useful function. Systems engineering ensures that all likely aspects of a project or system are considered and integrated into a whole. It involves discovering real problems, identifying the most probable failures and finding solutions to these problems. This book elucidates the concepts and innovative models around prospective developments with respect to this field. There has been rapid progress in system engineering and its applications are finding their way across multiple industries. As this field is emerging at a rapid pace, the contents of this book will help the readers understand the modern concepts and applications of the subject.

The information shared in this book is based on empirical researches made by veterans in this field of study. The elaborative information provided in this book will help the readers further their scope of knowledge leading to advancements in this field.

Finally, I would like to thank my fellow researchers who gave constructive feedback and my family members who supported me at every step of my research.

Editor

1

Natural Hazards: Systematic Assessment of their Contribution to Risk and their Consequences

Berg Heinz-Peter and Roewekamp Marina

Abstract

The significance of event scenarios from a variety of natural hazards —from seismotectonic over meteorological, hydrological up to biological ones — to all types of industrial facilitieshas been recognized in the near past and needs to be addressed systematically in the safety assessment. The most recent approaches for assessing the risk contribution from these hazards and their consequences start with a site-specific qualitative as well as quantitative screening of those individual hazards and event combinations with such hazards, which can be directly related, correlated, or occur independently during the mission time of another. In the second step, for those hazards and hazard combinations remaining with a non-negligible occurrence frequency, a detailed analysis of the facility-specific event scenario including interdependencies between the hazards to be considered, and the safety features and countermeasures in the facility being investigated is conducted in order to estimate the corresponding risk contribution and consequences.

Keywords: natural hazard(s), event combination(s), risk contribution, interdependencies, screening, countermeasures

1. Introduction

Natural hazards frequently cause disturbances for different types of critical infrastructures and, therefore, have a substantial impact on the safe and reliable operation of the respective infrastructures such as telecommunication, processes, and energy industry. Furthermore, the set of natural hazards can strongly affect the different types of transport infrastructures such as road, rail, waterways, and aviation.

In case of the electrical power industry, the impact, in particular of seismotectonic, hydrological, and meteorological hazards, is diversified. Examples are the energy production and the transmission and distribution lines of the suppliers where strong winds like tornados result in a disruption of the production (e.g., in case of wind turbines) or of the distribution lines due to trees fallen on overhead lines. An expected low water level may require the shutdown of a nuclear power plant because of potential core cooling problems.

Hazards can arise not only individually but often occur together with other events or hazards. The experience has shown that a variety of combinations of different types are possible. If and how frequently such hazard combinations do occur at a nuclear facility site depends on the site characteristics but also on the facility to be investigated and its design against various events.

In particular, for combinations of natural hazards with other events, the operating experience of the more recent past has shown that at least some of the huge amount of theoretically possible hazard combinations cannot be excluded to occur in principle. Some of these combinations represent—like individual hazards—low-frequency, high-damage events, others are more frequent, but the damage potential is much lower (so-called high-frequency, low-damage events).

For systematically considering all hazard combinations having the potential to impair the safe operation of an industrial facility, but enabling the analyst to exclude non-negligible combinations as well, the entire set of hazards, which can be anticipated at the site of the facility being analyzed, needs to be identified. In the second step, the individual hazards have to undergo a qualitative and quantitative screening process. In the third step, hazard combinations have to be identified starting from those individual hazards identified and not screened out by qualitative arguments. For these hazard combinations, again the screening has to be performed.

2. Different types of hazards and hazard combinations

When considering those hazards which may impair the safe operation of an industrial facility, in principle, two types of hazards have to be distinguished: internal and external hazards.

Internal hazards are those occurring under the responsibility of the operator of the industrial facility on the site of the corresponding installations (e.g., one or more industrial plants).

External hazards are those ones occurring independent of the facility being analyzed, off-site, and out of the responsibility of the plant operator. External hazards may result from natural causes—so-called natural hazards—or maybe induced by humans—so-called man-made hazards. Natural hazards can be further subdivided into different classes of hazards corresponding to the types of phenomena covered.

types of hazards in order to enable the analyst to perform a complete screening of hazard combinations.

2.1. Systematic binning of hazards

In **Table 1**, an overview of the different classes of internal and external hazards is given. **Tables 2–10** provide for all hazard classes mentioned in **Table 1** the binning of individual hazards to the different hazard classes.

I. External hazards

1. *Natural hazards*

Class A: Seismotectonic hazards

Class B: Flooding and other hydrological hazards

Class C: Meteorological hazards

Class D: Extraterrestrial hazards

Class E: Biological hazards

Class F: Geological hazards

Class H: Natural fires

2. *Man-made hazards* (Class Z)

II. Internal hazards (Class I)

Table 1. Overview of hazard classes, from [1].

Hazard	Type of individual seism tectonic hazard
A1	Earthquake (vibration ground motion (including long duration)
A2	Vibration ground motion induced or triggered by human activity
A3	Surface faulting (fault capability)
A4	Liquefaction, lateral spreading
A5	Dynamic compaction (seismically induced soil settlement)
A6	Permanent ground displacement subsequent to earthquake

Table 2. Class A hazards according to [2].

Hazard	Type of individual hydrological hazard
B1	Tsunami
B2	Flash flood by local extreme precipitation
B3	Flooding by melting snow
B4	Flooding by extreme precipitation outside the plant boundary
B5	Extreme groundwater increase
B6a	High water level due to obstructions in the course of the river
B6b	Low water level due to obstructions in the course of the river
B7a	High water level by natural changes in the course of the river
B7b	Low water level by natural changes in the course of the river
B8	Flooding by high fresh water waves due to volcanism, land, or snow slide
B9a	High water level with wave formation due to failure of water control or retention systems (e.g., damns, dykes, etc.)
B9b	Low water level with wave formation due to failure of water control or retention systems (e.g., damns, dykes, etc.)
B10	Seiche
B11	Tidal bore (running extremely river-up)
B12	Tidal high water, spring tide
B13	Storm-induced waves and monster waves
B14	Storm surge
B15	Corrosion resulting from contact with salt water
B16	Instability of coastal areas (of rivers, lakes, oceans) by erosion due to strong water flows or sedimentation
B17	Water flotsam (mud, debris, etc.)

Table 3. Class B hazards according to [2].

Hazard	Type of individual meteorological hazard
C1	Precipitation, snow pack
C2a	High air temperature
C2b	Low air temperature
C3a	High ground temperature

Hazard	Type of individual meteorological hazard
C3b	Low ground temperature
C4a	High cooling water temperature
C4b	Low cooling water temperature
C5a	High humidity
C5b	Low humidity
C6	Extremes of air pressure
C7	Drought
C8	Low ground water
C9	Low seawater level
C10	Icing
C11	White frost, rime
C12	Hail
C13	Permafrost
C14	Recurring soil frost
C15	Lightning
C16	High wind
C17	Tornado
C18	Waterspout
C19	Snowstorm
C20	Sandstorm
C21	Salt spray
C22	Wind-blown debris
C23	Snow avalanche
C24	Surface ice
C25	Frazil ice
C26	Ice barriers
C27	Mist, fog

Table 4. Class C hazards according to [2].

Hazard	Type of individual extra-terrestrial hazard
D1	Coronal mass ejection, solar flare
D2	Meteorite fall

Table 5. Class D hazards according to [2].

Hazard	Type of individual biological hazard
E1	Marine/river/lake growth
E2	Crustacean/mollusk growth
E3	Fish, jellyfish
E4	Airborne swarms, leaves
E5	Infestation
E6	Biological flotsam
E7	Microbiological corrosion

Table 6. Class E hazards according to [2].

Hazard	Type of individual geological hazard
F1	Subaerial slope instability
F2	Underwater landslide, and so on
F3	Debris flow, mud flow (including seismically triggered events)
F4	Natural ground settlement
F5	Ground heave
F6	Karst, leeching of soluble rocks (limestone, gypsum, anhydrite, halite)
F7	Sinkholes
F8	Unstable soils
F9	Volcanic hazards close to the volcano source
F10	Volcanic hazards far away for the volcano source
F11	Methane release
F12	Natural radiation
F13	Pole reversal (polar motion)

Table 7. Class F hazards according to [2].

Hazard	Type of individual natural fire hazard
H1	Wildfire

Table 8. Class H hazards according to [2].

Hazard	Type of individual man-made hazard
Z1	Industrial accidents: explosions
Z2	Industrial accidents: releases of hazardous substances
Z3	Industrial accidents: missiles
Z4	Accidental consequences of military facilities
Z5	Accidental military releases of hazardous substances
Z6	Accidental consequences of military activities
Z7	Ship accidents: direct impact
Z8	Ship accidents: Collisions with SSC
Z9	Ship accidents: Releases of solid or liquid substances
Z10	Transportation accidents: direct impact
Z11	Transportation accidents: explosions
Z12	Transportation accidents: releases of hazardous substances
Z13	Pipeline accidents: fire or explosion
Z14	Pipeline accidents: releases of hazardous substances
Z15	Accidental aircraft crash in the airport area
Z16	Accidental aircraft crash in air lanes/corridors
Z17	Satellite crash
Z18	Drone crash
Z19	Off-site excavation and construction work
Z20	External grid stability
Z21	Industrial impurity of high voltage insulations (of switchgears, etc.)
Z22	Electromagnetic interference (EMI)
Z23	Underground high-voltage Eddy currents (off-site)
Z24	Flooding due to man-made failure of water control or retention systems
Z25	Man-made fire (off-site)
Z26	Log jam (e.g., by driftwood)
Z27	Bore by water management activities
Z28a	High water level by building structures (wave breakers, moles, languets)
Z28b	Low water level by building structures
Z29	Man-made ground settlement

Table 9. Class Z hazards according to [2].

Hazard	Type of individual internal hazard
I1	Internal fire
I2	Internal flooding
I3	Component failure (including high energy faults)
I4	Pipe breaks (whip/jet effects, flooding)
I5a	Heavy load drop/falling objects
I5b	Collapse of structural building elements
I6	On-site collision of vehicles
I7	Internal explosion
I8	Multi-unit impact
I9	Electromagnetic interference (EMI)
I10	Missiles
I11	Release of hazardous substances
I12	On-site excavation and construction
I13	Underground high-voltage Eddy currents (on-site)

Table 10. Class I hazards according to [2].

For a systematic assessment of the contribution of natural hazards to risk, only the hazard classes A to H have to be considered as initially occurring individual hazards.

2.2. Categories of hazard combinations

When combining hazards with other anticipated events, three different categories of combinations need to be distinguished (see also [1]):

Related events:

- **Category 1:** Consequential (or subsequent) events:

  The events are causally related. An initial event, for example, an external hazard, results in another consequential event, for example, an internal hazard. Typical examples are seismic and consequential internal explosion and/or fire, internal fire and consequential internal flooding, external flooding and consequential high energy arcing fault (HEAF) of a component and subsequent fire.

- **Category 2:** Correlated events:

  Two or more events, at least one of them representing a hazard, do occur as a result from a common cause. The common cause can be any anticipated event including external hazards. The two or more events correlated by this common cause could even occur simultaneously[1].

[1] "Simultaneous" here does not mean that the events occur exactly at the same time but that the second event occurs before the previous one has been completely mitigated.

Typical examples are electromagnetic interference (EMI) as common cause for a station black-out (SBO) and an internal fire as two correlated events or Tsunami as common cause for external flooding, internal flooding, and internal fire as three correlated events.

- **Category 3:** Unrelated events:

  An initial event, for example, a (external or internal) hazard occurs independently from but simultaneously[1] to a hazard without any common cause. Typical examples are external flooding and independent internal fire or explosion, seismic event, and independent internal fire.

For each category of event combinations with hazards involved those combinations, which can occur site-specifically and according to the design and protection features of the facility or plant to be analyzed, have to be identified and undergo a systematic screening. In the frame of the assessment of the contribution of natural hazards to risk, some hazard classes cannot be combined with natural hazards (hazard classes A to H) depending on the category of combinations. This limits the amount of principally possible combinations significantly.

3. Hazards screening

For limiting further detailed analyses only to those hazards and hazard combinations, which can occur at the site and in the facility under investigation, a systematic screening is needed. In Germany, a clearly structured, systematic approach for hazards identification and screening has been developed in the recent past by GRS for probabilistic risk assessment of nuclear power plant sites with respect to hazards [2]. This approach uses for the collection and processing of generic as well as site- and plant-specific information needed for screening and detailed analysis an analytical tool called *Hazards Library*. Based on the information and data available in general and the plant under investigation, a step-wise screening with a qualitative and a quantitative screening step, first for individual hazards and, based on the results of the qualitative screening, afterwards also for hazard combinations, is done. The screening can be performed semi-automatically based on questions to be answered for the qualitative screening and applying preselected quantitative criteria for the quantitative screening. A schematic overview of the screening approach is given in **Figure 1**.

3.1. Individual hazards screening

The first step is the identification of those hazards, which cannot be directly excluded as practically impossible for the site being analyzed. For the assessment of natural hazards, in this first step, the hazard classes Z and I can be excluded as non-natural hazards.

For a given site, for example a riverine site in central Europe far from any coastal and/or tidal influence in an area with relatively high seismicity and no volcanic history, a variety of natural hazards can be excluded.

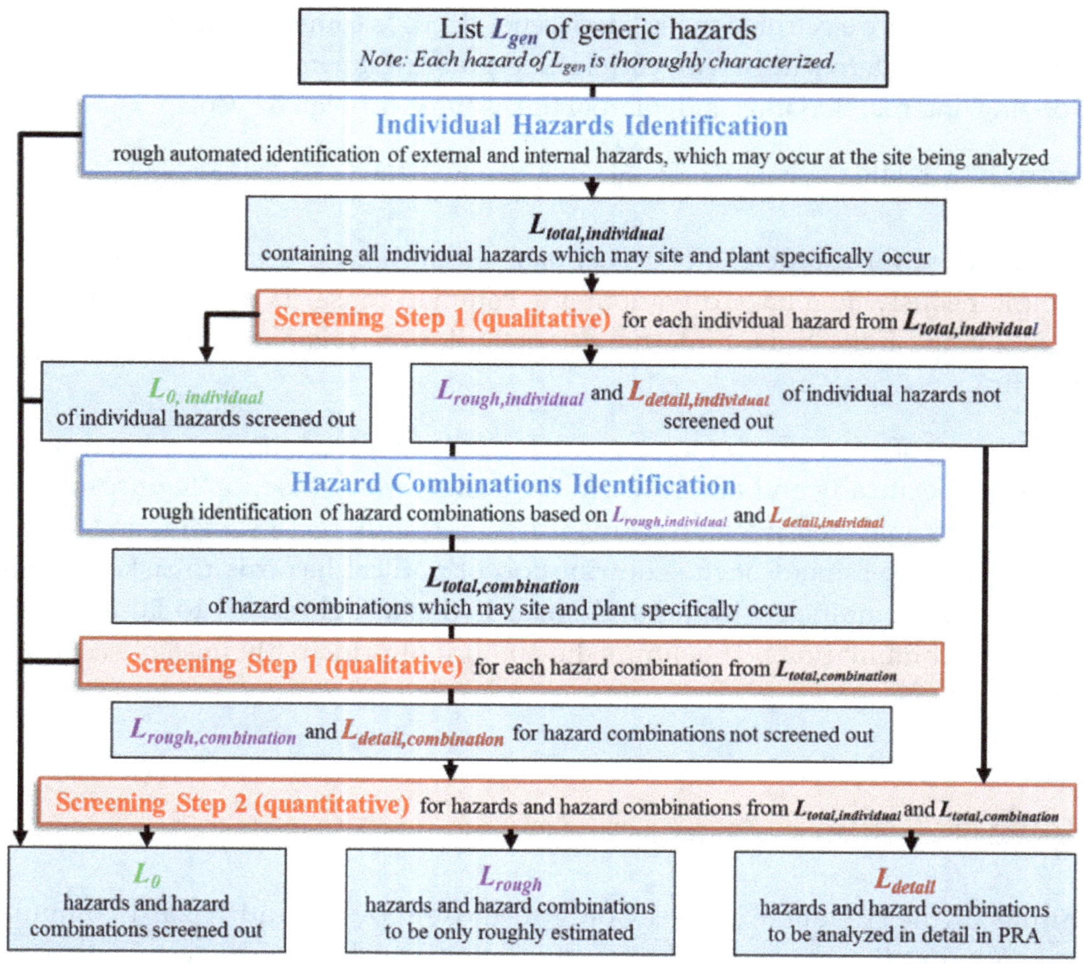

Figure 1. Overview of the stepwise approach for screening of hazards and hazard combinations, from [1].

3.1.1. Qualitative screening of individual hazards

The qualitative screening of individual hazards, which is mainly based on information available for the site being analyzed and from relevant operating experience, provides a list of site-specific remaining individual hazards, which cannot be physically excluded. Some hazards can be easily screened out from further analysis because of the general conditions not being met at the site, such as hurricane or tropical cyclone, which do only occur in areas with tropic or sub-tropic climate, or sandstorms, which cannot be assumed based on results of detailed analyses (e.g., for siting and design of building structures) being available for the site ground. In case of a plant site on rock, several hazards such as sinkholes can be easily screened out.

As a result, the following individual hazards remain for the site being investigated after qualitative screening:

Seismotectonic hazards: A1, A3, A5;

- Hydrological hazards: B2, B3, B4, B6a, B8, B9a;
- Meteorological hazards: C1, C2b, C3b, C4b, C5a, C10, C11, C12, C14, C15, C16, C19, C22, C24, C25, C27;

- Biological hazards: E6;
- Geological hazards: F6.

For these, individual natural hazards remaining after the qualitative screening, the second, quantitative screening step needs to be carried out.

3.1.2. Quantitative screening of individual hazards

The quantitative screening of individual natural hazards needs predefined quantitative criteria for screening out hazards by occurrence frequency or damage frequency. Such criteria are either available in the national or international regulation (e.g., for nuclear power plants, quantitative screening criteria by the regulatory bodies in charge of nuclear oversight are available), or conservative (pessimistic) cut-off criteria have to be defined for the facility to be investigated based on best practices.

For those hazards, for which the quantitative screening step needs to be carried out, the ranges of their occurrence frequencies have to be conservatively estimated. These are compared to cut-off frequency value corresponding to the screening criterion applied by the analyst.

Depending on the design of the facility with its protection measures and the corresponding safety margins for those hazards not screened out by frequency, a decision needs to be taken; one must decide for which hazards a rough risk estimate is sufficient and for which a detailed probabilistic analysis is needed.

For this purpose, the design requirements (national or international ones, for example, by the European Community) and their implementation at the site, for which the risk assessment shall be carried out, together with the site- and plant-specific boundary conditions and precautionary provisions against hazards impact need to be considered.

In case of the facility, for which the screening approach has been verified, the design against natural hazards such as external flooding covers events occurring once in 10,000 years corresponding to an occurrence frequency of 10^{-4} per year. Less frequent events as well as events with an occurrence frequency close to the design threshold but a damage probability of more than one order of magnitude lower can be screened out quantitatively.

The screening of hydrological (Class B) hazards for the reference site provides the result that B6, B8, and B9a can be screened out and only B2 "flash flood (torrent) by local extreme precipitation", B3 "flooding by melting snow" and B4 "flooding by extreme precipitation outside the plant boundary" remain for more detailed risk assessment. With respect to meteorological hazards (Class C), only C16 "high wind" remains at least for a rough analysis. Individual biological hazards are also screened out by frequency.

Those individual hazards screened out are stored in a list $L_{0,\text{individual}}$, those remaining after screening have to be considered for risk assessment and are stored, depending on their damage frequencies either in a list $L_{\text{rough,individual}}$ for only rough risk estimates or in a list $L_{\text{detail,individual}}$ for detailed analyses.

3.2. Hazard combinations screening

For a comprehensive Hazards probabilistic risk assessment according to the state-of-the-art, hazard combinations have to be included in the analyses. Since the number of hazards in L_{gen} and the resulting combinations is much too high to consider all combinations from the beginning in a generic manner, the screening of the hazard combinations starts from those hazards ($\boldsymbol{L}_{\mathbf{total,individual}}$) which cannot be qualitatively excluded at the nuclear power plant site being analyzed, and this in turn significantly reduces the hazards' screening effort.

In order to limit the analytical effort, at least for related hazards in a first step only first order combinations are qualitatively as well as quantitatively screened. For those combinations not screened out, potential second order combinations are identified and screened out. If there are still combinations remaining after screening, this process is repeated for the next order of combinations as long as there are combinations not yet screened out.

3.2.1. Qualitative screening of hazard combinations

From the list of individual hazards which may occur at the plant under investigation, different types of hazard combinations (consequential, correlated, and unrelated ones) involving the remaining individual hazards after qualitative screening for rough or detailed analysis (stored in $\boldsymbol{L}_{\mathbf{rough,individual}}$ and $\boldsymbol{L}_{\mathbf{detail,individual}}$) are identified. For these site- and plant-specific hazard combinations identified, the qualitative and quantitative hazards screening steps have again to be carried out (cf. **Figure 1**).

3.2.1.1. Category 1 combinations of consequential hazards

As already mentioned, for screening of causally related event combinations of natural hazards with other hazards, combinations of man-made or internal hazards with consequential natural hazards can be excluded.

For the site being investigated, only few individual natural hazards of the classes A, B, C, E, and F remain after the qualitative screening. For these, physically possible category 1 combinations have to be identified and screened out. As an exemplary result from [3], the following combinations with hydrological (Class B) hazards remain for quantitative screening (**Figure 2**):

3.2.1.2. Category 2 combinations of correlated hazards

For the qualitative screening of event correlations in the first step, the common causes have to be identified. These have to be systematically correlated to the different consequential events by phenomena. Typical correlations are possible between different hydrological hazards induced by precipitation (C1) as common cause. Examples of results from [3] are provided hereafter in **Figure 3**.

The final result of the qualitative screening of related hazards for the reference site analyzed is the following:

The longer duration external flooding hazards B3 and B4 resulting from snow melt or extreme precipitation can induce internal flooding hazards I2. All those external hydrological hazards with flooding potential not screened out individually besides B6a can in principle result in biological flotsam at the reference site.

- External flooding hazards from various origins and with differing duration B2, B3, and B4 can occur as correlated events from the same root cause or together with the meteorological hazard C1 (precipitation) correlated, for example, by extreme weather conditions. In addition, flash flood B2 and heavy rainwater flooding B4 can occur correlated to F1 (subaerial slope instability).

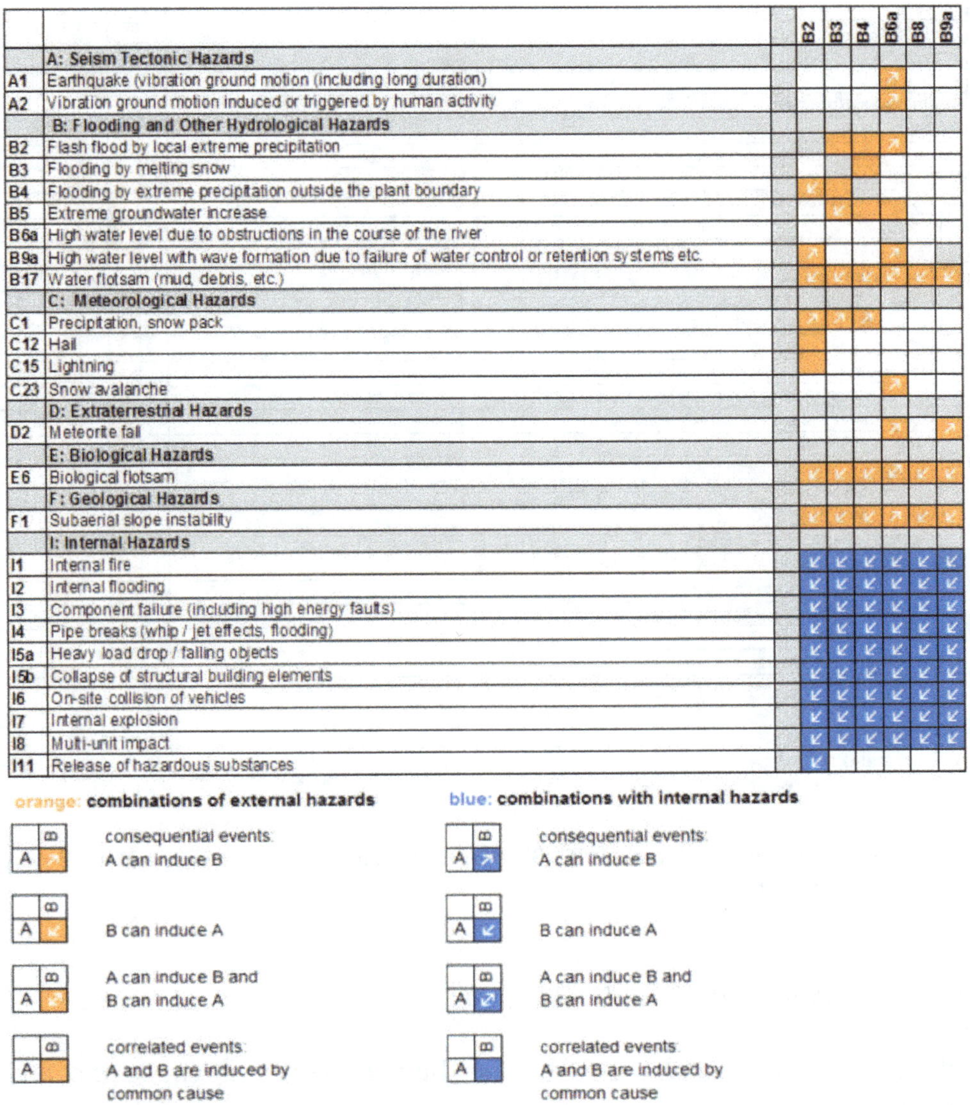

Figure 2. Result of the qualitative screening for combinations of those hydrological hazards not screened out at the site being investigated with other hazards.

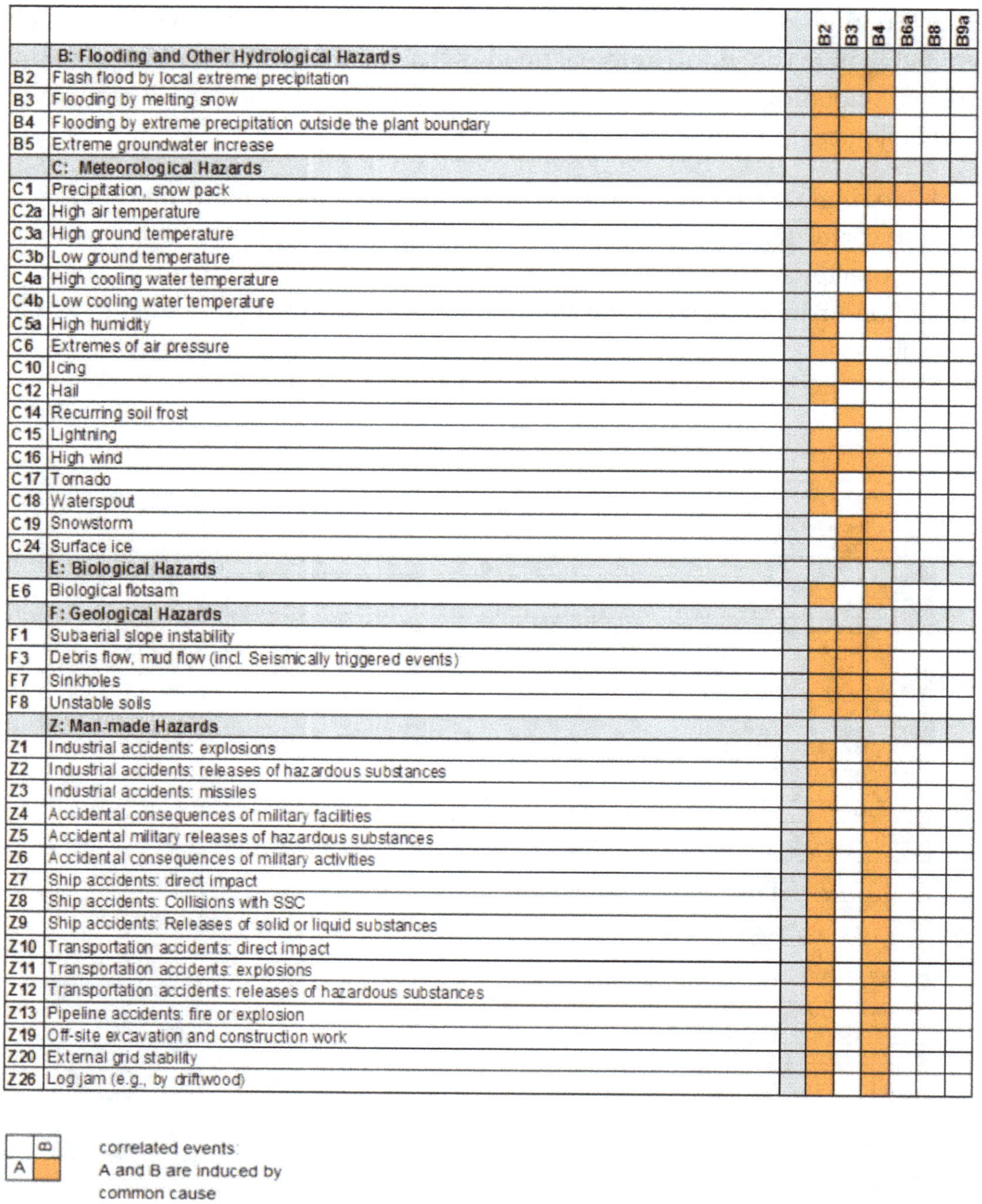

Figure 3. Correlations of the hydrological hazards B2, B3, B4, B6a, B8, and B9a not screened out qualitatively with other hazards due to a common cause, from [3].

3.2.1.3. Category 3 combinations of unrelated hazards

Hazards that occur independently of each other have no common cause and are unrelated. The simultaneous occurrence is in general highly unlikely and is therefore investigated on an international level mainly for hazards of longer duration. In the example of the hydrological hazards not screened out qualitatively (B2, B3, B4, B6a, B8, B9a) a broad majority of combinations with unrelated events is not possible or very unlikely.

In the first step, all those individual hazards not qualitatively screened out can be considered for this third category of combinations. This results in a relatively long list of category 3 combinations, for which qualitative screening is necessary.

3.2.2. Quantitative screening of hazard combinations

In the example of a German nuclear power plant site, given cutoff values from the German regulation [4] have been applied to the occurrence frequency and to the damage frequency.

The qualitative screening for category 1 combinations provides the following result for causally related combinations at the reference site with the hydrological hazards B2, B3, B4, B6a, B8, and B9a: Combinations of these hazards with E6 (biological flotsam) and F1 (subaerial slope instability) have been screened out quantitatively. Therefore, no category 1 combinations remain after the quantitative screening; higher order combinations are also not to be assumed.

The following category 2 combinations that remain after qualitative screening for the reference plant site have been analyzed: A meteorite fall (D2) can cause correlations of I2 (internal flooding) with B6a, B8, or B9a. I2 can also occur together with B6a, B8, or B9a as consequence of man-made explosions (Z1, Z4, Z6, Z11, or Z13). Resulting from a common cause such as a thunderstorm precipitation (C1) or F1 (subaerial slope instability) can be observed correlated with B2 or B4. All these correlations have been excluded quantitatively for the reference facility.

For category 3 combinations B2, B3, B4, B6a, B8, or B9a have to be assumed to occur independent from other hazards. Such combinations have only to be analyzed, if their occurrence frequencies exceed a given cut-off value under consideration of the durations of the individual hazards. According to this argumentation, in the example of screening for hydrological hazards for a given German site, only combinations of B2 with B3 or B4 finally remain for further detailed risk analysis.

It could be demonstrated that the remaining number of hazard combinations is significantly lower after qualitative and quantitative screening of hazard combinations.

4. Detailed analyses

The plant model for risk assessment of the facility under consideration needs to be extended by taking into account those hazards and hazard combinations remaining after screening. It has to be analyzed, which structural elements, plant operational components, or even complete systems maybe impaired in their required function (so-called initiating events, IEs). That also requires to extend the original list of risk-relevant functional unavailabilities, the so-called basic events (BEs) in the plant model by those ones related to the hazards and hazard combinations to be considered as well as by the corresponding failure dependencies. This requires another two analytical steps:

After identification of the potential hazard induced initiating events, these have also to be screened with respect to their significance for the facility. In this context, it is important to analyze within that screening step if and how far the identified initiating events from hazards do occur quasi simultaneously and need to be modeled as common cause initiating events.

- In a further step, the potential unavailability of structure, systems, and components depending on the impact by hazards needs further extension of the risk analysis model of the facility requiring for each hazard and hazard combination not screened out to identify those items which may functionally failing (so-called hazard equipment lists *HEL* as defined in [3]) and the corresponding failure dependencies (so-called hazards dependency list *HDL*, details see [3]). Again, for limiting the analytical effort, a reduction of these lists according to their risk significance by qualitative arguments and quantitative criteria is important.

As provided in more detail in [1, 3], the hazard equipment list for a single hazard H_k covers the entire number j of structures, systems, and components SSC_j identified to be vulnerable to H_k and for which their failure contributes to the risk induced by H_k:

$$\mathrm{H_kEL} = \{\mathrm{SSC_1}, \ldots, \mathrm{SSC_m}\}_{\mathrm{Hk}}.$$

In order to quantify the failure probabilities of the remaining structures, systems, and components vulnerable to the hazard $H_{k,}$ information from the facility being analyzed such as technical reliability of systems and components and other factors affecting the hazard-induced scenarios like human reliability in case of actions (e.g., for the remaining hydrological hazards B2, B3, and B4 and their combinations, temporary flood protection measures) have to be taken in a predefined period to prevent damage.

In a further analytical step, the dependencies among the failure characteristics of the vulnerable structures, systems, and components need to be investigated. Each dependency in this list $H_kDL = \{D_1, \ldots, D_n\}_{Hk}$ is characterized by a triple $D_k = \{A_k, S_k, c_k\}$ of parameters, which include the set of dependent structures, systems, and components S_k, the common characteristics of the elements of S_k (e.g., water level as cause for a flooding hazard-induced dependency) A_k, and a correlation factor c_k for the dependency strength. The hazard equipment lists and hazard dependency lists need to be generated based on the corresponding parameters to be estimated and are used for the qualitative plant model extension. For adequately modeling the dependencies between the structures, systems, and components and/or the hazards impact, the fault trees of the analytical risk analysis model need to be modified and multiplied for the different hazards to be considered. In addition, new elements of the fault trees have to be specified (see also [5]) within the database representing a probabilistic model of a plant system.

A schematic overview of the approach for the plant model extension by hazards is given in **Figure 4**.

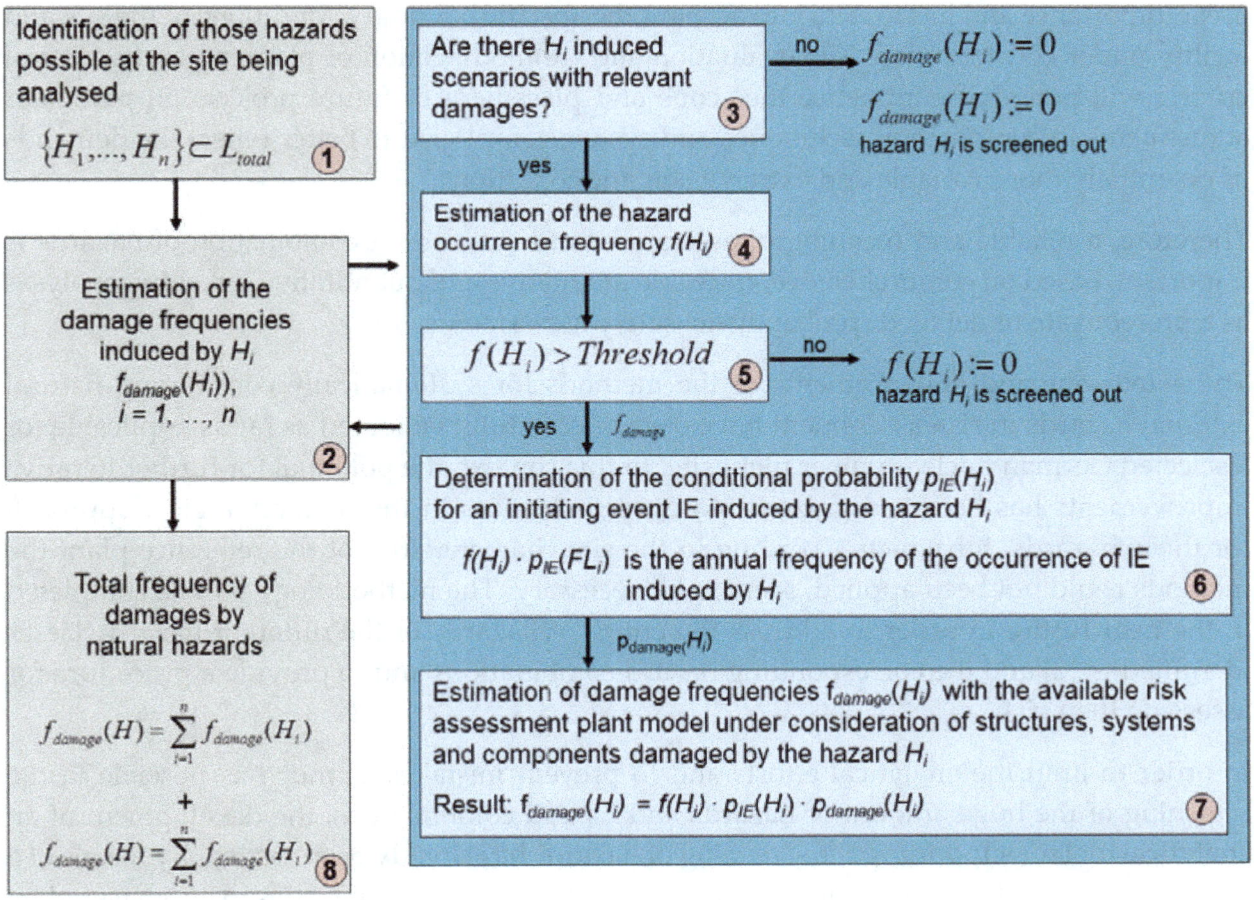

Figure 4. Extension of the model of the facility being analyzed for probabilistic risk assessment of hazards, adapted from [3].

The model extension also needs to take into account any countermeasures for preventing a risk-significant impact to the facility or mitigating the consequences of the hazards such that the damage to the facility remains non-negligible.

A typical example for preventive countermeasures is the timely implementation of rotatable bulkheads or stop logs as temporary means for protecting water ingress in case of flooding hazards. An example for mitigative measures in case of flooding events is the use of portable equipment to remove water from buildings with systems or components needed for safe operation of the facility such that their required function will not be inadmissibly impaired. In this context, the time and flooding scenario-dependent success paths including the manual actions to be taken have to be included in the probabilistic plant model considering also the human factor adequately in the corresponding HRA (human reliability analysis) model. As a result, additional end states (damage states) of the fault trees can be determined.

5. Conclusions and outlook

The systematic assessment of natural hazards including their contribution to risk and their consequences such as physical and operational impacts on critical infrastructures is still of great importance and has to take into account the specific boundary conditions of the site and facility under consideration. The evaluation and (re)modification of planning and technical criteria will potentially influence the scope and placement of future projects, in particular adjustments in construction techniques and systems employed to better reflect the demands of potentially more variable and extreme climatic conditions.

Therefore, a reliable and meaningful assessment of hazards and combination of hazards is important, based on comprehensive, traceable qualitative and quantitative screening analyses as a prerequisite of detailed (probabilistic) safety assessments.

The extensions and enhancements of the methods for systematically considering natural external hazards in risk assessment have been successfully validated as far as applicable for a selected German nuclear power plant site. In this context, the potential for further iterative improvements has been recognized. Moreover, advances in the methodological approach for those hazards, for which according to the site characteristics of the reference plant the methods could not been applied, seem to be necessary. The methodology will be completed in the near future in order to address the entity of hazards of the different hazard classes identified in [2] and the corresponding hazard combinations and to provide a procedure for assessing their risk.

In order to limit the analytical efforts and to prevent mistakes as much as possible in the screening of the huge amount of hazards and hazard combinations, the development of an analytical tool for supporting the screening of hazards has already been started. By means of a scroll down menu based on qualitative arguments formulated as questions to be answered by yes or no, such as "Is the site a tidal site?", various hazards can be directly screened out qualitatively. The menu offers to provide inputs on a generic or plant design-specific basis. The tool will offer, in a second step, to also semi-automatically perform the quantitative screening by selecting from a predefined menu of quantitative criteria, such as an occurrence frequency threshold value and apply these to those hazards or hazard combinations not qualitatively screened out. The tool will be as far as possible independent of database software products to enable any possible user to apply it without software restrictions. In addition, the output will be documented in simple, written text form as well as graphically.

For the detailed analyses in the frame of hazard risk assessment it is intended to advance the topological modeling methods provided by GRS in the analytical tool pyRiskRobot [5] for an as far as practicable automated integration of event combinations within particular natural hazards in the probabilistic plant models.

Acknowledgements

The authors would like to thank the German Federal Ministry for the Environment, Nature Conservation and Nuclear Safety (BMU) and the German Federal Ministry for Economic Affairs and Energy (BMWi) for funding and supporting the work presented in this chapter.

Author details

Berg Heinz-Peter[1*] and Roewekamp Marina[2]

*Address all correspondence to: bergheinzpeter@gmail.com

1 Bundesamt für kerntechnische Entsorgungssicherheit (BfE), Salzgitter, Germany

2 Gesellschaft für Anlagen-und Reaktorsicherheit (GRS) gGmbH, Köln, Germany

References

[1] Roewekamp M, Sperbeck S, Gaenssmantel G. Screening approach for systematically considering hazards and hazards combinations in PRA for a nuclear power plant site. In: Proceedings of ANS PSA 2017 International Topical Meeting on Probabilistic Safety Assessment and Analysis, September 24-28, 2017; Pittsburgh, PA, USA: On CD-ROM, American Nuclear Society, LaGrange Park, IL, USA; 2017

[2] Sperbeck S, et al. Analysehilfsmittel zur Bereitstellung von Informationen und Daten zur systematischen Durchführung von PSA für übergreifende Einwirkungen, GRS-A-3914, Gesellschaft für Anlagen- und Reaktorsicherheit (GRS) gGmbH, Köln, Germany; April 2018 (in German)

[3] Roewekamp M, et al. Methoden zur Bestimmung des standort- und anlagen-spezifischen Risikos eines Kernkraftwerks durch übergreifende Einwirkungen (Estimation of the Site and Plant Specific Risk of a Nuclear Power Plant from Hazards), Technical Report, GRS-A-3888, Gesellschaft für Anlagen- und Reaktorsicherheit (GRS) gGmbH, Köln, Germany; June 2017 (in German)

[4] Facharbeitskreis (FAK) Probabilistische Sicherheitsanalyse für Kernkraftwerke. Methoden zur probabilistischen Sicherheitsanalyse für Kernkraftwerke, Stand: August 2005, BfS-SCHR-37/05, Bundesamt für Strahlenschutz (BfS), Salzgitter, Germany; October 2005 (in German). http://doris.bfs.de/jspui/handle/urn:nbn:de:0221-201011243824

[5] Berner N, Herb J. Weiterentwicklung der Methodik zur automatisierten Integration übergreifender Einwirkungen in PSA-Modelle der Stufe 1, Technischer Fachbericht, GRS-454, ISBN 978-3-946607-36-6, Gesellschaft für Anlagen- und Reaktorsicherheit (GRS) gGmbH, Köln, Germany; March 2017 (in German). http://www.grs.de/publikation/grs-454

2

The Approach of Probabilistic Risk Analysis and Rationale of Preventive Measures for Space Systems and Technologies

Nikolay Paramonov

Abstract

This chapter is devoted to the probabilistic risk analysis of collision of satellites with space debris. The uncertainty and random space-time characteristics of dynamic space objects are researched for the rationale of preventive measures for space systems and technologies. The proposed approach is illustrated by analyzing space debris and their distribution on satellite orbits. The actuality is confirmed by many dangerous convergences of controlled satellites and fragments of the old objects that have been discarded and transformed into uncontrolled debris. The research demonstrates a possibility of probabilistic modeling allows calculating preventive measures for avoiding collisions.

Keywords: probabilistic risk, calculating, space systems, control of space, modeling, control system, crucial situations

1. Introduction

Space systems are essential to human progress. They are an integral part of everyday life and the development of science. Now, there are thousands of satellites that provide remote sensing of the Earth in near-earth space, assist navigation, provide accurate weather forecasts, etc. The use of space systems involves the ability to control satellites. The control of the space objects includes the ability to change their orbit and orientation. One of the difficult tasks in objects control is the problem of evading uncontrolled objects or space debris, an example of which is shown in **Figure 1** [1]. System of space monitoring is created in developed countries to solve these problems, and the main task is to track the trajectory of space objects and assess the risk of possible collisions. The technology of space control systems mostly reduced to the modeling

Figure 1. Space debris.

and prediction of space objects motion. This chapter gives examples of probabilistic risk analysis of collisions with space debris and preventive measures to avoid such collisions.

Analysis of the complex systems for control in the space branch requires a large volume of simulation. The mathematical model of a complex system is created using the principles of the functional association of the models of elements and subsystems into a single program complex of the implemented algorithms. This complex accomplishes an imitation of the processes for the entire variety of input conditions and current states of the real system [2].

There are some questions that arise in the possible risk analysis of space objects collision, including collision with space debris. Methods of risk analysis and predictive preventive measures are based on probabilistic modeling. The questions of choosing the modeling method are the most determining while analyzing risks and substantiating security space systems.

The index of efficiency **R** is the mathematical expectation of functional Y, which is determined on a set of functions

$$Z(t, \omega) \in Z$$

The output processes of a system in a single implementation characterize the function $Z(t,\omega')$ with a fixed value $\omega = \omega'$. A single implementation is a random interval of time $t \in [0, T]$ of system functioning. Assume that Ω is a space of elementary events ω with the possible measure P(A), where A is a random measurable subset Ω. Then,

$$R = \int_{\Omega} \varphi(z(t, \omega))dP \tag{1}$$

Every complicated system realizes transformation of input signals into output ones. The system model makes the same.

Assume that X is a set of input signals $x(t, \omega_X) \in X$, Ω_X is a space of elementary events ω_X with probabilistic measure P_x, and that for every $\omega_X \in \Omega_X$, there is an input signal $x(t, \omega_X) \in X$.

The system properties are described with a random operator $H(x(t, \omega_X), \omega')$. There is a probabilistic measure P_H in the set Ω_H. Elements of this set are $\omega_H \in \Omega_H$.

By definition, a random operator $H(x(t, \omega_X),\ \omega_H)$is a set of nonrandom operators $H(x(t, \omega'_X),\ \omega'_H)$, defined for each $\omega'_H \in \Omega_H$. It implements the mapping of the set X to the set Z for all $\omega_x \in \Omega_x$in the set Z. It means that every implementation of $Z(t, \omega')$ is a result of transformation of an input signal $x(t, \omega'_X)$ by a nonrandom operator $H(x(t, \omega'_X),\ \omega'_H)$. In the operator form, the process of transformation is

$$Z(t, \omega') = H(x(t, \omega'_X),\ \omega') \tag{2}$$

It follows from Eq. (2) that for a known structure of a random operator H, elements $\boldsymbol{\omega}$ from the set Ω are generated by the elements $\omega_X \in \Omega_X$and $\omega_H \in \Omega_H$, and for every (ω_X, ω_H), there is only one point ω of space Ω. This correspondence allows the integral Eq. (1) to be written this way:

$$R = \int_{\Omega=[\Omega_X \times \Omega_H]} \varphi\{H[(x(t),\ \omega_X),\ \omega_H]\} dP_X dP_H \tag{3}$$

Under real conditions, the structure of a random operator H and the probabilistic measures P_x, P_H are got by basis of a priory information I and information Z, which are obtained in natural tests of elements and the whole system. An example of such tests is the archives of ballistic situations that were created by the space control systems of Russia and the US.

The problem of calculating R can be regarded as a statistical problem of the synthesis of decision rules W provided estimates $\widehat{\mathbf{R}}$ with certain preassigned properties. The task determines the ultimate goals that include calculating indicators of efficiency of the complicated technical systems, risks of usage, and possible preventive measures.

In the operator form, the evaluation operation R can be written as:

$$\widehat{R} = W(Z, I, H) \tag{4}$$

While choosing rules W, it is required that the calculations related to finding $\widehat{R}$ by Eq. (4) are technically implementable, and the properties of the estimates $\widehat{R}$ satisfy the conditions of maximum achievable accuracy. In the problems of assessing the characteristics of complex technical systems, these requirements are usually decisive.

The features of the space systems tests consist in the fact that they can be carried out in the conditions of the regular functioning of the system in a very limited volume. Experiments on the system in crucial situations and operating modes usually requires considerable effort and material costs, and it is sometimes associated with a risk of system failure, which is unacceptable for space systems. That is why while evaluating the system's efficiency indicators, it is necessary to consider that the sample $\tilde{Z}(t,\omega_1), \ldots, \tilde{Z}(t,\omega_n)$ reflects the results of the tasks performed by the system only under normal operating conditions. In other words, the direct use of the sample $\tilde{Z}(t,\omega_1), \ldots, \tilde{Z}(t,\omega_n)$ to determine the characteristics of the system over a wide range of its operation, including critical modes, is practically impossible.

Calculation of system performance indicators for normal conditions of its operation is carried out in several stages:

a. a number of values $\varphi(\tilde{Z}(t,\omega_1)), \ldots, \varphi(\tilde{Z}(t,\omega_n))$are calculated;

b. statistical properties of random variables $\varphi(\tilde{Z}(t,\omega_i))$ are determined, relying on known laws of distribution of measurement errors;

c. the a priori information about the value of R is expressed as the a priori density of the distribution P (R);

d. a criterion for the optimality of the W estimates $\widehat{R}$ is chosen.

In the presence of the aforementioned information, it is possible to combine a priori information with the real information obtained in the process of carrying out field experiments to obtain estimates.

In calculating the integrals in Eq. (3) by simulation modeling methods, it is necessary:

a. to develop a model that allows the generation of processes $Z(t,\omega) = H((x(t),\omega_X),\ \omega_H)$for different $(\omega_X,\ \omega_H)$;

b. to determine the method of setting up experiments on the model, provided that the probabilistic measures P_x, P_n are given;

c. to develop algorithms for processing simulation results;

d. to build a plan for conducting experiments and processing their results.

However, in modeling real systems, the estimated situation proves to be much more complicated, because the structure of the random operator H and the probability measures P_x, P_n are determined as a result of the complex processing of a priori information and information obtained from field testing of the elements and the entire system as a whole. The limitation of real statistics and a priori information usually leads to the fact that both the operator H and the measures P_x, P_n will contain errors, which in the general case are of a probabilistic nature.

It follows that in assessing the performance indicators of complex systems and their risks, there will be components due to errors in the definition of the operator H and errors in calculating the probability measures P_x and P_n. In addition, modeling errors arise from inaccuracies in the implementation of the operator H on computational means and the limited amount of statistical data obtained by experimenting on the model.

Collectively, the aforementioned errors determine the total modeling error, which in general will be a random quantity consisting of deterministic and random components.

Further examples of probabilistic risk analysis and justification of preventive measures for space systems and technologies will be illustrated by using examples of the risk of collision of spacecraft with space debris.

The basis for assessing the risk of collisions in space is trajectory measurements.

2. Factors of uncertainty in processing trajectory measurements

In applied mathematics, the development of methods for processing trajectory measurements occupies a special place because of their complexity. General methodological problems in this area have not been solved in many respects due to the possibility of obtaining subjective conclusions or the use of excessive formalization, which makes it difficult to extract practically useful results.

An important role in solving problems associated with the development of methods for processing trajectory measurements is played by the concept of randomness. The concept of randomness is a certain type of uncertainty, characteristic of frequently observed events.

The methods used to solve problems related to the processing of trajectory measurements can be divided into sections corresponding to those deductive theories whose apparatus is used to solve problems from other fields of knowledge. If the studied problem, the solution of which provides the development of new methods for processing trajectory measurements, can be represented as a certain set of objects related by relations, then it is always possible either to find a suitable formalism to solve the problem or to create a new mathematical structure suitable for solving the problem.

It should be noted that the question of whether it is possible to single out objects forming a population in such a problem area as the development of new methods for processing trajectory measurements that allows them to be interpreted as sets connected by relations is a matter of not applied mathematics, but a part of the science of developing and testing of complex space systems.

The limited knowledge of the processes and phenomena studied does not allow the creation of absolutely accurate models of elements, including space systems and technologies. In addition, it is impossible to carry out an infinite number of experiments on the system model because of real limitations. For these reasons, the accuracy of the estimates obtained will be determined by the reliability of the information on the structure and parameters of the models being created, and also by the errors caused by the imperfection of the methods used for setting and processing the experiments carried out on the model.

If the accuracy of calculating the indicators will be sufficient for practical purposes, then the application of simulation methods can be considered justified. In practice, a series of control checks is carried out for this purpose, the main purpose of which is to establish a measure of proximity between the real and simulated processes. If, according to the results of the comparison, it turns out that for all input conditions, the differences in the simulation of real processes do not exceed some given critical values, then the model is considered adequate for the system under analysis. Otherwise, the system model needs to be improved.

In the process of modeling, it is possible to use various assumptions and coarsening, as a result of which the mathematical formulation of the investigated problem can only be its approximate reflection. When carrying out formalizations, there always arises the question of the adequacy of the resulting mathematical description of problems associated with the development of methods for processing trajectory measurements. If the solution of the content problem is not complicated and requires a numerical solution, an adequacy check can be made by experimental calculations using the initial data for which the desired results are obtained experimentally. The calculated results should differ from the reference data by no more than the task of assessing the risks of collision of space vehicles or carrying out preventive measures eliminating the danger of such a collision allows.

2.1. Analysis of the risk of collisions of space appliances

The area where most artificial earth satellites operate is very extensive; its volume is about 10^{12}–10^{13} km^3.

The density of artificial Earth satellites and space debris can be estimated by numbers in the order of 10^{15}, and this number is constantly increasing. For example, in 2007, China tested an anti-satellite missile, sending it to one of its old satellites, adding about 3500 extra fragments in the area between 160 and 2000 km above the surface. This is a very large amount of objects and they must somehow be taken into account. The general picture of the contamination of near-Earth space is clearly shown in **Figure 2** [3]. The orbital information on more than 20,000 space objects (SO) of more than 10 cm in size has been fixed and regularly updated.

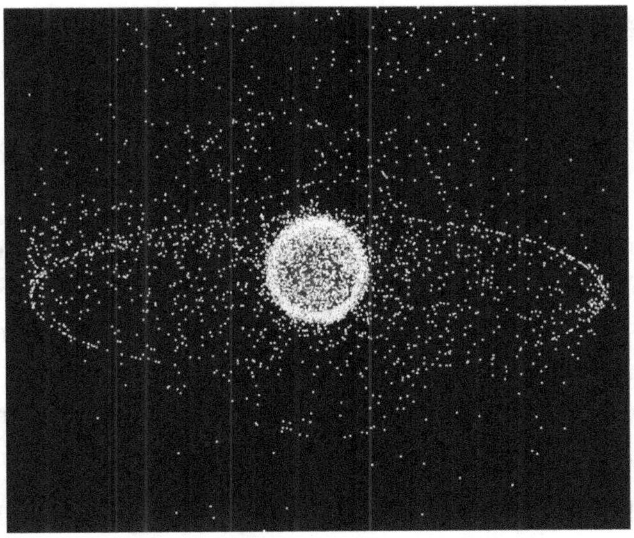

Figure 2. Earth space.

Simulation of the motion of such satellites reduces to obtaining, as a rule, systems of ordinary differential equations of motion of an object and their integration by one or another method. As a result, the dependence of the motion parameters on time is obtained for the given initial conditions. These equations are a form of representation of the laws of dynamics and kinematics and can be supplemented by equations of control.

By managing the catalogs of space objects (and such catalogs exist both in Russia and in the US), one can assess the mutual position and carry out the forecast of their movement. In particular, it is possible to assess the dangerous convergence and even collision of space vehicles. The main method for determining the motion of space objects is modeling.

Simulation of the motion of controlled space objects can be performed with different goals determined by the specific content of the tasks being solved. Among the tasks that involve the use of motion models for controlled objects, let us dwell on the identification of parameters and SO states based on the results of their measurements. Such a task is the basis for justifying the control actions in order to evade the collisions of space vehicles.

Any of the models used in solving the problem presented earlier is a mathematical idealization of real motion. Therefore, in modeling, of course, a special question arises about the adequacy of the mathematical description of the real movement of an object. The adequacy of the model is directly dependent on the degree of confidence in the a priori data, the completeness of their accounting for modeling, and the accuracy of the model's reproduction on a computer.

Requirements for accuracy in modeling can be considered to be dictated by the content of the problem being solved.

The a priori data on motion include the laws of kinematics and dynamics, the parameters (for example, the traction force, the nature of its variation in time and the time of the engine, the aerodynamic coefficients), the characteristics of the surrounding space (for example, the model of the atmosphere, the gravitational potential), the characteristics of the interaction of the object with the surrounding medium. Sometimes it becomes possible to use the data about the software path (program parameters) of the movement of the object. The a priori information may include the law governing the apparatus (for example, the method of parallel approach) and the features of its implementation by the control system (lag, other management errors). These data can include both deterministic and random parameters with known distribution laws.

The most reliable are usually the laws of dynamics and kinematics of the motion of objects. The validity of kinematic constraints such as "the linear velocity vector of the object is the first time derivative of the vector of its position" is beyond doubt. At the same time, the use of the dynamics relations should be carried out taking into account that they are sufficiently close to the real movement only under the conditions of their formulation. In many practical applications, the object is idealized as a material point located at its center of mass.

If by the conditions of the solution of the problem this assumption is relatively rough, then the degree of confidence in describing the trajectory of the movement of the object as a material point can turn out to be low. If necessary, the object can be considered as a system of rigid bodies and clarify the description. If it is required to take into account more subtle effects that have a noticeable effect on the components of the object's motion that are of interest, for example, on orientation in space, then the movements of the fuel components in the supply lines and the like can be modeled.

The degree of confidence in the mathematical model is determined not only by the nature of the idealization of the object (material point, aggregate of rigidly bound bodies, etc.), but also careful consideration when modeling individual factors affecting the movement of the apparatus. The adequacy of the model also depends on the accuracy of knowledge of the parameters appearing in the formulas (ballistic coefficient, aerodynamic coefficients). These values are determined, as a rule, experimentally and are therefore known with some errors.

As criteria of comparison and estimation of models of movement of SO, it is expedient to apply those or other modeling functional errors.

Denote $\lambda(t)$ is the real change in the motion parameter on the time interval, $[O, T]$, $\lambda_M(t)$ are the process simulating this motion. Then, the absolute deviation of the model motion from the real is a function of time $\Delta\lambda_M(t) = \lambda_M(t) - \lambda(t)$, describing the modeling errors in time. As the values that generalize such errors, extreme, averaged and confidence indicators can be used.

By extreme exponents, we mean the largest (least) values of some characteristics of a function $\Delta\lambda_M(t)$, for example, a module $\Delta\lambda_M(t)$:

$$\alpha_1[\lambda_M(t)] = \max_{t\in[O,T]} |\Delta\lambda_M(t)|$$

When the average is implemented, this or that characteristic of the function $\Delta\lambda_M(t)$ is averaged over a time interval $[O, T]$. An example of averaged index of modeling quality is the average square error of modeling:

$$\alpha_2[\lambda_M(t)] = \int_O^T [\Delta\lambda_M(t)]^2 dt.$$

The introduction of confidence indicators in quality is associated with the random nature of the modeling errors caused, for example, by the presence of random components in the movement of the object. The appearance of such components is due to fluctuations in the properties of the environment surrounding the SO, the parameters of various elements of the flight control system, and so on.

Confidence indicators are complex and combine a confidence interval and a confidence probability.

The confidence interval is the range of values of some characteristic β of the function $\Delta\lambda_M(t)$: $B = [\beta_{\min}, \beta_{\max}]$.

The confidence probability P is the probability that the calculated value of the characteristic $\beta = \beta[\Delta\lambda_M(t)]$ in the confidence interval B.

The confidence level of the modeling quality is the interval B, in which the error characteristic falls with a confidence probability.

In modeling, the influence of the atmosphere, the rotation of the earth, reactive forces, etc., can be taken into account to some extent. To simplify the model, a number of factors that determine the movement of the object, but are insignificant, are combined and replaced in the noise component model ("useful noise"). Useful noise is a random amount included in the model.

The resulting solution of the equations is the deterministic basis of the simulated motion. It can be supplemented by random specially imitated components, which are introduced additively, multiplicatively or additively-multiplicatively.

2.2. Differential equations of the spatial motion of space systems

Let us consider the system of equations of translational and rotational motions of SO. The first subgroup of equations characterizes the displacement of the center of mass, and the second characterizes the orientation of the object in space.

In doing so, we will use the normal terrestrial coordinate system. For simplicity, we assume that the acceleration of gravity is constant in magnitude and direction, Coriolis acceleration is absent; the curvature of the Earth is neglected, the wind is not taken into account.

The simplest way is that the velocity of the translational motion of the center of mass of the rocket relative to the Earth is described in the projections on the axis of the trajectory coordinate system $Ox_k\, y_k\, z_k$, since $v_{x_k} = v$, $v_{y_k} = v_{z_k} = 0$ in this case. Then,

$$\dot{V} = \sum F_x/m; \dot{V}_{w_{z_k}} = \sum F_{y_k}/m; \dot{V}_{w_{y_k}} = -\sum F_{z_k}/m,$$

where $\sum F_x$, $\sum F_{y_k}$, $\sum F_{z_k}$ are the projections on the axis of the specified coordinate system of the resultant force acting on the center of mass of the object; ω_{y_k}, ω_{z_k} are the projections of the angular velocity of rotation of the trajectory coordinate system $Ox_k\, y_k\, z_k$ relatively fixed system $Ox_g y_g z_g$ on the axis of the trajectory coordinate system. To determine them, we use the relations that reveal the relationship of these components of the angular velocity with the orientation angles of the trajectory coordinate system with respect to the normal one:

$$\omega \mathrm{x}_k = \dot{\Psi} \sin\theta; \omega_{y_k} = \dot{\Psi} \cos\theta; \omega_{z_k} = \theta$$

Here θ and Ψ are the angles of the path and the slope of the trajectory. As a result of the substitution, we obtain a system of equations of the form

$$\dot{V} = \sum F x_R/m, \dot{V} = \sum F y_R/m, V\dot{\psi}\cos\theta = \sum F z_R/m$$

In the right-hand side of the equations, we include the components of the traction force, gravity, and control forces.

When obtaining an expression for the aerodynamic force, we use the velocity coordinate system $0x_a y_a z_a$, and then from it, we proceed to the trajectory system. Projections of this force on the coordinate axis of the last system are represented in the form

$$RAx_R = -x_R = -x_a;$$

$$RAy_R = y_R = y_a \cos\gamma_a - z_a \sin\gamma a;$$

$$RAz_R = z_R = y_a \cos\gamma_a + z_a \sin\gamma_a$$

where γ_a is an angle of rotation of the velocity system relative to the trajectory system.

Further equations that take into account the features of the motion of the SO will lead to a system of equations of motion.

Solving this system, we find all the characteristics of the motion of the rocket or SO:

$$V(t), \theta(t), \psi(t), x_g(t), y_g(t), r(t), \vartheta(t), \gamma(t), \alpha(t), \beta(t), \gamma_a(t), \varpi_x(t), \varpi_y(t), \varpi_z(t), m(t), z_g(t), \psi(t).$$

Naturally, the initial conditions for integration must be given.

With a rigorous theoretical approach to the solution of the problem of modeling, it is obviously impossible to separate the equations describing only the translational motion of the center of mass or only the rotational motion of the relative center of mass, and the equations of longitudinal and transverse motion. The relationship between translational and rotational movements is manifested through so-called cross-links.

The probability that the calculated values of the parameters of two SO movements $\beta = \beta[\Delta\lambda_M(t)]$ in the confidence interval is estimated. In each of them, one can predict the risk of a dangerous convergence.

The same task is directly related to the definition of collision risk with space debris (CD). For probabilistic modeling of collision risks with space debris, special programs are used, for example:

Model SDPA [4] is a semi-analytic stochastic model for medium- and long-term forecasting of technogenic SG larger than 1 mm in low Earth orbits (LEOs) and geosynchronous Earth orbits (GEOs), for constructing the spatial distribution of concentration and velocity characteristics, as well as estimating collision risk [5].

The model uses summary data on SO of various sizes (including space debris without "binding" them to a specific source of pollution).

The measurement errors are estimated on the basis of averaging of the last measurements of the orbital parameters. In calculations of dangerous convergence, errors are calculated in several models, for example, in the orbital coordinate system, in the elements of the orbit, and in models of direct integration. The error matrix is used to calculate the collision probability.

If we consider the problem in posing the risk of collision of an uncontrolled SO with a controlled spacecraft (SC), then it is necessary to consider the process of mutual proximity in three-dimensional space and in the picture plane.

In this three-dimensional space, the spacecraft structure is considered as a sphere of a given radius. The region of possible position of the SO is represented in the form of an ellipsoid whose parameters and orientation are determined by the total error matrix

$$C_{SC+SO} = C_{SC} + C_{SO}$$

In the picture plane, the SV and the SO region are represented as a circle and an ellipse, respectively.

The problem reduces to the search for the probability of hit of a random vector whose density is given by the ellipsoid of scattering errors into the sphere of a given cone in **Figure 3**, where T_{dc} is the time of dangerous convergence, and V and D_v are the speed.

For example, the approach of a SO to an ISS is considered safe if $P_c < 10^{-5}$. For the value of P_c in the range between 10^{-4} and 10^{-5}, the collision risk is high enough, therefore, when planning the control of the spacecraft, necessary maneuvers should be provided for the purpose of evasion.

To assess the characteristics of the collision risk, an archive of dangerous convergence (ADC) between all objects in the catalog is maintained in the Information and Analytical Center for Near-Earth Space Monitoring, taking into account the data of the catalog of the American USSSN; they are available on the Internet [6, 7].

The ADC gathers all potentially dangerous convergences between all cataloged SOs. "Potentially dangerous" means either convergence of two SOs to a distance less than a given distance

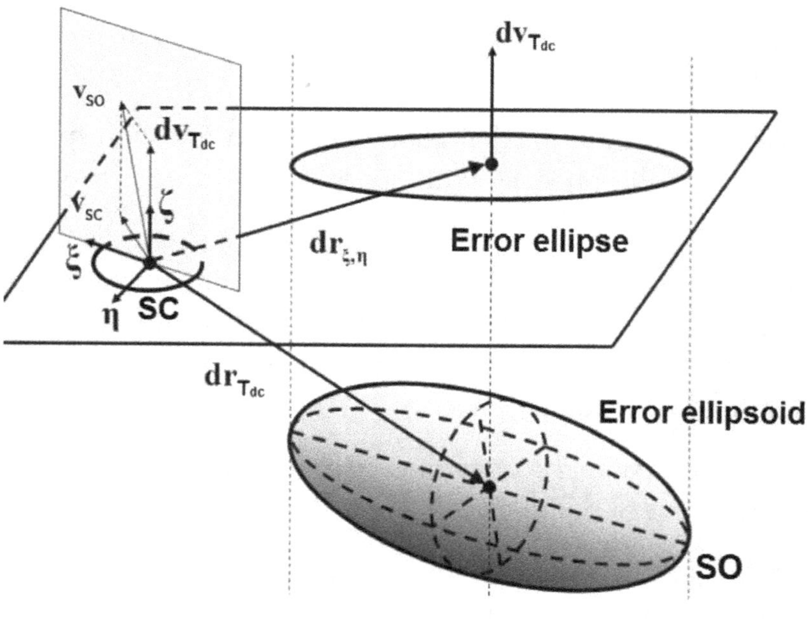

Figure 3. Geometric interpretation of the relative location of spacecraft and space objects in outer space at the time of dangerous convergence.

Δ, or a closer distance to a greater probability with a collision probability p_c greater than the threshold p_{min}. Such convergences are about ≈15,000 per day. The archive is more than 20 years old. For each convergence, the following characteristics are stored in the ADC: the convergence time, trajectory and non-trajectory parameters of the objects convergence, the residuals at the moment of convergence and their probabilistic characteristics, the probability of collision. Briefly, the algorithm for the supporting ADC is described in the book [8]. There, a method for evaluating various risk characteristics using ADC is also described. For more detail, see Ref. [9].

2.3. Modeling of space debris

At present, there are several models of the objects fragmentation at hypersonic collision. Review of the probabilistic models of space debris is represented in monograph [5]. The main result of collision of two objects with masses M1 and M2 becomes formation of the large number of fragments of various shapes, masses, and dimensions. The following features are used to describe the effect of the collision:

- *Nf* (>*m*), *Nf* (>*d*) — the number of fragments with mass more than *m*, or dimensions more than *d*. This is one of the fundamental characteristics. Some assumption about fragments shape and weight are used to recalculate the mass values to dimension values;
- *A*/*m*(*d*) — the ratio of the square of the typical cross section to mass for fragments of different sizes. This parameter is related to the difference of shapes and materials of the colliding objects; it is necessary in the analysis of the evolution of SD to calculate the deceleration of fragments in the atmosphere;
- *p*(Δ*V*) — the statistical distribution of the incremental speed of fragments by their size and direction. As a result of collision, some of the energy goes to changing speed of fragments, which leads to the spread of SD in some part of interplanetary space.

Experiments show that hundreds and even thousands of space debris objects formed in collisions with the satellites. In 2009, the collision of a communications satellite Iridium with Russian satellite Kosmos-2251 resulted in about 600 shards that flew at an altitude from 500 to 1300 km [10].

Space debris poses a great danger to functioning spacecrafts because of the large relative velocities of convergence (**Table 1**). In recent years, collisions with space debris have killed several spacecrafts.

Given the fact that simulation of motion of space debris is performed under conditions of essential indeterminacy of the input data and using the processing algorithms of random

Kalashnikov	KAMAZ with cargo	Space debris
Bullet weight: 7.9 g	Mass: 14.5 t	Mass of fragment: 40 g
Muzzle velocity: 715 m/s	Velocity: 90 km/h	Relative speed:15 km/s
Kinetic energy: 2 kJ	Kinetic energy: 4.5 MJ	Kinetic energy: 4.5 MJ

Table 1. Comparative analysis of the kinetic energy of objects [3].

events in large part, methods of probabilistic risk analysis and justification of preventive measures of damage reducing become most common for cosmic systems and technologies.

The movement of each element of the system of space objects can be divided into two components. First, the orbital object moves on a trajectory that can be represented in the elliptical form in the general case in the current time, which oriented in space in a certain way (osculating orbit). Second, the trajectory of the orbital object changes over time (generally, form and orientation are changing). Meanwhile, trajectories of motion of orbital objects change much slower than the position of orbital objects on these trajectories. Therefore, it is proposed to model changes of trajectories and identify the parts of the trajectories for the current moment in time, which are located from each other at a dangerous distance, from the point of view of possibility of collisions of orbital objects (nodes of mechanical conflicts). In other words, to simulate the nodes of mechanical conflicts, speed changing of which corresponds to speed changing of trajectories. For orbital objects, trajectories of which form a node conflict, it is necessary to determine the time intervals of their movement through the node conflicts without a significant investment of time (on the dangerous part of trajectory). Hence, the method of modeling system of orbital objects is based on the method of modeling the nodes of mechanical conflicts and the method of determining the time intervals of movement of the orbital object through a node of conflicts.

Tasks of the analysis of conflicts of orbital objects can be divided into two classes. In the first class, there are the tasks, where it is possible to analyze only the risk of collisions and not to predict specific orbital conflicts. Their solution is based on the consideration of the altitude-latitudinal density distribution of the orbital objects at a specific point in time.

In the second class, there are the tasks that demand prediction of orbital collisions. This prediction boils down to the prediction of convergence of pairs of orbital objects at a dangerous distance, from the point of view of their collision at possible deviations of objects from their calculated trajectories (these can be called dangerous or conflict convergences). In many cases, it is sufficient to predict only dangerous convergence of the orbital objects, and not to simulate the effects of conflicts, which change the trajectories of the colliding objects and form new orbital objects. Such tasks are solved when it is necessary to predict dangerous collisions for spacecraft, which can make the maneuver to avoid collisions. The task of prediction of dangerous convergence can be used as a base for models of near-Earth space contamination by orbital objects. The direct deterministic method is the most common. It is based on the formation of an archive of dangerous convergences of all possible pairs of orbital bodies at a specified time interval, which is included in the considered set of orbital objects (for each dangerous convergence, the passing time interval, the geometric characteristics of convergence and the probability of collision are determined).

The traditional method to predict dangerous convergence is based on modeling the movement of objects and analyzing the current distance between them. There is a difficulty in this method. The relative speed of orbital objects can be more than 10 km/s. Meanwhile, the convergence at a dangerous distance of several kilometers lasts less than 1 s. Therefore, the prediction of dangerous convergences requires modeling with a correspondingly small time step. At larger sizes of sets of orbital objects, it leads to significant time consumption.

An effective way to solving this problem is the implementation of the prediction of dangerous convergences in several stages. Each stage is the check of the possibility of dangerous convergence based on some rule or the simplified method of prediction.

There are three stages of checking the possibilities of dangerous convergence implemented for a given pair of orbital objects. In the first stage, the overlapping region of heights above the Earth's surface, where their trajectories pass is checked. The second stage is based on the fact that the conflict between orbital objects is possible only when their trajectories intersect. It is assumed that the orbital object cannot deviate from the position on the calculated trajectory more than a certain distance. Hence, in each moment of time, the orbital object may be within a sphere, which has the center of the calculated trajectory, and radius is R_{cr}. A pair of sections of trajectories will be called a node of the mechanical conflicts, if that trajectories are located on the distance $L < L_{cr}$ from each other. If for a pair of trajectories of orbital objects there is defined a node of conflict, then the condition of the second stage is fulfilled. The third stage is based on the fact that the conflict convergence is possible during simultaneous motion of orbital objects on segments of trajectories, which form a node of conflict. In the third stage, the time intervals of orbital objects' motion through the node of conflicts are defined. If these time intervals overlap each other, then dangerous convergence is possible, and its probability can be calculated.

Assume [11] that the set of the cataloged orbital objects is considered as a multi-element mechanical system. There are some quasiregular components in the movement of the elements of this system. Meanwhile, the interactions of the elements of the system are not taken into account. Such restrictions allow the allocation of the node of conflict at the current time, which is formed by the dangerous parts of the trajectories of orbital objects k and l. This node of conflict restricts the dangerous part of the trajectory of each of these orbital objects. Considering the regularity of the motion parameters of the objects allows to simulate space debris as a combination of deterministic and probabilistic models.

3. Summary

Probabilistic modeling is an important method of risk analysis and justification of preventive measures for space systems and technologies.

Methods of calculating ballistic trajectories and assessment of collision risks with space debris are based on conversion of inaccurate source data, results of which are random variables. Therefore, the risk of collision can be specified as a probability measure.

Preventive measures for controlled spacecraft are reduced to change of trajectory, which could prevent or reduce this risk.

4. Conclusion

This section contains examples of the probabilistic risk analysis of collision of satellites with space debris.

It is shown how to predict the dangerous convergence between space objects and to justify preventive measures to reduce collision risk, solving tasks of modeling with random parameters.

The approach to constructing a probabilistic mathematical model of a complex system based on the principles of functional integration of the models of elements and subsystems in a single integrated software and implemented algorithms to perform the simulation processes for different input conditions and current state of the real system is given.

Modeling the motion of such satellites typically boils down to the obtainment of systems of ordinary differential equations of object motion and their integration by any method. The result is a dependence of the parameters of motion from time under given initial conditions. These equations are the form of representation of the laws of dynamics and kinematics, and can be supplemented with the equations of control.

Considering the catalogs of space objects (such catalogs exist in Russia and the United States), it is possible to estimate their relative position and to forecast their movement. In particular, it is possible to assess the threat of convergence and even the collision of spacecrafts.

Author details

Nikolay Paramonov

Address all correspondence to: paramonov_n_b@mail.ru

Moscow Technological University Mirea, Russia

References

[1] Melrae Pictures, Space Junk 3D [Online image]: Retrieved January 11, 2017 from http://www.spacejunk3d.com/

[2] Paramonov NB, Tokarev DA. Preliminary simulation of systems. Herald of MSTU MIREA. 2015;**4**(9):165-170

[3] Kozoriz FI, Skornyakov VA. Assessment of csollisions in the approach of the ISS to the observed objects. Lesnoy vestnik. 2009;**2**:164-167

[4] Space environment (natural and artificial). Model of spatial and time distribution for space debris in LEO. GOST-25645.167-2005

[5] Nazarenko AI. Modeling of Space Debris. Moscow: IKI RAS; 2013. 216 p

[6] Space track catalog of objects [Internet]. 2017. Available from http://www.space-track.org [Accessed: January 11, 2018]

[7] CelesTrack [Internet]. 2017. Available from http://celestrack.com[Accessed: January 11, 2018]

[8] Agapov V. Space debris, Book 1. Methods of observation and models of space debris, Chapter 3, Moscow: Fizmatlit, 2013. 248 p

[9] Khutorovsky ZN, Kamensky SY, Boikov VF, Smelov VL. The risk of collision of space objects at low altitudes, Collisions in near-Earth space, Collection of scientific works, RAS, Institute of Astronomy, 1995

[10] Zverev PS, Dovgal VM. Method and algorithm of recognition of artificial Circum-terraneous orbital objects and "dust" for support of safety of space flights. Vestnik VGTU. 2010;**6**(4):105-109

[11] Labutkina TV, Petrenko AN. New aspect of design of multiple-unit system of orbit objects. Vestnik NTU "HPI". 2013;**19**:60–65

3

Periodic Monitoring and Recovery of Resources in Information Systems

Alexey Markov, Alexander Barabanov and Valentin Tsirlov

Abstract

This section deals with the issues of business continuity and recovery after disasters. The authors analyzed standards, laws, and regulations pertaining to the parameters of periodic monitoring and recovery in information systems. This section includes mathematical models of resources and environment periodic monitoring as well as periodic backup and recovery after interruptions or disasters. The work demonstrates that the well-known deterministic periodic monitoring and backup models do not take into account stochastic peculiarities of ergatic systems to the full extent. The authors developed new stochastic models of restricted monitoring and backup that allow taking into consideration resources constrains and random factors of information systems operation. The notion of Bernoulli stream has been introduced. This section suggests the criteria for selecting deterministic or stochastic monitoring and backup models and their combinations. A solution of direct and reverse task of the calculation of control and monitoring procedures frequency is offered. This section also provides a methodology for information system stability management, considering periodic monitoring, rollback, and recovery in case of interruption.

Keywords: business continuity, backup, rollback, recovery, regular procedures, limited stochastic control, Bernoulli flow, stochastic models, deterministic models, periodic inspection, stochastic redundancy

1. Introduction

Basic business continuity planning and disaster recovery procedures include periodic monitoring (control) of resource integrity and periodic backup [1–4].

Requirements for periodic monitoring and backup established by current regulatory documents are briefly described subsequently.

2. Parameters of periodic monitoring and recovery in information systems

2.1. Periodic monitoring parameters

The main parameters of periodic monitoring and recovery in protected information systems (ISs) are provided as follows:

- frequency of monitoring (internal monitoring) of security functions operability of information security controls used in the information systems;
- frequency of external (external audit) of security functions operability of information security controls applied in the information systems; and
- update frequency of the information system parameters and characteristics relating to the information security (change of passwords, update of the information security controls decision rules or signatures).

The results of completed analysis are shown in **Table 1**.

Name of document	Frequency of internal monitoring	Frequency of external monitoring	Frequency of parameters update
ISA 62443–3-3:2013	+	+	+
ISO/IEC 27001:2013/ISO/IEC 27002:2013	+	+	+
PCI DSS	+ (6 months)	+ (6 months)	+ (90 days)
Australian Government Information Security Manual. Controls[1] (Australia)	+	+	+ (90 days)
The IT-Grundschutz Catalogs[2] (Germany)	+	+	+
Cyber Essentials Scheme Requirements for basic technical protection from cyber attacks[3] (Great Britain)	—	—	+
Information Security Provisions in Federal Information Systems[4] (Russia)	+	+	+ (180, 120, 90, and 60 days)
Requirements for Information Security in Process Control Systems (Russia)	+	+	+ (180, 120, 90, and 60 days)
NIST SP 800–53[5]/NIST SP 800-63B[6] (USA)	+	+	+

[1]https://www.asd.gov.au/publications/Information_Security_Manual_2017_Controls.pdf
[2]https://download.gsb.bund.de/BSI/ITGSKEN/IT-GSK-13-EL-en-all_v940.pdf
[3]https://www.gov.uk/government/uploads/system/uploads/attachment_data/file/647,619/requirements_archived.pdf
[4]http://nvlpubs.nist.gov/nistpubs/SpecialPublications/NIST.SP.800-63b.pdf
[5]https://nvd.nist.gov/800-53/Rev4
[6]http://nvlpubs.nist.gov/nistpubs/SpecialPublications/NIST.SP.800-63b.pdf

Table 1. Requirements for the periodicity of control.

2.2. Periodic backup parameters

In practice [1, 2, 4, 5], the main parameter defining the frequency of periodic information backup is the recovery point objective (RPO)—the maximum period of data loss occurring due to an information security incident. The value recovery time objective (RTO) is the period of the information system unavailability in case of the information security incident. The value of RPO defines the backup frequency (**Figure 1**).

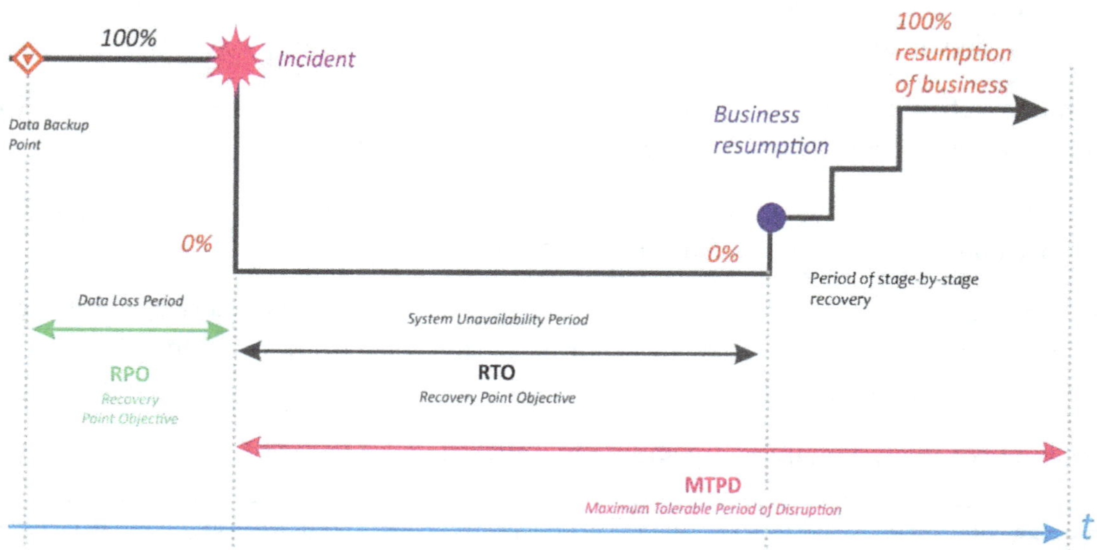

Figure 1. Diagram of the system operation and incident recovery.

Document name	Requirements for periodic backup	Quantitative values or calculation formulae
ISA 62443-3-3:2013	+	–
ISO/IEC 15408	+	–
ISO/IEC 27001:2013/ ISO/IEC 27002:2013	+	–
Australian Government Information Security Manual Controls	+	+
The IT-Grundschutz Catalogs	+	–
Cyber Essentials Scheme Requirements for basic technical protection from cyber attacks	–	–
GOST R 56939	+	–
Information Security Provisions in Federal Information Systems	+	–
Requirements for Information Security in Process Control Systems	+	–
NIST SP 800–53/NIST SP 800–34	+	–
Framework for Improving Critical Infrastructure Cybersecurity[1]	+	–

[1]www.nist.gov/sites/default/files/documents/cyberframework/cybersecurity-framework-021214.pdf

Table 2. Requirements for backup frequency.

The analytical review of regulatory documents and methodologies defining the requirements for information security relating to periodic backup and recovery is shown in **Table 2**.

As the completed review shows (**Tables 1** and **2**), there are clear requirements for periodic monitoring and backup though their main parameters are defined either by expert judgments or by management order.

Considering high subjectivity of such decisions, it is reasonable to develop mathematical models for the calculation of periodic monitoring and backup parameters.

3. Mathematical models of periodic monitoring and backup

As noted earlier, the basic mechanism for providing functional stability to information systems (ISs) is systematic monitoring and backup against possible failures. There are two key approaches to arranging monitoring in IS. The first one relates to the occurrence of a certain event of the computation process (message processing, initiating an exchange among processes, system program call, etc.) [6]. This approach's drawbacks are the difficulty in identifying a set of controlled events of the computation process and the potential for unlimited growth of control points. The latter makes the approach hard to apply during IS normal operation, given the specified resource and task-time restrictions.

The second approach involves a periodic check of the system at predetermined intervals [7–10]. This is consistent with time schedules and allows the existing resource restrictions to be taken into consideration, but fails to fully reflect the stochastic nature of the occurrence of various errors and irregularities. Furthermore, a number of subjective factors make it impossible, in the first place, to organize periodic control in ergatic systems at strictly specified intervals. There is another approach, however, that takes into account the stochastic external factors of IS functional stability, given the specified time and economic constraints.

Under ISO/IEC 15408–1:2009,[1] monitoring covers not only SW (assessment object) but also the operational environment. Let us consider stochastic and deterministic models of the earlier procedures.

3.1. Periodic resource monitoring models

Let us conditionally present the IS software (SW) operating process as alternating flows of errors $I(y)$, normal operation recovery $I(z)$, failures $I(Q)$, and SW/environment control (**Figure 2**).

Being mutually alternative, the flows of failures and normal system operation recovery result from the flow of errors and are shifted with respect thereto by the values $Q(t)$ and $z(t)$. The maximum of these values determines the manifestation of a respective flow. Assuming the recovery time to be instantaneous, the normal operation recovery flow may be considered part

[1]ISO/IEC 15408–1:2009: IT—security techniques—evaluation criteria for IT security.

of the control flow. In this case, the task of providing SW functional stability comes down to that of optimizing restricted control that meets the condition $z(t) < Q(t)$.

Let us consider the SW life cycle period t, having regard to the conducted inspection control of repeatable accuracy. Because the period t far exceeds the control time, let us assume the latter to be instantaneous. Then, the SW repeatable accuracy is characterized by the probability $P(\hat{z} < Q) = F\hat{z}\ (Q)$ that the irregularity/vulnerability/error detection time $\hat{z}$ within the inter-control interval is not longer than the permissible SW life cycle period Q, where there is an irregularity. A periodic control fragment is shown in **Figure 3**.

Let us consider the flow of irregularities (errors and vulnerabilities) to be the simplest one with the density of interval $\hat{y}$ distribution among them:

$$g_{\hat{y}} = \lambda e^{-\lambda y} \tag{1}$$

where λ is the intensity of irregularities.

Let us define a stochastic model for the detection of irregularities. In this case, control is undertaken a certain number of times with equal probability and independently of one another. Thus, the limited flow formed by all the control points is **Bernoulli's flow** with the density of interval $\hat{T}$ distribution among the control points [11]:

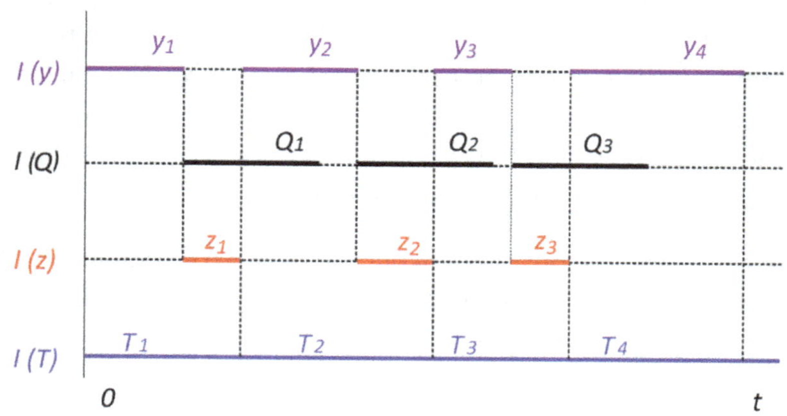

Figure 2. Flows of errors, failures, recovery, and control.

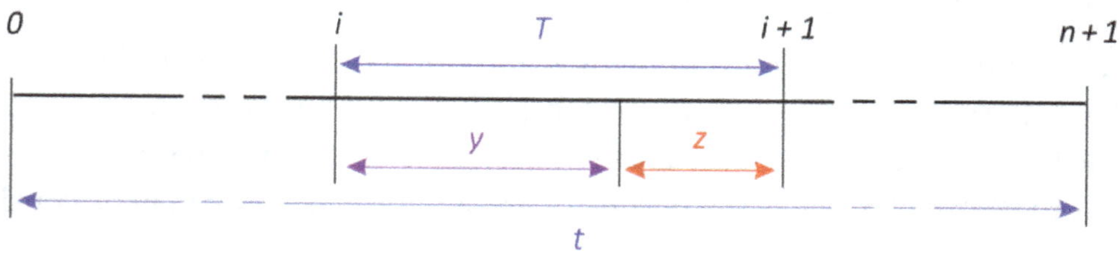

Figure 3. A fragment of the inspection control of an information security tool.

$$f_{\hat{T}} = \mathrm{n/t}(1 - T/t)^{n-1} \quad (2)$$

where n is the number of control points.

The delay time $\hat{z} = \hat{T} - \hat{y}$ is a function of two stochastic variables and has the following distribution function:

$$F_{\hat{z}}^{s} = \iint_{S)} \frac{n}{t\left(1 - \frac{T}{t}\right)^{n-1}} \lambda e^{-\lambda y} dT dy;\ (t > 0, n > 0) \quad (3)$$

Having defined the integration limit (**Figure 4**), we obtain the following:

$$\mathrm{F}_{\hat{z}}^{\mathrm{s}} = \int_0^{\mathrm{t-z}} \left(\int_{\mathrm{y}}^{\mathrm{y+z}} \mathrm{n/t}(1 - \mathrm{T/t})^{\mathrm{n-1}}\, \lambda \mathrm{e}^{-\lambda \mathrm{y}} \mathrm{dT} \right) \mathrm{dy} + \int_{t-z}^{t} \left(\int_{y}^{t} n/t(1 - T/t)^{n-1}\, \lambda e^{-\lambda y} dT \right) dy \quad (4)$$

After simplifying (Eq. (4)), we have the following formula:

$$F_{\hat{z}}^{s} = \lambda / t^{n} \left(\int_0^{t-z} e^{-\lambda y}((t-y)^{n} - (t-z-y)^{n}) dy + \int_{t-z}^{t} e^{-\lambda y}((t-y)^{n}) dy \right) \quad (5)$$

Having expanded the formula integrands as a power series, we obtain an approximate value of the distribution function that is the basic computational ratio:

$$F_{\hat{z}}^{s} = \lambda \sum_{i=0}^{n} \sum_{j=0}^{r} \sum_{l=1}^{n+j+1} (-1)^{i+j+l+1} C_n^i C_{n+j+1}^l \frac{\lambda^j t^{j+1-l} z^l}{j!(1+j+i)} \quad (6)$$

where r is the number of iterations.

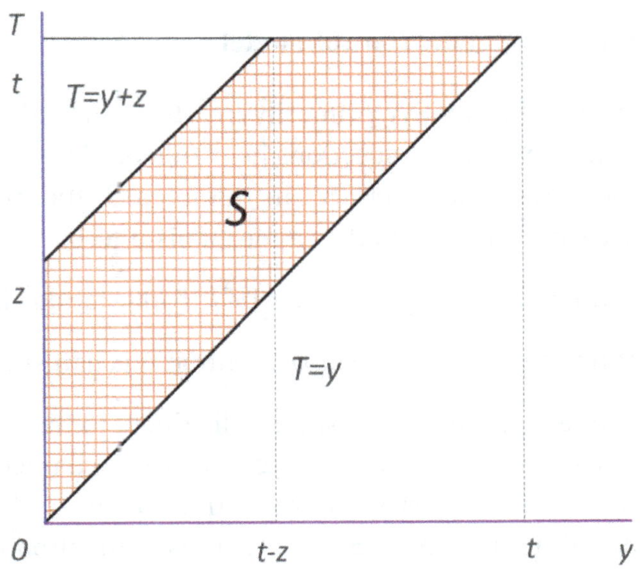

Figure 4. Domain of integrating the irregularity detection time delay interval.

In order to compare stochastic and deterministic models, let us elaborate on the latter. The deterministic model's control points form a regular flow with a constant value of the interval $T = t/(n + 1)$ and the irregularity detection time distribution density:

$$g_{\hat{z}} = \lambda e^{-\lambda(T-z)}; \quad (0 < z < T) \tag{7}$$

It can be shown that the expression for the distribution function in the deterministic model is as follows:

$$F^{d}_{\hat{z}} = e^{-\lambda T}\left(e^{\lambda z} - 1\right); \quad (0 < z < T) \tag{8}$$

Comparison of the expressions for the models (Eqs. (4) and (8)) suggests that the models under review conform to the process of detecting SW repeatable accuracy disturbances (at specified values λ, Q, t, and n):

$$F^{s}_{\hat{z}}(z) \lesseqgtr F^{d}_{\hat{z}}(z) \tag{9}$$

The irregularity detection probability P_n for the SW life cycle period t can be presented as follows:

$$P_n = (n + 1) \cdot F_{\hat{z}} \tag{10}$$

where n is the number of inspection control points $(n > 0)$, $F_{\hat{z}} = \max\left(F^{s}_{\hat{z}}(z), F^{d}_{\hat{z}}(z)\right)$.

A review of the models discussed earlier showed an advantage of the stochastic model, given a small number of inspection control points. Conceptually, it can be accounted for by the fact that even with a small number of random points of SW characteristics control, there is always a likelihood that an irregularity is detected once it has occurred, whereas in the case of the deterministic model, the inspection period may not be less than the specified value.

3.2. Operational environment periodic control model

The control of restrictions imposed on SW primarily involves inspecting SW environment and operation/production conditions. Such inspections help rule out irregularities (errors, vulnerabilities) concerning the SW front-end interface. In this regard, the procedures for detecting environment irregularities can be interpreted as a mechanism to prevent SW irregularities.

Environment control requirements are specified by ISO 15408 standards.

Let us consider IS operation where an SW error prevention mechanism is available.

When developing environment control models, we will adhere to the approach outlined in the previous section. We will assume SW repeatable accuracy to be characterized by the probability $P(\hat{z} < Q) = F\hat{z}(Q)$ that the preliminary control $\hat{z}$ time between the environment control point and a possible point of occurrence of SW characteristic disturbance does not exceed the permissible time Q. Let us define a stochastic model of environment irregularity control (**Figure 5**).

It can be shown that the preliminary control time is a function of two random values $\hat{z} = \hat{y} - \hat{T}$ and has the following distribution function:

$$F_{\hat{z}}^{s} = \iint_{(S)} \frac{n}{t\left(1 - \frac{T}{t}\right)^{n-1}} \lambda e^{-\lambda y} dT dy; \qquad (t > 0, n > 0) \tag{11}$$

where n is the number of environment control points and λ is the SW characteristic disturbance intensity.

Having defined integration limits (**Figure 6**) and simplified the expression, we obtain the following:

$$F_{\hat{z}}^{s} = \int_{0}^{z} f\hat{T}(T) e^{-\lambda T}\left(1 - e^{-\lambda T}\right) dT + \int_{z}^{t-z} f\hat{T}(T) e^{-\lambda T}\left(1 - e^{-\lambda z}\right) dT + \int_{t-z}^{t} f\hat{T}(T) e^{-\lambda T} dT - e^{-\lambda t}\left(\frac{z}{t}\right)^{n}, \tag{12}$$

Having expanded the formula integrands as a power series, we obtain an approximate value of the distribution function that is the basic computational ratio:

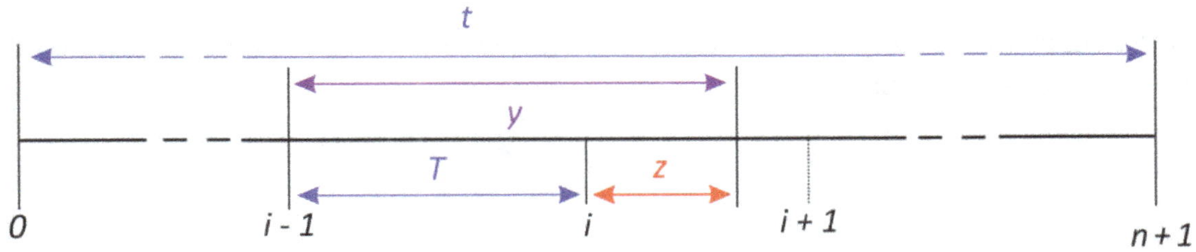

Figure 5. Operation of the system, with an irregularity error prevention mechanism available.

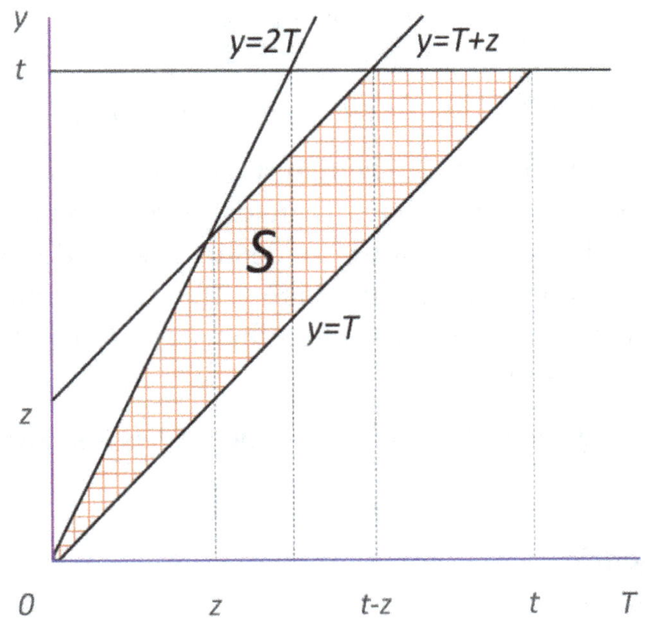

Figure 6. Domain of integrating the irregularity prevention time interval.

$$F_{\hat{z}}^{s} \approx n \sum_{i=0}^{r} \left(\sum_{j=0}^{n-1} b_1 b_2 \right) - e^{-\lambda t} \left(\frac{z}{t} \right)^n; \qquad (t > 0, n > 0) \tag{13}$$

where r is the number of iterations; $b_1 = (-1)^{-i+j} C_{n-1}^{i} \frac{\lambda^j}{(j! t^{i+1} (i+j+1))}$; $b_2 = t^{i+j+1} - z^{i+j+1}$ $(2^j - e^{-\lambda z})(t - z)^{i+j+1} e^{-\lambda z}$.

Let us compare the obtained stochastic model and the deterministic one. The deterministic model's control points form a regular flow with a constant value of the interval $T = t/(n + 1)$ and the following preliminary control time distribution density:

$$g_{\hat{z}} = \lambda e^{-\lambda(T+z)}; \qquad (0 < z < T) \tag{14}$$

Hence, the expression for the distribution function in the deterministic model will be as follows:

$$F_{\hat{z}}^{d} = e^{-\lambda T} \left(1 - e^{-\lambda z}\right); \qquad (0 < z < T) \tag{15}$$

By comparing computational model expressions at specified values λ, *Q*, *t*, and *n*, we obtain a criterion to choose a model:

$$F_{\hat{z}}^{s}(z) \lesseqgtr F_{\hat{z}}^{d}(z) \tag{16}$$

The probability P_n of irregularity prevention for SW life cycle period t can be presented as follows:

$$P_n = (n+1) \cdot F_{\hat{z}} \tag{17}$$

where n is the number of control points $(n > 0)$, $F_{\hat{z}} = \max\left(F_{\hat{z}}^{s}(z), F_{\hat{z}}^{d}(z)\right)$.

Comparative analysis of stochastic and deterministic models showed the former's effectiveness with a small number of control points. Therefore, when managing system information security by numerical methods, it is possible to identify preferred models (stochastic, deterministic, or combined) that bolster confidence in SW. This gives an effect akin to introducing structure redundancy, that is, a special type of redundancy—**stochastic**—the use of which is unlikely to result in higher costs [11].

An example of comparing deterministic and stochastic models is shown in **Figure 7**.

3.3. Periodic backup models

The previous subsections dealt with deterministic and stochastic SW control models. When tackling comprehensive tasks of providing IS operational reliability and security, it is important to ensure information safety in case of incidents. This can be achieved by developing an incident management system.[2]

Apart from control models, this work also investigates backup and recovery models.

The backup mechanism is intended to recover a system's normal operation in case of a failure or an incident, such a recovery starting from the last backup time (**Figure 8**).

The backup mechanism control task boils down to developing a checkpoint (CP) setting model that minimizes the mathematical expectation of the program operation delay time, given the restrictions on the total SW operation time and the number of CP. The issues of minimizing the mathematical expectation of delay time by changing the CP setting frequency and the determined interval among checkpoints are discussed in [9].

Let us consider a situation when an interval is a random value.

If the failure flow of the computation process is regarded as simple, it can be shown that the delay time $\hat{z} = \hat{y} - \hat{T}$ is a function of two random values and has the following mathematical expectation:

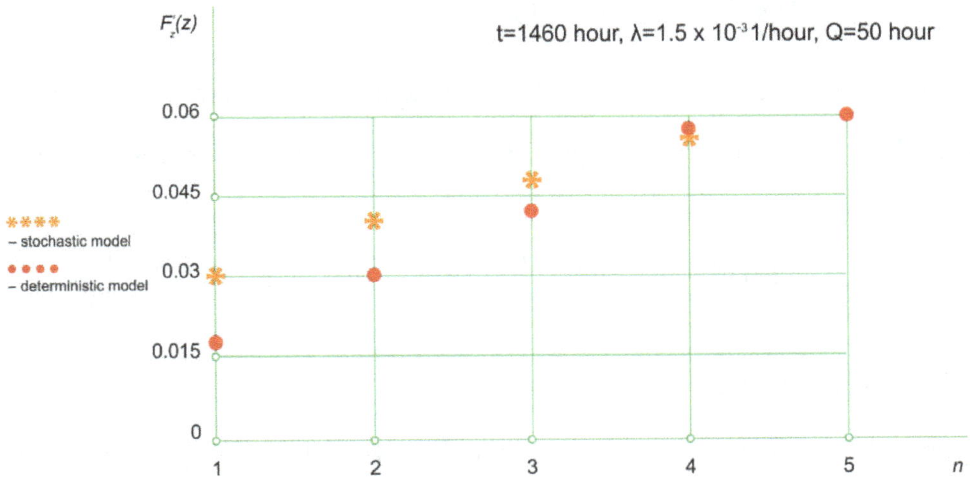

Figure 7. Irregularity prevention probability versus the number of preliminary control points.

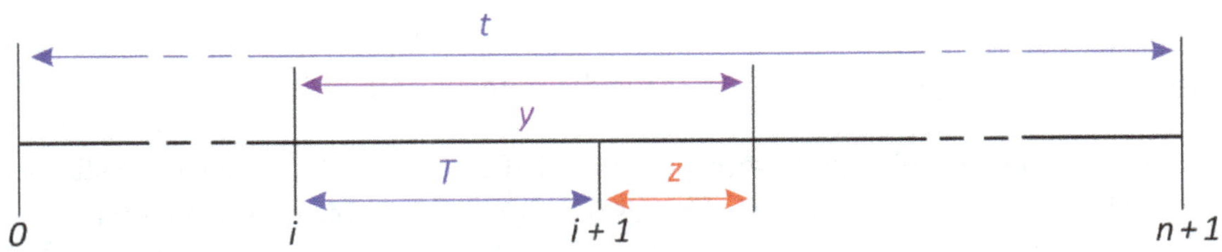

Figure 8. Program operation using a checkpoint mechanism.

[2] ISO/IEC TR 18044:2004 IT—security techniques—information security incident management.

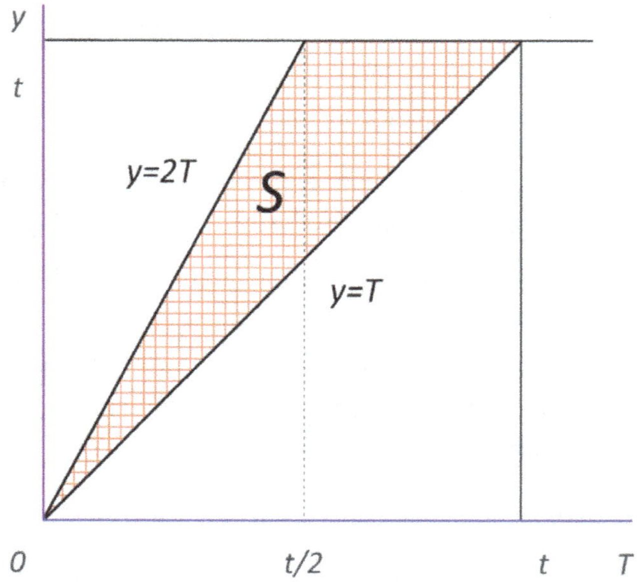

Figure 9. Domain of integrating the program operation delay time interval.

$$M_{\hat{z}}^{s} = \iint_{(S)} (y - T)\frac{n}{t}\left(1 - \frac{T}{t}\right)^{n-1} \lambda e^{-\lambda y} dTdy; \qquad (t > 0, n > 0) \tag{18}$$

where n is the number of environment control points and λ is system's failure intensity.

Having defined integration limits (**Figure 9**) and simplified the expression, we obtain the following:

$$M_{\hat{z}}^{s} = \int_0^t n/t(1 - T/t)^{n-1}\, \frac{e^{-\lambda T}}{\lambda}\, dT - \int_0^{t/2} n/t(1 - T/t)^{n-1}\, e^{-2\lambda T}\left(T + \frac{1}{\lambda}\right) dT - b_1, \tag{19}$$

where $b_1 = \frac{e^{-\lambda t}}{2^n}\left(\frac{1}{\lambda} + nt/(2n + 2)\right)$.

Having expanded the integrands as a power series, we obtain an approximate value of the mathematical expectation of delay time:

$$M_{\hat{z}}^{s} \approx n \sum_{i=0}^{n-1} \sum_{j=0}^{r} (-1)^{j+i} C_{n+1}^{i}\, \frac{\lambda^j t^j b_2}{j!} - b_1, \tag{20}$$

where $b_2 = \frac{1}{(\lambda(i+j+1))} - \frac{t}{2^{i+2}(i+j+2)} - \frac{1}{2^{i+1}\lambda(i+j+1)}$, r is the number of iterations.

In order to compare the obtained stochastic model (Eq. (20)) and the deterministic one, we consider the latter in more detail. The deterministic model's checkpoints form a regular flow with a constant value of the interval $T = \frac{t}{n+1}$. The delay time distribution density will be as follows:

$$g_{\hat{z}} = \lambda e^{-\lambda(T+z)}; \qquad (0 < z < T). \tag{21}$$

It can be shown that the expression for the mathematical expectation of delay time in the deterministic model is as follows:

$$M_{\hat{z}}^{d} = \frac{e^{-\lambda T}}{\lambda}\left(1 - e^{-\lambda T}(\lambda T + 1)\right); \qquad (0 < z < T) \tag{22}$$

By comparing the expressions (Eqs. (20) and (22)), we obtain a criterion allowing a model to be chosen at specific values λ, t, and n:

$$M_{\hat{z}}^{s}(z) \lesseqgtr M_{\hat{z}}^{d}(z) \tag{23}$$

Considering the CP setting time and restart to be instantaneous, we obtain a total SW operation time model, given the availability of the CP mechanism:

$$t'(n) = t + (n+1)M_{\hat{z}} \tag{24}$$

where t is the SW operation time, n is the number of checkpoints, and $M_{\hat{z}} = \max\left(M_{\hat{z}}^{s}(z), M_{\hat{z}}^{d}(z)\right)$ is the mathematical expectation of the SW operation delay time in case of failure.

Here is an example using the department archive data for the first half of 2017. The database (DB) was inspected seven times over this period. The inspections revealed 12 errors, all of which were corrected by standard methods, with the relevant entry made in the administrator log. The following error parameters were calculated:

- the average time between errors $M_{\hat{z}} = 43.83$ h;
- the error intensity $\lambda = 0.022$ 1/h;
- the average quadratic deviation $\delta_{\hat{z}} = 30.04$ h; and
- *Cramér-von Mises criterion* (goodness of fit) $k(n) = 0.55$.

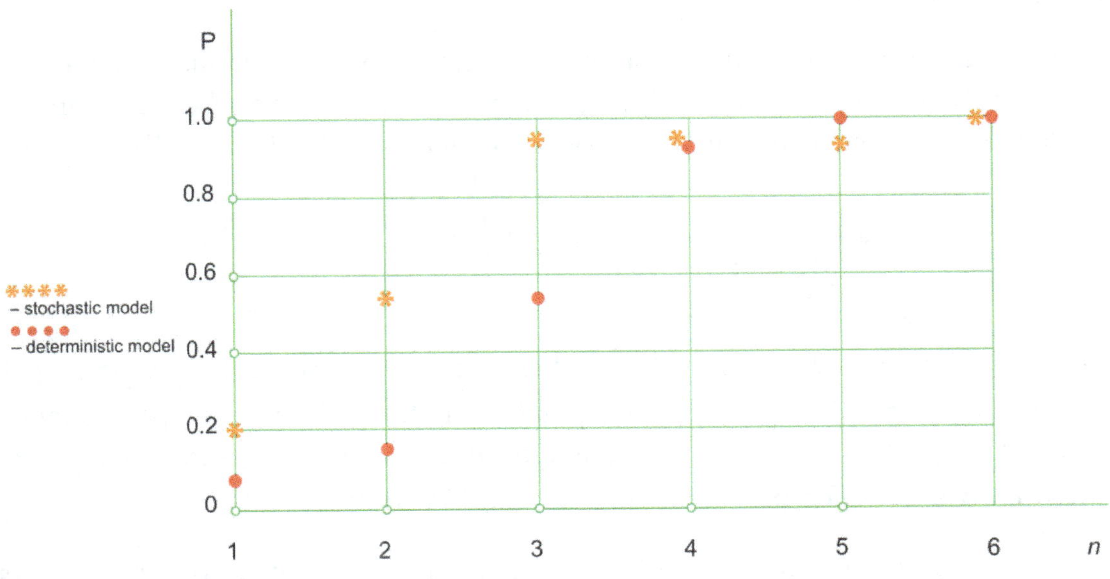

Figure 10. DB error detection probabilities versus the number of control points.

Control type	Number of control points	Man hours	Recovery probability
Conventional	7	42	1.0
Deterministic	4	24	0.99996
Stochastic	3	18	0.99997

Table 3. An example of control parameter calculation results.

This allowed the error flow to be considered a stationary Poissonian flow. A study of the electronic archive DB operation in 2000–2017 showed that the restriction on the correctable error detection time Q was more than 1 month.

The DB recovery probabilities calculated by the formulas (Eqs. (20) and (22)) and their dependence on the number of control points for the first half of 2010 ($t = 1052$ h) are shown in **Figure 10**.

Taking into account the electronic archive availability requirements, it is advisable to use only three control points when applying the stochastic model (**Table 3**).

Thus, the practical solutions offered in this work allow for the stochastic nature of DB errors. This permits the desired error detection model to be chosen at a specified DB and electronic archive parameters.

4. System functional stability management

In general, IS periodic control involves performing a number of standard procedures:

- software error control;
- operational environment error control; and
- backup in case of failure.

Choosing a strategy and the number of control/backup points helps manage the system's stability, integrity, and accessibility levels [12]. For example, considering the earlier procedures, one can define the system availability ratio (operational availability factor [13]):

$$R = \left(\frac{t}{\left(t + M^r_{n_r}\right)}\right)\left(p + (1-p)\left(P^e_{n_e} + \left(1 - P^e_{n_e}\right)P^p_{n_p}\right)\right) \tag{25}$$

where t is the task solution time, $M^r_{n_r}$ is the mathematical expectation of the program operation delay time in case of n_r being the backup points, p is the SW error-free performance (SW efficiency), $P^e_{n_e}$ is the error prevention probability in case of n_e environment control points, and $P^p_{n_p}$ is the error detection probability in case of n_p SW control points.

In the above formula, p is the SW failure-free performance probability; error prevention probability—$P^e_n = (n_e + 1)\cdot F\hat{z}$; error detection probability—$P^p_n = (n_p + 1)\cdot F_{\hat{z}}$; availability factor $R' = \left(\frac{t}{(t+M^r_{n_r})}\right)$; $M^r_{n_r} = t + (n_r + 1)M_{\hat{z}}$.

The constraining factor (Eq. (25)) is the SW dependability cost index defined as the cost of standard procedures:

$$C\left(n_e, n_p, n_r\right) = C_e\, n_e + C_p\, n_p + C_r\, n_r \tag{26}$$

where C_p is the cost of one SW error detection control event; C_e is the cost of one environment control event for error prevention; and C_r is the cost of setting one checkpoint.

The SW operation security management task comes down to optimizing the availability factor, with the constraining factor (Eq. (26)). The following two optimization tasks can be defined:

1. **Direct task:** Using partial redundancy of the number of standard procedures, ensure that the SW security index is at least equal to the specified index R_{rg}, with a minimum possible cost of standard procedures in general, that is

$$\min\left\{C\left(n_e, n_p, n_r\right) \mid R\left(n_e, n_p, n_r\right) > R_{rg}\right\} \tag{27}$$

2. **Reverse task**: Using partial redundancy of the number of standard procedures, ensure that the cost of all standard procedures does not exceed the specified value C_{rg}, with a maximum possible SW reliability index, that is

$$\max\left\{R\left(n_e, n_p, n_r\right) \mid C\left(n_e, n_p, n_r\right) < C_{rg}\right\} \tag{28}$$

4.1. Direct optimization task

Analysis (Eq. (25)) showed that R is a nondifferentiable monotone increasing function that is strictly convex upward.

In order to solve optimization tasks, therefore, it is advisable to employ sequential search methods.

Let us assume the value of incremental difference $\Delta R(, n_r)/\Delta C$ to be an enumeration criterion. Let us determine an enumeration step in accordance with the dichotomy rule. In this case, the computational scheme for solving the direct task can be presented as follows:

1. Define a set of initial values of the number of standard procedures:

   $N_o = \{n_0{}^i, n_0{}^j, n_0{}^k\}$, where $i,j,k \in \{e,p,r\}$.

2. Calculate the initial value R:

   $R_0 = R(N_o)$.

3. If $R_0 > R_{rg}$, perform the following operations:

3.1. Find a set of search interval values

$L_0 = \{ L_0^i, L_0^j, L_0^k\}$, where $L_0^l = n_0^l - n_{lw}^l$; n_{lw}^l is the lower boundary n^l.

3.2. Obtain the set of possible values of the number of standard procedures

$N_1^i = \{n_1^i, n_0^j, n_0^k\}$, where $n_1^i = n_0^i - \|L_0^i/2.\|$

3.3. Find another set of values of the number of standard procedures

$N_1 = N_1^i$, where i is the index of the standard procedure conforming to the minimization condition:

$\min(R_0 - R(N_1^i)/(C^i\|L_0^i/2\|))$, where $i,j,k \in \{e,p,r\}$.

4. If $R_0 < R_{rg}$, perform the following operations:

4.1. Find a set of search interval values

$L_0 = \{ L_0^i, L_0^j, L_0^k\}$, where $L_0^l = n_u^l - n_0^l$; n_u^l is the upper boundary n^l.

4.2. Obtain three sets of possible values of the number of standard procedures

$N_0^i = \{n_1^i, n_0^j, n_0^k\}$, where $n_1^i = n_0^i + \|L_0^i/2\|$.

4.3. Find another set of values of the number of standard procedures

$N_1 = N_1^i$, where i is the index of the standard procedure conforming to the maximization condition:

$\max(R(N_1^i) - R_0 - /(C^i\|L_0^i/2\|))$, where $i,j,k \in \{e,p,r\}$.

5. Increase the iteration index

$\tau = \tau + 1.$

6. Calculate the value R

$R_\tau = R(N_\tau).$

7. Find a set of search interval values

$L_\tau = \{ L_\tau^i, L_{\tau-1}^j, L_{\tau-1}^k\}$, where $L_\tau^i = \|L_{\tau+1}^i/2\|$.

8. If $R_0 > R_{rg}$, perform the following operations:

8.1. Obtain a set of possible values of the number of standard procedures

$N_\tau^i = \{n_{\tau+1}^i, n_\tau^j, n_\tau^k\}$, where $n_{\tau+1}^i = n_\tau^i - \|L_\tau^i/2\|$.

8.2. Find another set of values of the number of standard procedures

$N_\tau = N_{\tau+1}^i$, where i is the index of the standard procedure conforming to the minimization condition:

$\min(R_0 - R(N_{\tau+1}{}^i)/(C^i\|L_\tau^i/2\|))$, where $i,j,k \in \{e,p,r\}$.

If $N_{\tau+1} = N_{\tau-1}$, withdraw from the procedure.

9. Otherwise ($R_\tau < R_{rg}$), perform the following operations:

9.1. Obtain the set of possible values of the number of standard procedures:

$N_{\tau+1}{}^i = \{n_{\tau+1}{}^i, n_\tau{}^j, n_\tau{}^k\}$, where $n_{\tau+1}{}^i = n_\tau{}^i + \|L_\tau{}^i/2\|$.

9.2. Find another set of values of the number of standard procedures

$N_{\tau+1} = N_{\tau+1}{}^i$, where i is the index of the standard procedure conforming to the maximization condition:

$\max(R(N_{\tau+1}{}^i - R_0)/(C^i\ \|L_\tau^i/2\|))$, where $i,j,k \in \{e,p,r\}$.

9.3. If $N_{\tau+1} = N_{\tau-1}$, record the value $R_{\tau+1} = R\ (N_{\tau-1})$ and withdraw from the procedure.

10. Proceed to item 5.

The period of this computation scheme can be reduced as follows:

- by specifying the effective initial values, for example, by using personnel's experience (knowledge) or statistically accumulative tables;
- by reducing the calculation of standard procedure indices to their calculation only as per deterministic models. This is acceptable with a great number of standard procedures (more than 5–20) when stochastic models are less effective than deterministic ones.

4.2. Reverse optimization task

The reverse task can be solved using the branch-and-bound procedure. In this case, the computation scheme will be as follows:

1. Specify a cost-ordered set N of initial values of the number of standard procedures

$N_\tau = \{n_\tau{}^1, n_\tau{}^2, n_\tau{}^3\}$, $C^1 \geq C^2 \geq C^3$,

which meets the normalization requirement:

$0 \leq C_{rq} - \sum_{i=1}^{3} (n\, C^i) \leq C^i;\ i = \overline{1;3}$,

where τ is the ramification index;

2. Calculate the maximum value R by directed enumeration n^2at the fixed value $n^1 = n_\tau^1$ and the initial value $n^2 = n_\tau^2$:

$R(N_\tau) = \max(R \mid n^1 = n_\tau^1)$,

where N_τ meets the normalization requirement;

3. Calculate the maximum value R by directed enumeration n^2at the fixed value $n^1 = n_\tau^1 + 1$ and the initial value $n^2 = n_\tau^2$:

$R(N_{\tau+1}) = \max(R \mid n^1 = n^1_\tau + 1)$,

where $N_{\tau+1}$ meets the normalization requirement;

4. Calculate the maximum value R by directed enumeration n^2 at the fixed value $n^1 = n^1_\tau - 1$ and the initial value $n^2 = n^2_\tau$:

   $R(N_{\tau-1}) = \max(R \mid n^1 = n^1_\tau - 1)$, where $N_{\tau-1}$ meets the normalization requirement;

5. If $R_\tau = \max(R_{\tau-1}, R_\tau, R_{\tau+1})$, withdraw from the computation scheme;
6. If $R_{\tau+1} = \max(R_{\tau-1}, R_\tau, R_{\tau+1})$, let $\tau = \tau + 1$, perform item 3 and proceed to item 5;
7. If $R_{\tau-1} = \max(R_{\tau-1}, R_\tau, R_{\tau+1})$, let $\tau = -1$ and proceed to item 4.

In practice, there may be a task of calculating indices not for the SW functional stability (dependability) system in general but for a part thereof (checkpoint or error prevention/detection mechanisms). This means a transition from multidimensional to unidimensional task interpretation, which helps substantially simplify computational procedures. Thus, solving a partial reverse optimization task boils down to a single calculation of a specific index when $n = C_{rq}/C$.

When solving a direct task, the effectiveness of the computation scheme can be additionally improved by adjusting the variable change interval, for example, by defining the next variable value in accordance with a distribution law, and so on.

5. Conclusion

Thus, in this section, we have considered stochastic and deterministic models of SW periodic monitoring and backup, which allow for time and computational/data resource constraints. Representing monitoring and backup points as a restricted Bernoulli's flow helps obtain random time intervals with the preset number thereof and, accordingly, allow for the effect of stochastic external factors on the system operation process.

Comparative analysis of stochastic and deterministic models showed the former's effectiveness with a small number of control and backup points. Therefore, when managing IS stability by numerical methods, it is possible to identify preferred models (stochastic, deterministic, or combined) which enhance IS functional stability. This gives an effect akin to introducing structure redundancy, that is, a special type of redundancy (stochastic), the use of which is unlikely to result in higher costs. The application of stochastic models in engineering systems can be facilitated by using a random-impulse generator that forms random-restricted Bernoulli's flows [11].

A similar approach was taken as a basis to solve the problem of efficiency assessment of the diagnostic mechanism for data array failures. Apart from the IS resource control and backup domain, the above-stated results can be of use in assessing the cost-effectiveness of control measures and mechanisms being implemented in various engineering and management systems. For better use of stochastic models, it is possible to use a random pulse generator (e.g., [14]).

Author details

Alexey Markov[1]*, Alexander Barabanov[2] and Valentin Tsirlov[2]

*Address all correspondence to: mail@cnpo.ru

1 Bauman Moscow State Technical University, Moscow, Russia

2 NPO Echelon, Moscow, Russia

References

[1] Bird L, Higgins D, editors. Good Practice Guidelines 2013—Global Edition: A Guide to Global Good Practice in Business Continuity. 3rd ed. The Business Continuity Institute; 2013. 115 p

[2] Engemann KJ, Henderson DM. Business Continuity and Risk Management: Essentials of Organizational Resilience. Rothstein Associates; 2011. 370 p

[3] Pompon R. IT Security Risk Control Management: An Audit Preparation Plan. 1st ed. Apress; 2016. p. 311

[4] Stewart JM, Chappie M, Gibson D. CISSP: Certified Information Systems Security Professional Study Guide. 7th ed. Sybex; 2015. p. 1104

[5] Cummings D. Dataflow-Based Rollback Recovery in Distributed and Multi-Core Systems: A Novel Software Approach for Building Highly Reliable Distributed and Multi-Core Systems. Müller: VDM Verlag Dr; 2009. 236 p

[6] Garcia E, Antsaklis PJ, Montestruque LA. Model-Based Control of Networked Systems. Cham: Birkhäuser; 2014. 382 p. DOI: 10.1007/978-3-319-07803-8

[7] Getta JR. Discovering irregular periodic patterns in audit data. In: 2016 2nd IEEE International Conference on Computer and Communications (ICCC); 14–17 October 2016; Chengdu, China: IEEE; 2016. 16867467. DOI: 10.1109/CompComm.2016.7924671

[8] Huai L, Qiushi L, Jianxin H, Tongzhou J. A novel fault-tolerant scheduling algorithm for periodic tasks of distributed control systems. In: 2009 Chinese Control and Decision Conference (CCDC '09); 17–19; June 2009; Guilin, China: IEEE; 2009. p. 1584–1588. DOI: 10.1109/CCDC.2009.5192227

[9] Kostogryzov A, Nistratov G, Nistratov A. Some applicable methods to analyze and optimize system processes in quality management. In: Aized T, editor. Total Quality Management and six Sigma. InTech; 2012 Chapter 7. DOI: 10.5772/46106. Available from: https://www.intechopen.com/books/total-quality-management-and-six-sigma/some-applicable-methods-to-analyze-and-optimize-system-processes-in-quality-management

[10] Yin JY, Guo G-C, Wu Y-X. A Hybrid Fault-Tolerant Scheduling Algorithm of Periodic and Aperiodic Real-Time Tasks to Partially Reconfigurable FPGAs. In: 2009 International

Workshop on Intelligent Systems and Applications (ISA 2009); 23–24 May 2009; Wuhan, China: IEEE; 2009. p. 1–5. DOI: 10.1109/IWISA.2009.5072624

[11] Markov AS, Kernozhitsky VA. Economically effective data bases diagnostics method. Advances in Modeling and Analysis B: Signals, Information, Data, Patterns. 1995;**33**(3):5-12

[12] Markov AS. Paradigma ogranichennogo stohasticheskogo kontrolya [paradigm of limited stochastic control]. Izvestiya Instituta inzhenernoy phiziki. 2012;**1**(23):15-19 (In Rus.)

[13] Ushakov IU. Probabilistic Reliability Models. Wiley; 2012. 248 p

[14] Random pulse generator. Patent SU 840856. USSR; 1981; G06F 1/02; Bull. 23

4

Probabilistic Modelling in Solving Analytical Problems of System Engineering

Anatoly Lepikhin, Vladimir Moskvichev and Nikolay Machutov

Abstract

This chapter provides some aspects to probabilistic modelling in solving analytical problems of system engineering. The historically developed system of the formation of scientific bases of engineering calculations of characteristics of strength, stability, durability, reliability, survivability and safety is considered. The features of deterministic and probabilistic problems of evaluation of the characteristics of strength, stiffness, steadiness, durability and survivability are considered. Probabilistic problems of reliability, security, safety and risk assessment of engineering systems are formulated. Theoretical bases and methods of probabilistic modelling of engineering systems are stated. The main directions of solving the problems of ensuring security and safety according to the accident risk criteria are determined. The possibilities of probabilistic modelling methods in solving the problems of strength, reliability and safety of engineering systems are shown in practical examples.

Keywords: engineering system, multi-level concept, probability, modelling, safety, survivability, security, safety, risk

1. Introduction

Sustainable development of social systems and the natural environment is determined by the state and prospects of the development of engineering and technology. Modern engineering and technology are created on the basis of the achievements of fundamental scientific research. Particular importance is the development of fundamental foundations of mechanics, which is the basis for the design and produce of engineering systems. New machines and structures are creating, based on achievements of construction mechanics, theories of elasticity, plasticity and strength of materials. Multivariance of design and engineering solutions to engineering problems and increase of uncertainties associated with the manifestation of complex combinations of dangerous natural, technical and social factors in the creation and operation of technical objects require the application of new approaches. These approaches will increasingly be based on a combination of traditional deterministic and developing statistical and prospective probabilistic methods of modelling, calculation and testing. Of particular importance is the development of methods of statistical mechanics, probabilistic fracture mechanics, reliability theory and safety theory of engineering systems [1, 2].

At the present time, a multi-level concept has been developed to ensure the safe operation of engineering systems (**Figure 1**). This concept includes specific stages, requirements, criteria, calculated parameters and directions of development. Each higher level is created and developed on the achievements of the lower levels. At the first stages, the methodology of modelling engineering systems and the calculation and experimental validation of operability were based on deterministic methods, with elements of statistical analysis (stages I–III). Understanding the role of random factors in the disruption of operability led to the use of probabilistic methods of modelling and analysis (stages IV, V). At the end of the twentieth century, operability analysis of complex engineering systems began to use parameters of safety S and risk R of disasters. These parameters take into account natural, technical and social hazards (stages VI, VII). On this basis, by the end of the twentieth century, a complex of interconnected multi-level deterministic and probabilistic requirements to engineering systems and their parameters was formed: "strength $R_\sigma \rightarrow$ stiffness $R_\delta \rightarrow$ steadiness $R_\lambda \rightarrow$ durability $R_{N,\ \tau} \rightarrow$ reliability $P_{P,\ R} \rightarrow$ survivability $L_{l,\ d} \rightarrow$ safety S". Each stage in the development of fundamental research and requirements in this structure corresponds to a certain practical result in the design, creation and operation of engineering systems: "indestructibility - preservation of size and shape - durability - fault tolerance - survivability - risk of disasters". Risk is considered as a quantitative probabilistic measure of safety.

The basic equation for determining these characteristics of engineering systems can be written in the following form [1, 2]:

Years	Stages	Characterictics	Criterias	Parameters
2030-2016	VIII	*Security*	*Acceptable risk*	$Z(\tau)$
2000	VII	*Risk*	*Acceptable loses*	$R(\tau)$
1990	VI	*Safety*	*Acceptable hazards*	$S(\tau)$
1980	V	*Survivability*	*Stability of damages*	$L_{d,l}$
1970	IV	*Reliability*	*Fault tolerance*	$P_{P,R}$
1960	III	*Durability*	*Operation life time*	$R_{N,\tau}$
1940	II	*Stiffness Steadiness*	*Saving of form*	R_δ R_λ
1920	I	*Strength*	*Indestructibility*	R_σ

Figure 1. Structure of system for ensuring operability of engineering systems.

$$\{R, S, L_{l,d}, P_{P,R}, R_{N,\tau}, R_\delta, R_\Lambda, R_\sigma\} = \Psi\Big\{\varphi_Q(Q, N, t, \tau); \varphi_\sigma(\sigma_y, \sigma_b, E, \nu, m, \psi, K_{1c}); \varphi_A(\alpha_\sigma, l, A)\Big\} \quad (1)$$

where Ψ is a generalized function of technical state; $\varphi_Q(.)$ is loading functional that takes into account load parameters Q, number of cycles N, temperature t, time τ of loading; $\varphi_\sigma(.)$ is functional of physical and mechanical properties of structural materials, taking into account the yield strength σ_y, ultimate strength σ_b, fatigue limit σ_r, modulus of elasticity E, Poisson's ratio ν, hardening ratio m, ultimate deformation ψ, critical stress intensity factor K_{1c}; $\varphi_A(.)$ is a functional of constructive forms, taking into account cross sections area A, lengths l of the defects, and stress concentrators α_σ.

Expression (1) can be considered for limiting states, under which the engineering system ceases to meet the requirements of operation, and for admissible states, determined by the system of safety factors n.

The modern stage of research of engineering systems takes into account the largest man-made accidents and disasters of nuclear, hydraulic and thermal power engineering objects, transport systems and in chemistry objects from twentieth to twenty-first centuries. Taking this into account, stages VII–VIII consider the protection of technical objects based on according risk criteria. The defining equation of this new direction of the engineering methodology of probabilistic modelling, calculation and experimental justification for security Z becomes the functional of the following form:

$$Z(\tau) = F_Z\{R, S, L_{l,d}, P_{P,R}, R_{N,\tau}, R_\delta, R_\Lambda, R_\sigma\} \quad (2)$$

The probabilistic characteristics play a decisive role in the structure of the functional (1) and (2). Therefore, for their analysis, further development of probabilistic modelling methods of engineering systems is necessary.

2. Theoretical foundation of probabilistic modelling for engineering systems

2.1. Statement for probabilistic modelling problems of engineering systems

The peculiarity of the above multi-level concept (**Figure 1**) ensuring operability of engineering systems in the form (1) is that each of the stages I–VIII considers its own, specific, calculating situations (**Figure 2**). At each stage, special fundamental problems of the mechanics of solids are solved:

- boundary problems for stress determining in the most loaded elements, in cross sections A and local volumes $V(x, y, z)$

$$\{\sigma_{ij}, e_{ij}\} = F_\sigma\{Q, A, V(x, y, z)\}$$

- experimental problems of obtaining metal deformation diagrams (equations of state)

$$\{\sigma_{max}, e_{max}\} = F_{\sigma,e}\{\sigma_{ij}, e_{ij}\}$$

- experimental problems of estimating the criterial values of stresses and deformations corresponding to the achievement of the conditions for breaking strength (fracture)

$$\{\sigma_c, e_c\} = F_c\{\sigma_b, e_f\}.$$

The nominal stresses σ_n and deformations e_n, local stresses σ_l and strains e_l, fracture stresses σ_f and deformations e_f, as well as actual and critical dimensions of technological and operational defects l and l_f are used as the determining parameters. The values of fracture stresses and deformations characterize the limiting states and are determined taking into account the loading regime in terms of the number of cycles N and time τ, the fatigue diagrams $(\sigma_f - N_f)$ and $(e_f - N_f)$; long-term strength $(\sigma_f - \tau_f)$ and $(e_f - \tau_f)$; fracture toughness $(\sigma_f - l_f)$ and $(e_f - l_f)$.

The basic characteristics of strength R_σ, stiffness R_δ, steadiness R_λ, and durability $R_{N,\tau}$ are considered for design situations when the values of all parameters are in the deterministic limits established by the project:

$$\{R_\sigma, R_\delta, R_\lambda\} = F_R\{(\sigma_n, e_n); (\sigma_l, e_l); (\sigma_f, e_f)\} \tag{3}$$

$$R_{N,\tau} = F_{N,\tau}\{(N,\tau); (R_\sigma, R_\delta, R_\lambda)\} \tag{4}$$

If statistical properties are taken into account for combinations $(\sigma_n, e_n); (\sigma_l, e_l); (\sigma_f, e_f)$, the characteristics of strength, stiffness and resource can be determined with use quantile of probability

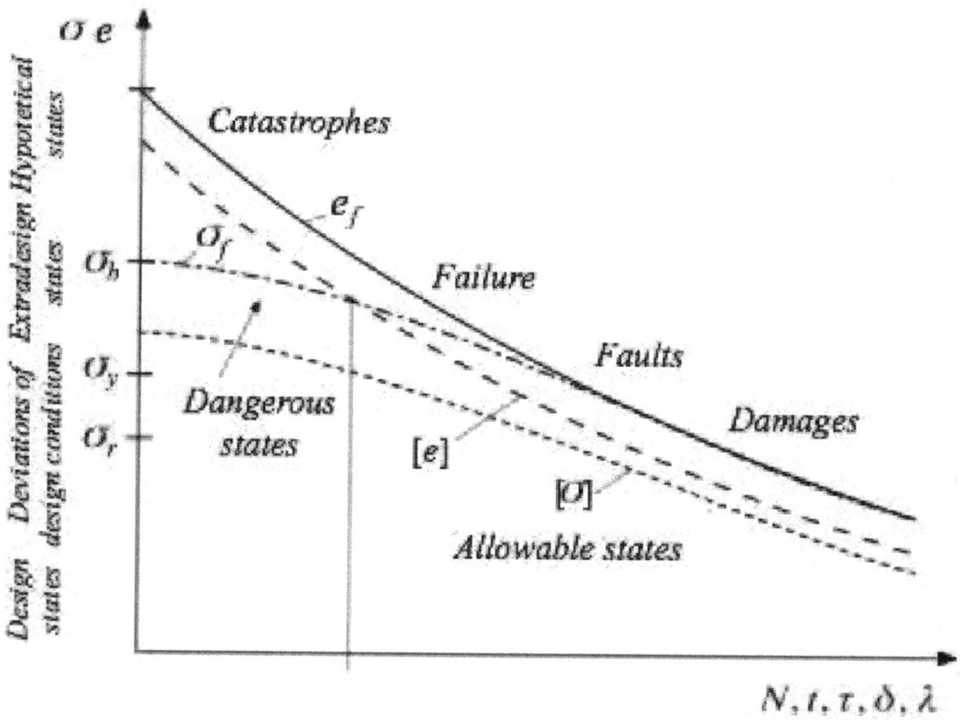

Figure 2. State diagram of engineering systems.

U, given by margin factors for nominal and local stresses n_σ and $n_{\sigma l}$, nominal and local strains n_e and n_{el}, destroying stresses $n_{\sigma f}$, and deformations of n_{ef}.

$$\{R_\sigma, R_\delta, R_\lambda\}_U = F_R\{(\sigma_n/n_\sigma, e_n/n_e); (\sigma_l/n_{\sigma l}, e_l/n_{el}); (\sigma_f/n_{\sigma f}, e_f/n_{ef})\} \quad (5)$$

$$\{R_{N,\tau}\}_U = F_{N,\tau}\{(N/n_N, \tau/n_\tau); (R_\sigma, R_\delta, R_\lambda)\} \quad (6)$$

As probabilistic modelling methods evolved, problems (3) and (4) were considered in a probabilistic formulation, when the basic parameters are given by probability distribution functions. In this case, by the methods theory of probability and reliability theory, one can obtain diagrams of limiting states in the coordinates $(\sigma_f, e_f) - (N, \tau, \delta, \lambda, l)$ for different probabilities P of their realization (1%, 50%, 99%) (**Figure 3**).

The reliability of engineering systems is determined in the presence of probability distribution functions of the basic parameters of operability. In a general case, reliability is estimated by the given probabilistic properties P of the characteristics of strength, stiffness, steadiness, durability:

$$P_{P,N,\tau} = F_P\{P|\,(Q, N, \tau); (\,R_\sigma, R_\delta, R_\lambda, R_{N,\tau})\} \quad (7)$$

The survivability of engineering systems is considered for design situations and beyond design situations with considering damage accumulation processes D. In engineering, practice damage is characterized by the sizes of the technological and operational defects L or scalar measure of accumulated damage d:

$$D = \{d, l\}; d = F_d\left\{\frac{N}{N_f}, \frac{\tau}{\tau_f}, \frac{l}{l_f}, \frac{\delta}{\delta_f}, \frac{\lambda}{\lambda_f}, \frac{\sigma}{\sigma_f}, \frac{e}{e_f}\right\}; l = F_l\{l_i(N, \tau), i = 1, m\} \quad (8)$$

Survivability can be estimated in deterministic and probabilistic formulation by using expression:

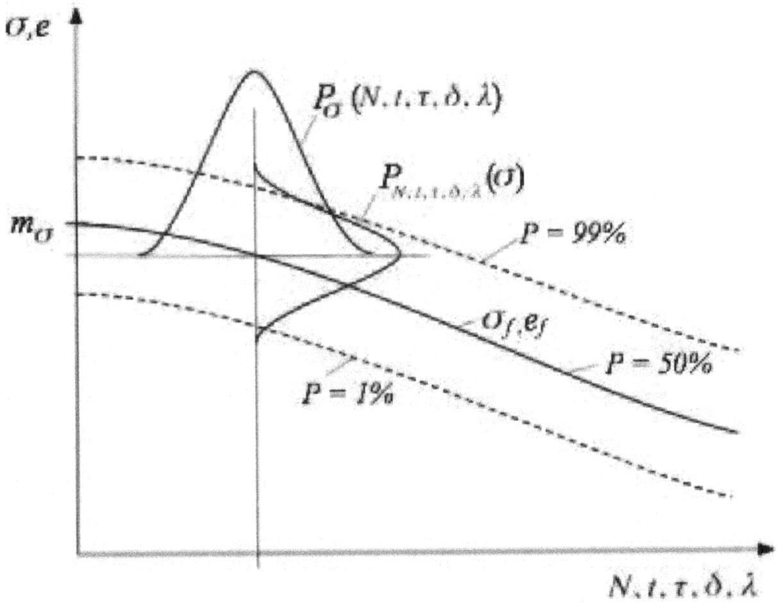

Figure 3. Probabilistic diagrams of limiting states.

$$L(\tau) = \{l_i(N, \tau), i = \overline{1, m}\} \tag{9}$$

In the probabilistic formulation, solution of problem (8) consists of obtaining probability distribution function of the survivability $F(L_{d,l})$ for given probabilistic properties of accumulated damage. Here it should be noted that at stresses above the yield point, the calculation of damages in the stress values $d = \sigma/\sigma_f$ is significantly more complicated. Therefore, the damage is calculated in terms of relative deformations, $d = e/e_f$.

From these positions, in the analysis of survivability, new calculation cases are considered such as deviations from design situations, beyond design situations, and hypothetical situations that characterize the transition from failures to accidents and disasters (**Figure 2**).

Currently, national and international programs to ensure the safety of engineering systems, engineering infrastructures, and natural environment (Rio-1992, Johannesburg-2002, Kobe-2005, Hyogo-2015) focus attention on security characteristics S. Quantitative assessments of safety characteristics are based on complex analysis of reliability and survivability of engineering systems [2, 3]:

$$S(\tau) = F_S\{P_{P,N,\tau}, L_{d,l}\} \tag{10}$$

The safety is the ability of the engineering systems to remain operative in damaged states and fracture states. In engineering practice, quantitative safety characteristics have become associated with the risks of accidents and disasters. Risk in quantitative form is defined as a function of the probabilities P_f of accidents and catastrophes and the associated losses U_f:

$$R(\tau) = F_R\{P_{P,N,\tau}, L_{d,l}; U_f\} = F_c\{P_f, U_f\}, P_f = F_f\{P_{P,N,\tau}, L_{d,l}\} \tag{11}$$

It is important to note that the probabilities P_f are estimated for beyond design and hypothetical situations, with the extreme values of Q^{extr} operation parameters and extreme strength and resource characteristics, not envisaged by the project:

$$P_f = F_P\{P|\ (Q^{extr}, N, \tau); (\ R_\sigma, R_\delta, R_\lambda, R_{N,\tau})^{extr}\} \tag{12}$$

The presented analysis shows that probabilistic models and probabilistic methods acquire an increasingly important role in ensuring the operability of engineering systems.

2.2. The development of traditional probabilistic methods

The development of theoretical foundations' probabilistic approaches to the analysis of the operability of engineering systems covers a significant historical period (stages I–VI, **Figure 1**). The first studies in this direction were carried out by M. Mayer (1926), N.F. Hotsialov (1929), and W. Weibull (1939). In these studies, the significant variation of strength characteristics for structural materials was shown, and the idea of introducing safety factors was proposed. Essential development of these studies was the work of N.C. Streletsky (1935). In his studies, the strength characteristics of materials (σ_f, e_f) and load parameters (σ_n, e_n) were considered as random variables. The further development of this approach was made by A.R. Rzhanicyn (1947). In his works, the relationship between safety factor n_σ and reliability P was established.

Theoretical basis for calculating the reliability of structures in form (7) was formulated for case of two random variables: the load q and the strength r (**Figure 4a**).

In the 1960s and 1970s, Polovko et al. developed methods for probabilistic calculation of fatigue life $R_N\tau$ of and reliability of machine parts $P_{P,N,\tau}$ according to expressions (5) and (6). These methods have been used to calculate the probability diagrams of fatigue characteristics of aircraft, transport, and other equipment.

According to F. Freudenthal (1956) and M. Shinozuki (1983), the reliability problem was formulated for a finite number of random variables $X = \{x_i, i = 1, n\}$ (geometrical parameters, material properties, loads, environmental factors, etc.). In this case, it was assumed that probabilistic properties of random variables are determined by joint probability distribution function $f(X)$. Reliability is determined by computing a multi-fold integral on given security region Ω_S:

$$P(X) = P\{X|G(X) > 0\} = \iiint\limits_{\Omega_S} f(X)dX, \Omega_S = \{X|G(X) > 0\} \tag{13}$$

Difficulty in computing this probability has led to the development of various approximation methods: the first-order reliability method (FORM) (Hasofer, Lind, 1974) and the second reliability method (SORM) (Tvedt L., 1988). These methods are widely used in engineering practice [4, 5]. The main drawback of these methods is that they relate to cases when changing random parameters in time does not exceed the limits of their statistical variability characterized by function $f(X)$.

The further development of probabilistic methods was obtained in the 1970–1980s of the twentieth century on the basis of three conceptual ideas [6]. The first idea was that the external conditions of operation structure and its reaction to these conditions are random processes. Therefore, probabilistic methods for calculating structures should be based on methods of the theory of random functions. The second idea was that the failure of the structures in most cases is a consequence of the accumulation of damage (d, l). These damages, reaching a certain value, begin to interfere with the normal operation of structures. The third idea was that the main indicator of reliability should be probability staying of system parameters in a certain permissible region. Violation of normal operation (accident)

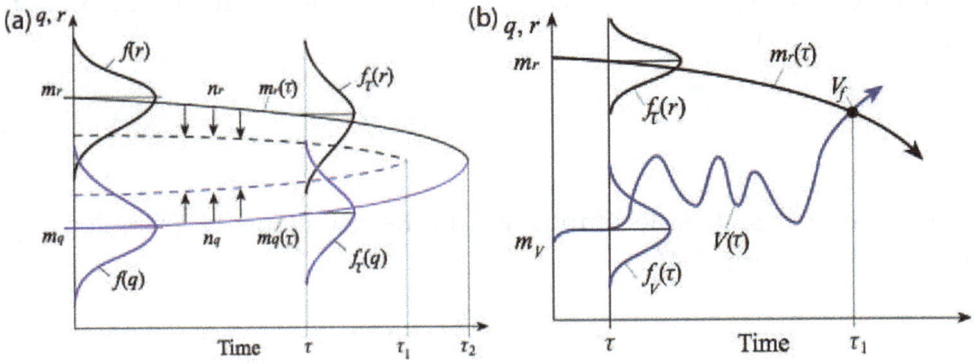

Figure 4. Failure models for calculating reliability.

of the system was interpreted as an output of parameters from this area (**Figure 4b**). Taking these ideas into account, reliability was determined in the form:

$$P\{X(\tau)\} = P\{V(X,\tau) \in \Omega_S, \tau \in [0,\mathrm{T}]\} \tag{14}$$

Thus, the interpretation of reliability as the probability of the fulfilment of some inequality connecting random variables gave way to a more adequate and in-depth interpretation in form emissions of random functions from an admissible region [6–8].

In subsequent years, based on the achievements of reliability theory, extensive studies were carried out to substantiate and improve the normative design calculations using probabilistic methods and reliability theory. Research is carried out in nuclear engineering construction [9], aerospace technology [10], and other industries. An important role in this was played by the achievements of fracture mechanics, taking into account the presence of technological and operational defects and structural damage [11, 12]. The development of probabilistic models of fracture mechanics made it possible to create a concept and methods for probabilistic risk analysis of engineering systems [13, 14].

At the present time, probabilistic modelling and probabilistic methods of calculation have become an integral part of a wide class of problems of statics and dynamics of engineering systems, in which randomness plays an essential role and is introduced by the variations of their geometric and physical properties. To this class belong the problems of strength of micro-inhomogeneous materials, composite materials, and structures, including nanomaterials and microstructures. Significant progress in this direction is associated with the development of numerical methods of analysis and computational technologies [15, 16].

2.3. New directions for solving engineering problems of security and safety by risk criteria

Engineering systems, with rare exceptions, are complex structures of elements of different nature. The problems of probability modelling of such structures turn out to be multivariate and lead to ambiguous solutions. In conditions of complexity and statistical diversity of states of the engineering systems, the diversity of elements, the multiplicity of the mechanisms of catastrophes, it seems unlikely that an integrated comprehensive risk model will be constructed in the near future. A more promising direction can be development individual models of risk, based on the representation of the engineering systems in the form of a structure Σ, consisting of subsystems σ and elements e [13].

$$\Sigma = \underset{i}{\cup}\sigma_i\left(\underset{j}{\cup}e_{ij}\right), \quad i = 1, n, \quad j = 1, m. \tag{15}$$

These models must realize the decomposition of R-characteristics (risk-decomposition) of structure (13) in the next form [14].

$$R_\Sigma \rightarrow \{R_i\} \rightarrow \{R_{ij}\} \rightarrow \{R_{ijk}\} \tag{16}$$

where R_Σ is integral (system) risk, R_i is complex (subsystem) risk, R_{ij} is elemental risk, and R_{ijk} is criterial risk.

The final level in this expansion is criterial risk, which allows the connection of systemic risk with mechanisms of catastrophes.

When constructing a system of models, implementing decomposition (16), it is necessary to take into account the following problem features of the engineering systems as the objects of risk analysis. First, in most cases, we have to analyse situations that have not been seen before, since the coincidence of all circumstances of disasters is an almost impossible event. Second, the analysis is carried out under conditions of high uncertainty associated with both the random nature of external influences and processes in the elements of systems, and with the ambiguity of objectives and safety criteria, as well as alternatives to decisions and their consequences. Third, the analysis is performed with time limit. At the stage of analysis of design decisions, these restrictions are determined by the design time, at the stage of operation—by the time of response to an emergency or emergency situation.

These features make specific requirements for model representations, the computer, and the information base for risk analysis. The development of model representations and a computational technology is connected with the solution of a number of specific problems. The first task is to describe the engineering systems from the standpoint of integrity and hierarchy. The creation of a substantial and compact model with a large number of significant parameters belongs to the number of difficult tasks, even with the use of modern mathematical and computational technologies.

The second task is to formulate information support for risk analysis. This task has two aspects. The first aspect is related to the task of processing information. Information in the hierarchical system comes in the language of the level that is being analysed. For conclusions at a higher hierarchical level, generalization is required, and at a lower level, detailing this information is needed. In both cases, this translation is ambiguous. The second aspect is related to the need to construct hypotheses about the states of elements on base the available information. The reliability of such hypotheses depends on the level of completeness of information and its reliability.

The third task is connected with the choice of the risk criterion. It can be solved on the basis of an analysis or development of special indicators that have the necessary properties of indicators of limit states of engineering systems. This choice can also be ambiguous or multicriteria.

Finally, the fourth task is to create theory and methods for risk analysis at given parameters. This apparatus can be considered as a set of mathematical models that reflect the mechanisms of catastrophes in a given sequence of the process of risk analysis. Here it is necessary to take into account the accidental nature of the catastrophe event of the system and the possibility of a formalized description and measurement of the random parameters of the systems.

A separate and difficult task is modelling the processes of accumulation of damage. In the general case, it is necessary to consider multicriterial damage (MCD) for each element and multi-structural damage for system (MSD).

To take into account multifocal character of damages and their structural hierarchy, we use the principle of selective scale and select the hierarchy of scales $M = \{\cup M_i,\ i = 1, n\}$ on which damages develop. Each scale M_i is considered as internal for a given level and is analysed by appropriate methods. For example, for the scale level of structural elements can be used by the methods of fracture mechanics, and for the level of construction, by the methods of structural mechanics. It should be noted that if fracture of individual elements, caused by MCD, can be considered as independent events, then at structure level there is an agreed redistribution of loads, and formation of the focus of MSD should be considered as a cooperated process.

2.4. Statistical information for risk analysis and safety

The safety and risk analysis is carried out using statistical information on dangerous events and damages. The systematization of data on major natural disasters and man-made disasters is carried out at the international and national levels [2, 3, 17]. Statistical studies show that modern global, national, sectoral and object security problems are the result of centuries-old quantitative and qualitative transformations both in social development and in the system "nature-machine-human." The uneven growth of damage from major disasters creates a real threat to the economy not only of individual regions but also for the planet as a whole. The scale and consequences of natural disasters and man-made disasters today are very tangible not only for developing countries but also for technologically advanced countries. The total losses currently for developed countries are 5–10% of GDP or more. In value terms, the total losses exceed 350 billion dollars (**Figure 5**).

Extreme losses are also attributed to individual catastrophic events. The losses from the hurricane Katrina (USA, 2005) amounted to 140 billion dollars. The accident at the Sayano-Shushenskaya hydroelectric power station resulted in the death of 75 people and damaged over 7.5 billion Roubles. Tsunami and the accident at the nuclear power plant "Fukushima" (Japan, 2011) led to the death of 20,000 people and damage of over $ 300 billion.

Based on the analysis of statistical data and estimates, modern man-caused hazards are characterized by the following values [2]:

- in the frequency of occurrence of failures, accidents, and disasters (1/year): objects of technical regulation 10^1–10^0; hazardous industrial facilities 10^0–10^{-1}; critical objects 10^{-1}–10^{-2}; strategically important objects 10^{-2}–10^{-3};
- in economic losses (dollars): objects of technical regulation 10^3–10^5; hazardous industrial facilities 10^4–10^7; critical objects 10^5–10^9; strategically important facilities 106–1011;
- in the risks of failures, accidents, and disasters: objects of technical regulation 10^3–10^5; hazardous industrial facilities 10^4–10^6; critical objects 10^4–10^7; strategically important objects.

The given statistical data can serve as a basis for categorizing man-made hazards according to the levels of risks of accidents and catastrophes. Risk diagrams can be represented by a power law of the following type:

$$R(\tau) = C_u\{U(\tau)\}^m \tag{17}$$

where C_U is a coefficient that depends on the dimension of the coordinates; m is an indicator that depends on the type of object.

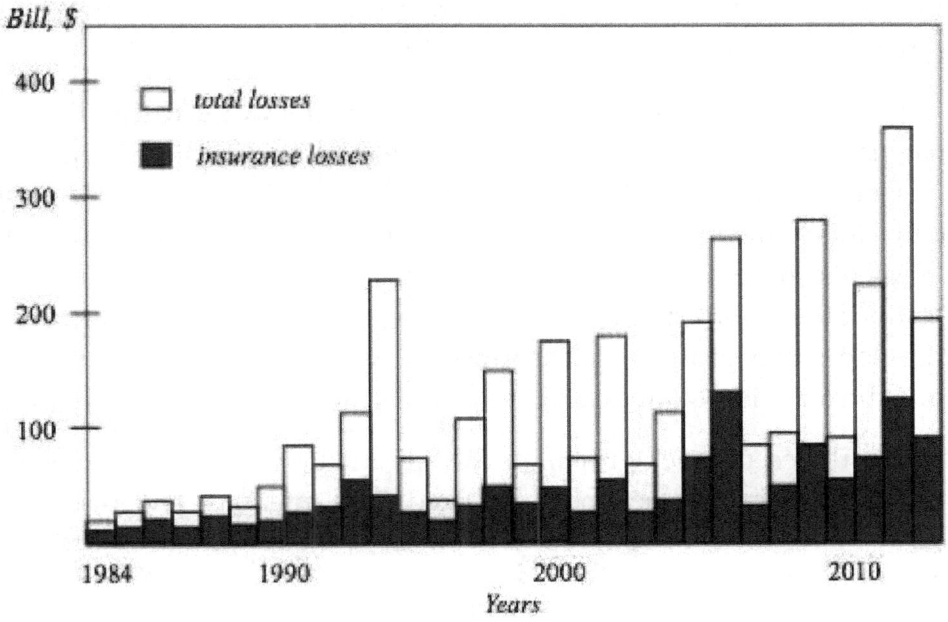

Figure 5. Losses from catastrophes of recent decades [3].

For natural disasters, natural-technogenic, and technogenic accidents and disasters, the value of m is in the range 0.3–1.0. For technogenic accidents and disasters, m = 0.55–0.60. The principal feature of distribution of losses according to the probabilities is that for critical and strategically important objects, large losses occur, leading to "heavy tails" of distributions.

The development and implementation of large infrastructure projects based on the achievements of science and technology not only dramatically increased opportunities in all areas of the world community but also created high risks of man-caused and natural-technogenic catastrophes at a global level. Modern engineering systems have destructive energy potential comparable to those of natural disasters. At the same time, the possibilities of parrying and localizing technogenic catastrophes are limited, despite the achievements of scientific and technological progress.

3. Solving engineering problems using probabilistic modelling

3.1. Probabilistic modelling of safe crack growth and estimation of the durability of structures

Crack growth up to a critical size under cyclic and long-term static loading is a rather complex process, which can be described by various crack growth equations. Methods for estimation of the lifetime of structures containing defects can be developed on the basis these equations. However, there are insufficient studies of the probabilistic aspects of crack growth, which greatly limit the opportunity for practical applications of these methods. To overcome this restriction, probabilistic models of the crack growth have been developed. This part presents the results, in a generalized manner, of these studies involving the probabilistic modelling of safe crack growth and the estimation of the durability of a structure [18, 19].

Probabilistic factors of crack growth are present both at the micro- and macro-levels of deforming materials. At a micro-level, these factors are the structural heterogeneity of materials and the heterogeneity of the stress-deformed conditions of local zones at the level of grain size. The important factors at the macro-level include the heterogeneity of intensely deformed zones of structural elements, the uncertainty of form, size, and orientation of cracks, and the dispersion in the evaluation of the cyclic crack growth resistance of materials. It is an extremely complex problem to develop probabilistic models of crack growth that reflect all levels of the process. Therefore, our main attention is directed to probabilistic models that handle macro-level factors.

Three models can represent crack growth: a discrete model with casual moments of time; a continuous model with casual increments at fixed time intervals; and discrete continuous model with casual increments of both types. In all cases, the conditions of irreversibility $\delta l_\tau \geq 0$ and kinetic conditions apply

$$\frac{dl}{d\tau} = \varphi(\Delta\sigma, l_\tau) \tag{18}$$

The problem of probabilistic modelling of the crack growth consists of the assignation of probabilistic features of trajectories $l(\tau)$, which adequately describe real processes. The problem of the probabilistic estimation of functions $f(l\,|\,\tau)$ and $f(\tau\,|\,l)$ on given probabilistic features of the trajectories. Modelling trajectories can be carried out on the basis of models of the theory of casual processes, empirical models, and probabilistic models of fracture mechanics.

The theory of casual processes offers a wide spectrum of models. Among the analytical models, it is possible to consider diffusive models as being the most respective. The use of diffusive models allows one to write the kinetic function in the manner of:

$$l(\tau) = a(l, \tau)d\tau + b(l, \tau)dw(\tau) \tag{19}$$

Modelling of processes with jumps requires that the kinetic function given by Eq. (19) must contain an additional component, that is:

$$l(\tau) = l(0) + \int_\tau a\{l(\tau), \tau\}d\tau + \int_\tau b\{l(\tau), \tau\}dw(\tau) + \int_\tau \theta(l, \tau)d\tau \tag{20}$$

The models represented by Eqs. (18)–(20) allow one to directly obtain the densities of the distribution of defects $f(l\,|\,\tau)$ or durability $f(\tau\,|\,l)$. In particular, Eq. (19) creates a diffusive durability distribution of kind:

$$f(\tau|l_f) = \Phi\left\{\frac{a\tau - l_f}{b\sqrt{a\tau/l_f}}\right\} + \exp\left\{\frac{2a^2}{b^2}\right\}\Phi\left\{-\frac{a\tau + l_f}{b\sqrt{a\tau/l_c}}\right\} \tag{21}$$

It is necessary to point out that such diffusive models present difficulties with respect to a physical interpretation of the parameters. So, if the physical sense of functions a, b, θ are understood, then the sense of component w remains unclear. Nevertheless, as will be shown later, the practical use of this approach gives rather efficient results.

Using a Monte Carlo method can be considered as another effective approach. The advantage of this method is the possibility to use determined forms of the equations of the crack growth with casual parameters. Let us consider an example of the kinetic equation. As is known, it has the form:

$$\frac{dl}{dN} = C(\Delta K)^m = C\left\{\Delta\sigma\sqrt{\pi l}\varphi(l)\right\}^m \tag{22}$$

If one accepts that parameters C, m, $\Delta\sigma$, l are casual variables with given probability distribution functions, use of the Monte Carlo method allows one to obtain casual realizations of the process trajectory $l(N)$ for a given number N of loading cycles:

$$l(N) = \left[\int_{l_0}^{l_f} l^{-m/2}(\varphi(l))^{m-1}\right]^{-1}\left\{C\sum_{j=1}^{N}\sigma_j^m\right\} \tag{23}$$

Distribution densities $f(l\,|\,\tau)$ or $f(\tau\,|\,l)$ in the case are a result of frequent realizations of the model expressed by Eq. (23). Obviously, the determination of specified probability distributions is a complex and labour-consuming task, and representative statistics on the parameters of cyclic crack resistance are not available at present.

The most productive approach is a creation of special probabilistic models, the parameters of which can be determined by means of fracture mechanics. Assuming the fundamental basis of the mechanics of the crack growth, a probabilistic kinetic model can be formulated in the manner of:

$$l(\tau) = \varepsilon_N P_N + \varepsilon_\tau p_\tau \tag{24}$$

Increments can be calculated by means of fracture mechanics. As a first approximation, it is possible to suppose that these values are the parts of the plastic zone at the crack front, where deformations reach the fracture state e_f. Using an energy notation of crack growth, the probabilities can be written as follows:

$$p_N = 1 - \exp\left\{-\left(\frac{W_N}{W_{fN}}\right)^\alpha\right\}, p_\varepsilon = 1 - \exp\left\{-(W_\varepsilon/W_{f\varepsilon})^\beta\right\} \tag{25}$$

Obviously, the values of W_N and W_ε are widely variable, while W_{fN} and $W_{f\varepsilon}$ only slowly change. It is possible to estimate values of W_N and W_ε by means of computational fracture mechanics.

The presented models contain initial defect sizes $l(0)$. Therefore, probabilistic models of the distributions of the defect sizes must be included into the main models. Statistical studies of defects in welded joints show that it is possible to use a two-parameter Weibull distribution as a basic probabilistic model for distribution of defect sizes:

$$f(l) = \frac{\gamma}{\theta}\left(\frac{l}{\theta}\right)^{\gamma-1} \exp\left\{-\left(\frac{l}{\theta}\right)^{\gamma}\right\} \tag{26}$$

An estimation of the parameters of Eq. (26) for structures of different types shows that parameter of form γ changes from 0.4 to 4.0. The second parameters θ changes over rather wide limits (from 1.5 mm to 25 mm and more).

These probabilistic models were used to study crack kinetics in the welded joints for high-pressure vessels and estimation of durability and reliability functions. The results of calculation reflect the character of the model trajectories over a range of dispersion in the function $l(\tau)$, which corresponds to those observed in laboratory experiments.

The empirical Paris-Erdogan model, in combination with the statistical simulation method, was used for modelling kinetic of crack growth in the weld joint of a pipeline in the nuclear reactor VVER-1000. Processing of the results of the probabilistic modelling of the crack kinetics allowed one to estimate reliability functions for a weld joint fracture as is shown in **Figure 6**.

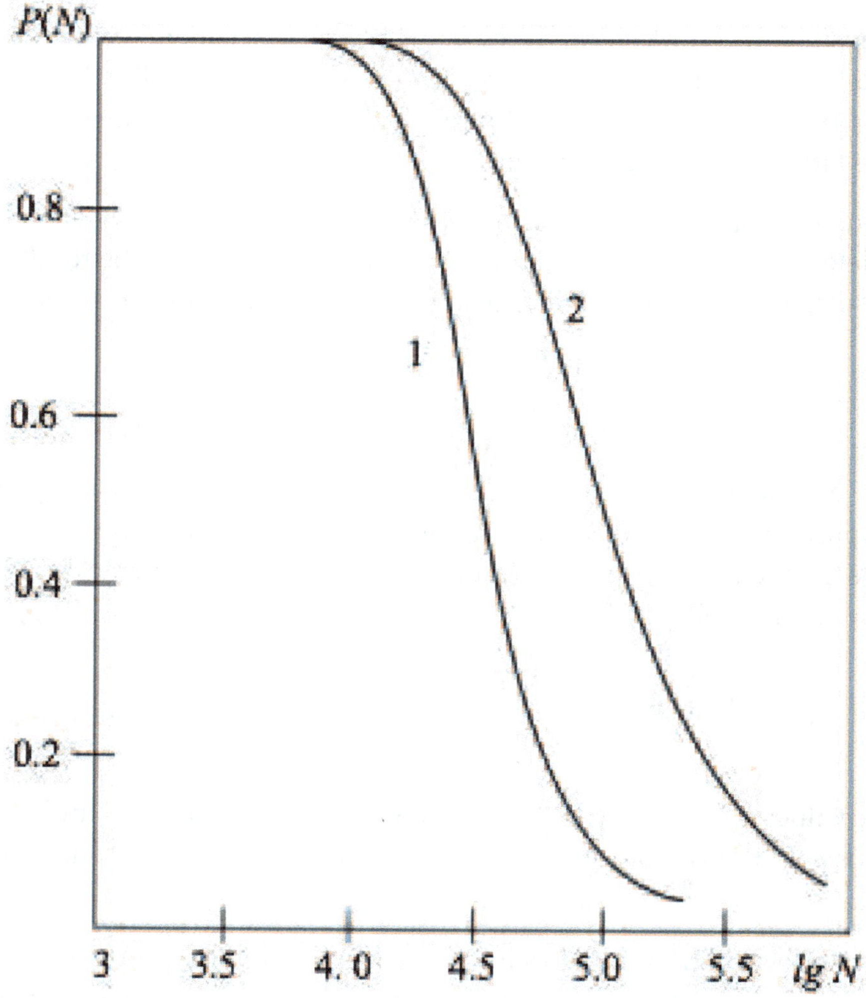

Figure 6. Reliability functions for hermetic breach criterion (1) and fracture criterion (2).

Producing complete probabilistic diagrams of integrity is a major prospect. These diagrams are based on information gained from the same model. The structure of complete probabilistic diagrams of integrity is shown in **Figure 7**. A complete probabilistic diagram of integrity presents itself as a number of sections of a surface connecting three parameters: probability *P*, safety factor *n*, and durability *N*. They allow one to estimate the durability and probability of its attainment. Additionally, there is a possibility for a decision of an inverse task—the definition of probability that for a chosen safety factor, a certain durability will be achieved. Thus, a complete probabilistic diagram of integrity presents itself as a number of sections of a surface connecting three parameters: probability *P*, safety factor, and durability.

3.2. Reliability and risk assessment of metal-liner composite overwrapped pressure vessels

Metal-composite pressure vessels (MCOPVs) have found a wide application in aerospace and aeronautical industries. Such vessels should combine the impermeability and high weight efficiency with enhanced long-term safety and durability. To meet these requirements, theoretical

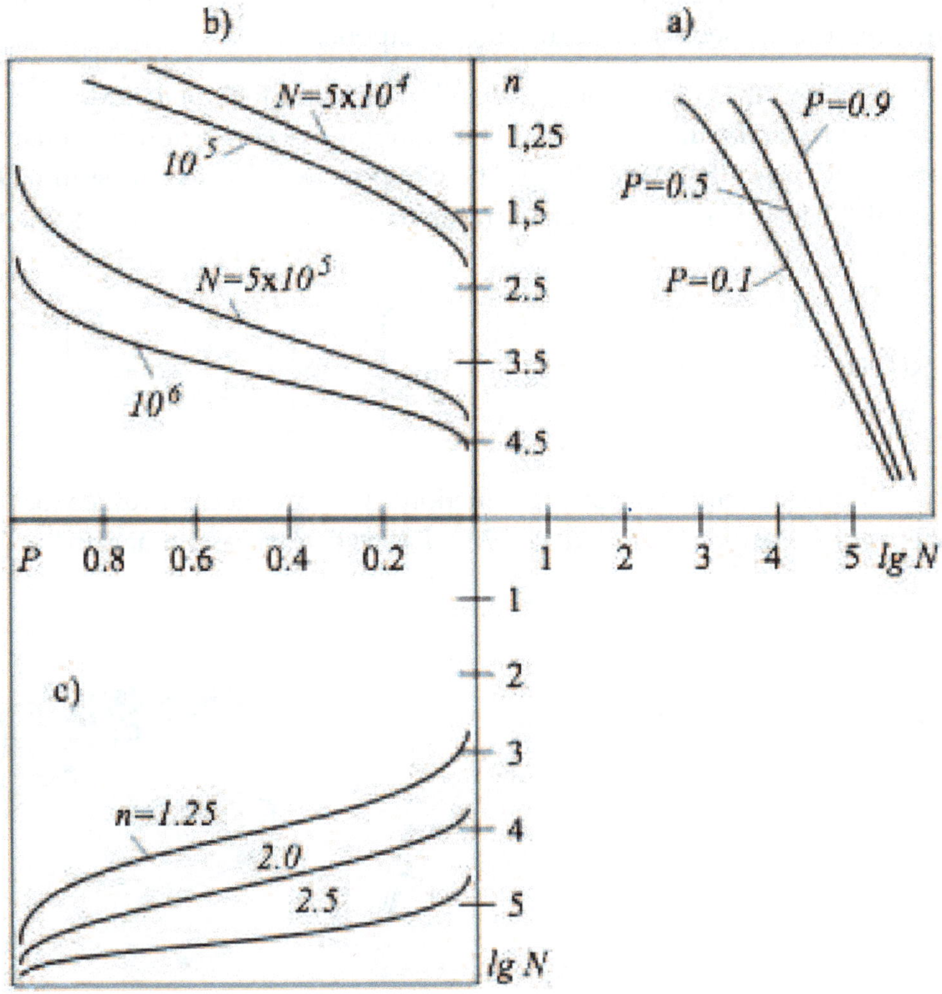

Figure 7. A complete probabilistic integrity diagrams: (a) probability of fatigue, (b) durability distributions, and (c) functions of equal safety factors.

and experimental studies on the mechanics of deformation and failure of MCOPVs are required [20].

Investigate reliability and risk fractures were based on results of numerical stress analysis and experimental tests' full-scale samples of MCOPV. The construction of MCOPV was having an axisymmetric ellipsoid-like shell of revolution with the minor to major diameter ration of about 0.6 (**Figure 8a**). The thin-welded liner was made of VT1–0 titanium alloy. The composite shell was formed by helical winding of IMS-60 carbon fibres impregnated with a polymer matrix. The stress analysis of MCOPV under internal pressure was performed using the finite element method. The calculations were carried out with finite element models developed to reflect all significant geometric and deformation characteristics of the composites vessel (**Figure 8b**).

Actual MCOPVs structures will exhibit a non-uniform distribution of stresses and deformation owing to a number of factors. These include the nuances of liner geometry and its interaction with the overwrap winding pattern, the relative stiffness of the liner to the overwrap, the liner-overwrap interface slips characteristics, and the presence of incompatible curvature changes. Load equilibrium in the bimaterial vessels requires that the total applied pressure be equal to the sum of the pressure carried by the individual components.

Taking this into account, the calculation of reliability function $R(P, \tau)$ included the evaluation of two components: the reliability $R(DM)$ at the beginning of service and the reliability $R(\tau, \sigma)$ during operation—$R(P, \tau) = R(DM)\times R(t, \sigma)$. The component $R(DM)$ was estimated by means of a conventional "load-strength" model, assuming the Gaussian law for load and strength values of MCOPV [20]:

$$R(DM) = Prob\{DM > 1\} = \Phi\left\{\frac{\mu_f - \mu_p}{\sqrt{s_f^2 + s_p^2}}\right\} = \Phi\left\{\frac{DM - 1}{\sqrt{V_f^2 DM^2 + V_p^2}}\right\} \quad (27)$$

where Φ is the standard normal distribution function; μ_f, μ_p are median values for P and P_f; V_f, V_p are coefficients of variations for P and P_f, $DM = P_f/P$ is design safety margins.

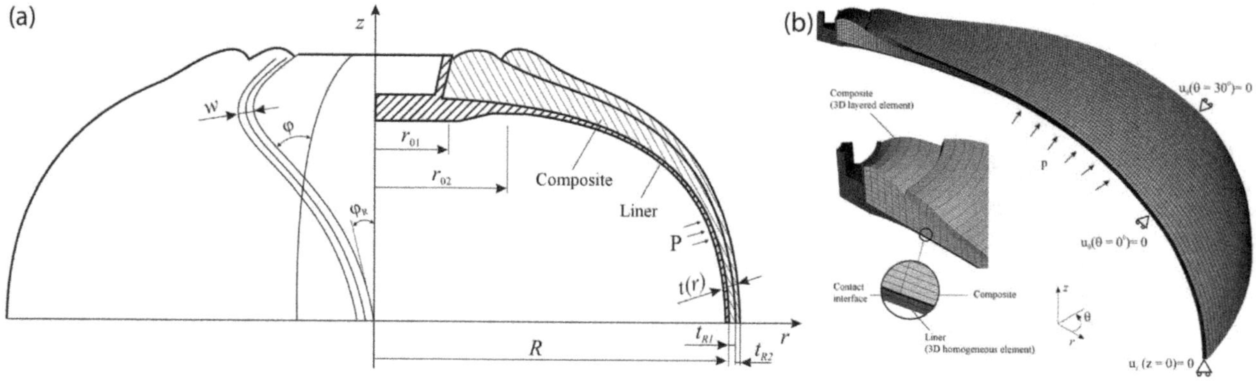

Figure 8. A metal-lined composite pressure vessel: calculation scheme (a), finite element model (b).

The Phoenix approach based on the Weibull reliability model was used to determine the term $R(\tau, \sigma)$. To account the influence of structural-mechanical heterogeneity of MCOPV, a reference measure M_0 was introduced. It is assumed that within M_0, the deformation of material is uniform. The probability of failure-free operation $R(\tau, \sigma)$ can be expressed as follows:

$$R(\tau,\sigma) = \exp\left\{-\frac{M}{M_0}\left[\frac{\tau}{\tau_c}\left(\frac{\sigma_P}{\sigma_f}\right)^{\alpha}\right]^{\beta}\right\} \quad (28)$$

where M is the total "scale" of MCOPV; τ is the time; τ_c is the characteristic (reference) time, which can be considered as time of failure during static test; σ_p is stress corresponding to operating pressure P; σ_f is stress at burst pressure; and α, β are statistic parameters.

Risk assessment was performed on the basis of an analysis of possible mechanisms of MCOPV destruction. The event of fracture of the MCOPV decays into two events: the destruction event of the liner and the event of destruction of the power composite shell. In the first case, the main parameter controlling the state of the liner is deformation, and in the second case, the stresses in the composite shell are the determining parameter:

$$\{\text{Risk}\} \rightarrow \begin{cases} \{\text{Leakage}\} \rightarrow P_f(\tau,\varepsilon) = P\{\tau|\varepsilon_p \geq \varepsilon_f\} \\ \{\text{Fracture}\} \rightarrow P_f(\tau,\sigma) = P\{\tau|\sigma_p \geq \sigma_f\} \end{cases} \quad (29)$$

Reliability functions $R(t,\sigma)$ for MCOPV in the orbit are shown in **Figure 9**. As can be seen from the figure, while ensuring the homogeneity of the properties of the composite sheath and

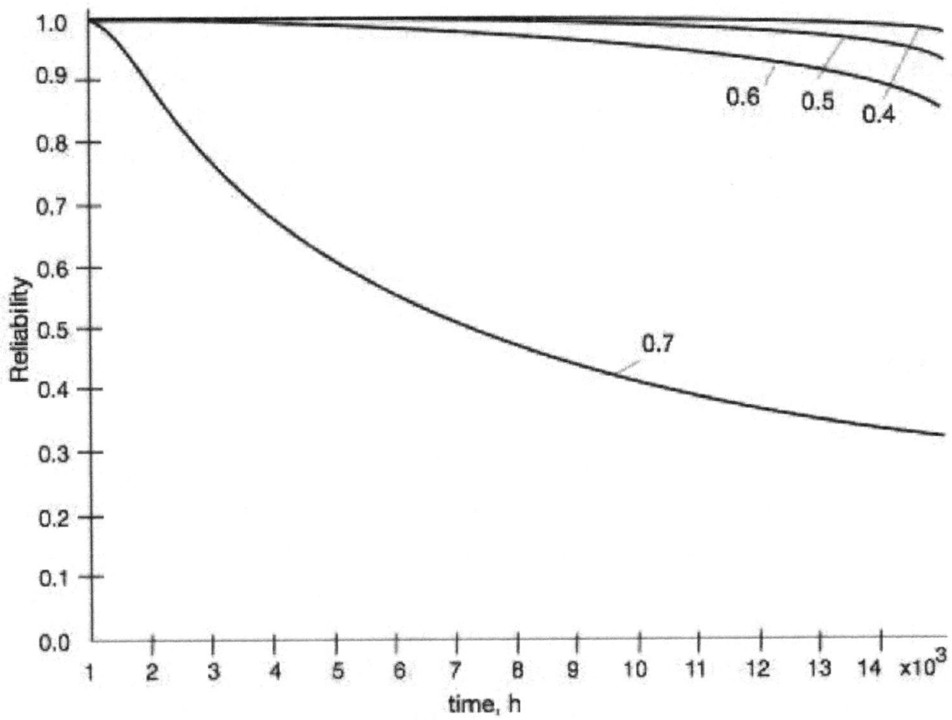

Figure 9. Functions of reliability depending on the time and the stress level σ_p/σ_c.

Type of destruction	Risk fracture of MCOPV	
	7-layer shell	9-layer shell
Leakage breach of liner in the test	1.3×10^{-6}	1.1×10^{-6}
Leakage breach of liner in the space (orbit)		
at the beginning work	1.1×10^{-8}	2.5×10^{-9}
at the end work	2.8×10^{-7}	6.3×10^{-8}
Fracture of MCOPV in the space		
at the beginning work	2.4×10^{-8}	4.7×10^{-9}
at the end work	1.8×10^{-6}	3.7×10^{-7}

Table 1. Calculated estimates of risk fracture of MCOPV.

operating stresses not exceeding 0.5 of the strength level of the composite material, the reliability of MCOPV is ensured at a level of at least 0.999 at the end of operation time in orbit. When the relative stresses level is increased to 0.6, high reliability is ensured only in the first 500 h of operation. The load level of more than 0.6 is unacceptable for the MCOPVs.

Quantitative risk assessments were performed for the most dangerous scenarios. The risk calculation was performed for the time moments 1000 h (the beginning work on the orbit) and 15,000 h (at the end work on the orbit). The calculation of the risk fracture of MCOPV was carried out according to the Phoenix approach, replacing the standard fracture stress of the composite σ_f by the numerical or experimental estimate of the actual value of the composite strength σ_c.

The results of the calculation are presented in the **Table 1**. The obtained risk assessments can be regarded as tentative, since they do not take into account possible processes of creep of liner and composite under load during the MCOPV operation. It should be noted that obtained values of the probabilities of destruction of the MCOPV belong to the class of extremely unlikely events. Therefore, the risk of destruction of the MCOPV in the orbit can be considered acceptable.

4. Conclusion

This chapter provided some aspects to probabilistic modelling in solving analytical problems of system engineering. The main tasks of engineering design are analysis of system operation from the moment of the conception and substantiation of the initial idea to the moment of writing off and wrapping the product so that it does not fail during the service period. When solving these problems, variability in the employed materials, loads, manufacturing process, testing techniques, and application inevitably arise. Probabilistic models and probabilistic methods proceed from the fact that various uncertainties are inevitable end essential features of the nature on an engineering system or design and provide ways of dealing with quantities whose values cannot be predicted with absolute certainty. Unlike deterministic methods, probabilistic approaches address more general and more complicated situations, in which behaviour of the engineering system cannot be determined with certainty in each particular experiment or a situation. Probabilistic models enable one to establish the scope and limits of the application of deterministic theories and provide a solid basis for substantiated and goal-oriented accumulation and the effective use of empirical data. Realizing the fact that probability of failure of engineering system is never zero, probabilistic methods enable one to quantitative assess the degree of uncertainty in various factors, which determine the safety of system and design on this basis a system with a low probability of failure.

Analysis of the largest man-caused and natural-technogenic catastrophes of recent years indicates the need to improve methods and means of ensuring the safety of the engineering systems. One of the main ways in this direction is to improve the historically established system of forming the scientific basis for engineering calculations of the characteristics of strength, stability, durability, reliability, survivability, and safety. Of decisive importance is the need to switch to a new methodological framework and principles for ensuring the safety of engineering systems by the criteria for risks of accidents and disasters. A special role in this direction is that security defines all the main groups of requirements for engineering systems: strength, rigidity, stability, reliability, survivability, and risk.

Author details

Anatoly Lepikhin[1]*, Vladimir Moskvichev[1] and Nikolay Machutov[2]

*Address all correspondence to: aml@ict.nsc.ru

1 Institute of Computational Technologies SB RAS, Krasnoyarsk, Russia

2 Mechanical Engineering Research Institute RAS, Moscow, Russia

References

[1] Makhutov NA. Strength and Safety. Fundamental and Applied Research. Novosibirsk: Nauka; 2008. 528 p

[2] Makhutov NA, editor. Strength, Resource, Survivability and Safety of Machines. Moskow: Librocom; 2008. p. 576

[3] Makhutov NA. Safety and Risks: System Research and Development. Novosibirsk: Nauka; 2017. 724 p

[4] Kapur KC, Lamberson LR. Reliability in Engineering Design. New York: John Wiley & Sons; 1979. 586 p

[5] Shuhir E. Applied Probability for Engineering and Scientists. New York: McGraw-Hill; 1997. 533 p

[6] Bolotin VV. Methods of Probability Theory and Reliability Theory in Calculations of Structures. Moscow: Stroiizdat; 1982. 351 p

[7] Timachev CA. Reliability of Large Mechanical Systems. Moscow: Nauka; 1982. 183 p

[8] Gusev AC, Cvetlitcky VA. Calculation of Structures under Random Actions. Moscow: Machinostroenie; 1984. 240 p

[9] Reactor safety study – an assessment of accident risk in US commercial nuclear power plant draft. WASH-1400. Washington: USAEC; 1974. 271 p

[10] NASA/SP-2010-580 System safety handbook. Vol. 1. Washington: NASA; 2011. 103 p

[11] Sih G. Mechanics of Fracture. Leyden Noordhoff International Publishing; 1973. 517 p

[12] Prowan JW, editor. Probabilistic fracture mechanics and reliability. Mart. Nijn. Publ; 1987. 467 p

[13] Shokin Yu I, Lepikhin AM, Moskvichev VV. Problems of mechanics of catastrophes and safety of technical systems/Risk management. In: Risk Sustainable development: Synergetic. Moscow: Nauka; 2000. 431 p

[14] Lepikhin AM, Makhutov NA, Moscvichev VV, Chernyaev AP. Probabilistic Risk Analysis of Technical System Constructions. Novosibirsk: Nauka; 2003. 174 p

[15] Contreras H. The stochastic finite-element method. Comp. Struct. 1980;**12**(N3):341-348

[16] Abdelal G, Abuelfoutouh N, Gad A. Stochastic finite element and satellite structure design. In: Finite Element Analysis for Satellite Structures. London: Springer; 2013. pp. 203-249

[17] Security of Russia. Legal, socio-economic and scientific-technical aspects. Risk analysis and safety issues: In 4 Parts. Part 4. Applied Questions Risks Analysis of Critical Objects. Moscow: Znanie; 2007. 816 p

[18] Lepikhin A, Moskvichev V, Doronin S. Statistical fracture modelling of weld joint for nuclear reactor components. Theoretical and applied fracture mechanics. 1998;**29**:103-107

[19] Lepikhin AM, Makhutov NA, Moskvichev VV, Doronin SV. Probabilistic modelling of safe crack growth and estimation of the durability of structures. Fatigue Fract. Engng. Mater. Struct. 2000;**23**:395-401

[20] Lepikhin AM, Burov AE, Moskvichev VV. Possibilities of the design estimates of the reliability of high-pressure metal-composite tank. Journal of machinery and reliability. 2015;**44**(N4):344-349

5

Probabilistic Model of Delay Propagation along the Train Flow

Vladimir Chebotarev, Boris Davydov and Kseniya Kablukova

Abstract

In this chapter, we propose a probabilistic model for train delay propagation. There are deduced formulas for the probability distributions of arrival headways and knock-on delays depending on distributions of the primary delay duration and the departure headways. We prove some key mathematical statements. The obtained formulas allow to predict the frequency of train arrival delays and to determine the optimal traffic adjustments. Several important special cases of initial probability distributions are considered. Results of the theoretical analysis are verified by comparison with statistical data on the train traffic at the Russian railways.

Keywords: train traffic, stochastic model, train delay propagation, probabilistic modeling, operative management

1. Introduction

The trains' movement is subject to a variety of random factors which leads to unplanned delays. This causes the scattering of the arrival times, hence, the inconvenience to passengers and consignees. Knowledge of the arrival times' distribution properties leads to the possibility of predicting the characteristics of the train traffic and making correct decisions on the transportation process management. This makes it possible to improve the punctuality of train traffic and save resources, in particular, electric power.

The properties of the arrival headways distributions allow us to estimate the probability of delays emergence and theirs characteristics, which are important from a practical point of view. Probabilistic modeling of the delay propagation process along the train flow is the main tool for solving this problem.

The models for the distribution of delays in a dense train flow are divided into two classes. These are deterministic and stochastic models. Stochastic models take into account the unpredictable nature of obstacles in the railway. A mathematical model, proposed in the present chapter, make it possible to determine the probability distributions of the arrival headways of two consecutive trains at the station. The distribution properties are analyzed for different scattering of input random variables (the primary delay and the initial headways). Comparison of theoretical distributions with real statistics of train traffic on the Russian railways is performed.

2. Literature review

A substantial volume of literature is devoted to study of the train delays effect on the railway functioning. Deterministic models for primary and knock-on delays description were proposed in [1, 2]. These models based on the application of graph theory allow adjust the train traffic schedule. However, such approach considering the different characteristics of train traffic (e.g., travel and dwell times, headways, etc.) as deterministic values does not take into account the uncertainties that arise in reality.

Stochastic modeling takes the influence of random factors (e.g., see [3–8]) into account. Authors of [7] determine a probabilistic distribution of the arrival times. The problem of finding a distribution of arrival train delays is examined in [8]. It should be noted that in these papers, special cases of primary delay distribution are considered. It is supposed in [8] that the random duration of the primary delay corresponds to some generalization of the exponential law. The paper [7] employs discretization of the delay distribution.

Some of the researchers have analyzed statistical data on deviations of the train arrival times from the planned ones. In particular, the papers [9–11] show that scattering of these deviations correspond to the exponential distribution.

3. Description of models and analysis of the arrival headways distribution

3.1. The first model

Trains follow one path one after another in one direction from station A to station B with the same average speed v_0. Let the total number of trains is n. The distance from the train j to the train $(j - 1)$ is denoted by $X_j + s_0$, where $j = 2, 3, \ldots, n$, $s_0 > 0$ is the minimal safe distance between trains, and $X_2, X_3, \ldots, X_n$ are the random variables (without any assumptions about their distributions). All trains have the same destination station.

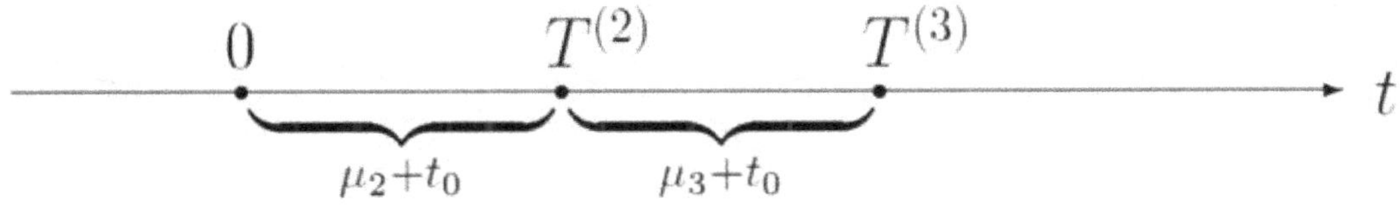

Figure 1. Departure times of trains 1, 2, and 3 from station A.

Let us also introduce the notations: $\mu_j = X_j/v_0$, $t_0 = s_0/v_0$. Suppose that train 1 departs from station A at the time $t = 0$. Then, the moment $T^{(m)}$ of departure train m can be found as (as shown at **Figure 1**):

$$T^{(m)} = \sum_{j=2}^{m} \mu_j + (m-1)t_0, \; m = 2, 3, \ldots, n \tag{1}$$

Assume that at some point in time, train 1 makes unplanned stop. The duration of this stop is random value τ. The subsequent trains suffer knock-on delays, when the value τ is large enough. Following train stops when the distance to the front train is reduced to s_0. It is assumed that as soon as the front train restore running, then the next one immediately follows it. The following problem is considered: to find out the probability distribution of the random arrival headway between the trains $(k-1)$ and k at the destination B (denote this headway as ν_k), assume that only the first train makes an unplanned stop. In other words, we need to find the (cumulative) distribution functions $W_k(t) = P(\nu_k < t)$, $k = 2, 3, \ldots, n$. Call this problem by the first problem.

3.2. The second model

Suppose that train 1 was delayed at station A at the moment $t = 0$ and waited for a random time τ. If $\tau < \mu_2$, then trains 2, 3, and so on, depart at the planned times: $T^{(2)}$, $T^{(3)}$, etc. If $\tau > \mu_2$, then train 2 will be delayed and will depart at the time $\tau + t_0 > T^{(2)}$. Train 3 departs according to the same rule depending on the delay time of train 2, and so on. In this formulation, ν_k is actual departure headway between the trains with numbers $(k-1)$ and k. It is required to determine the distribution functions $W_k(t)$ of random variables ν_k, $k = 2, 3, \ldots, n$.

Example 1. Let $n = 5$, $\mu_k = 2$, $k = \overline{2,5}$, $t_0 = 1$. The moments of planned departures of trains satisfy the equalities $T^{(k)} = 3(k-1)$, $k = \overline{1,5}$. **Figure 2** shows the process of headways ν_k forming, $k = \overline{2,5}$, depending on the six values of the interval τ. The dots represent real train departure times that result from the primary delay τ.

The basic model assumptions are follows: (1) only train 1 is exposed to primary delay τ. (2) $T^{(k)} - T^{(k-1)} > t_0$, $k = 2, 3, \ldots, n$.

Denote by $R^{(k)}$ the real departure time of the train with number k, which depends on τ and t_0.

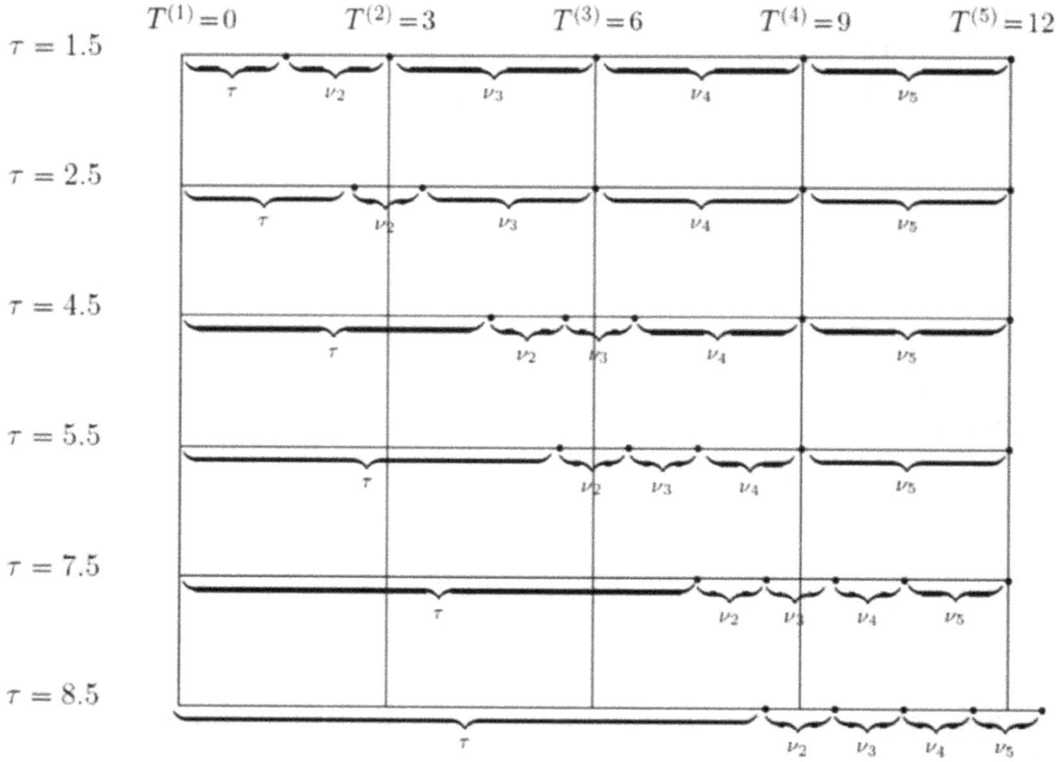

Figure 2. The headways ν_k for some values τ.

We suppose that the departure times of trains satisfy the following two rules. Let k be fixed, $2 \le k \le n$. The first rule: if $R^{(k-1)} \le T^{(k)} - t_0$, then $R^{(k)} = T^{(k)}$. The second rule: if $R^{(k-1)} \ge T^{(k)} - t_0$, then $R^{(k)} = R^{(k-1)} + t_0$. Obviously, $R^{(k)} \ge T^{(k)}$.

In what follows, we use the notation $I(x \in A) = \begin{cases} 1, \text{if } x \in A, \\ 0, \text{if } x \in R \setminus A, \end{cases}$ where A is an arbitrary set on the real line R.

Suppose that the total number of trains is equal to $n \ge 2$. Formally, we set $\nu_k = 0$ if $k > n$. Let us proceed to the formulation of the obtained results. We note that the proofs of the majority of the assertions are not given here due to the condition on the size. They take up a lot of space and will be published in our other work.

Theorem 1. 1. *If* $\tau < \mu_2$, *then* $\nu_2 = \mu_2 + t_0 - \tau$, $\nu_k = \mu_k + t_0$, $3 \le k \le n$.

2. *Let k be a fixed integer,* $2 \le k \le n$. If $\tau \ge \sum_{j=2}^{k} \mu_j$, then $\nu_2 = \ldots = \nu_k = t_0$.

3. If $\sum_{j=2}^{k} \mu_j \le \tau < \sum_{j=2}^{k+1} \mu_j$, then

$$\nu_{k+1} = I(k+1 \le n)\left[\sum_{j=2}^{k+1} \mu_j + t_0 - \tau\right], \qquad (2)$$

$$\nu_m = I(m \le n)(\mu_m + t_0), \; m = k+2, \ldots, n \tag{3}$$

Theorem 2. *Let $n \ge 2$. For any k, $2 \le k \le n$, the following formula holds*

$$W_k(t) = I(t > t_0)\left[\mathrm{P}\left(\mu_k < t - t_0, \tau < \sum_{j=2}^{k-1}\mu_j\right) + \mathrm{P}\left(\tau + t - t_0 > \sum_{j=2}^{k}\mu_j, \tau \ge \sum_{j=2}^{k-1}\mu_j\right)\right], \tag{4}$$

in particular,

$$W_2(t) = I(t > t_0)\mathrm{P}(\tau + t - t_0 > \mu_2) \tag{5}$$

Let us introduce the notations, $G(x) = \mathrm{P}(\tau < x)$, $\overline{G}(x) = \mathrm{P}(\tau > x)$. Note that $G(x) + \overline{G}(x) + \mathrm{P}(\tau = x) = 1$. We denote by $g(x)$ the density function of τ in the case when it is absolutely continuous.

Further, some corollaries of Theorem 2 are formulated.

Corollary 1. *Let μ_j, $2 \le j \le n$, be arbitrary positive numbers, then for $2 \le k \le n$*

$$W_k(t) = I(t_0 < t \le \mu_k + t_0)\,\overline{G}\left(\sum_{j=2}^{k}\mu_j - t + t_0\right) + I(t > \mu_k + t_0), \tag{6}$$

in particular,

$$W_2(t) = I(t > t_0)\overline{G}(\mu_2 - t + t_0). \tag{7}$$

Example 2. Let the primary delay τ have exponential distribution, that is,

$$g(t) = I(t \ge 0)\lambda e^{-\lambda t}, \; \lambda > 0 \tag{8}$$

As initial parameters, we take the following quantities.

$$\lambda = 0.4, \; t_0 = 3, \; \mu_2 = 5, \; \mu_3 = 6, \; \mu_4 = 10 \tag{9}$$

Graphs of the functions $W_2(t)$ from Eq. (7), $W_3(t)$ and $W_4(t)$ from Eq. (6) with the parameters (Eq. (9)) are depicted in **Figure 3**.

It should be noted that in this and the subsequent examples, we use the following measures for the values: μ_k, $T^{(k)}$, t_0, τ, τ_k, ν_k, T, b, $\mathrm{E}\nu_k$ (minutes, min); λ (1/min); $\mathrm{D}\nu_k$ (min^2). The product $\alpha\beta$ (as mean of μ_k), where α is a shape parameter, β is a scale parameter (in min).

Corollary 2. *Let $\mu_j = T$, $2 \le j \le n$, be a positive constant, then for $2 \le k \le n$*

$$W_k(t) = I(t_0 < t \le T + t_0)\overline{G}((k-1)T - t + t_0) + I(t > T + t_0), \tag{10}$$

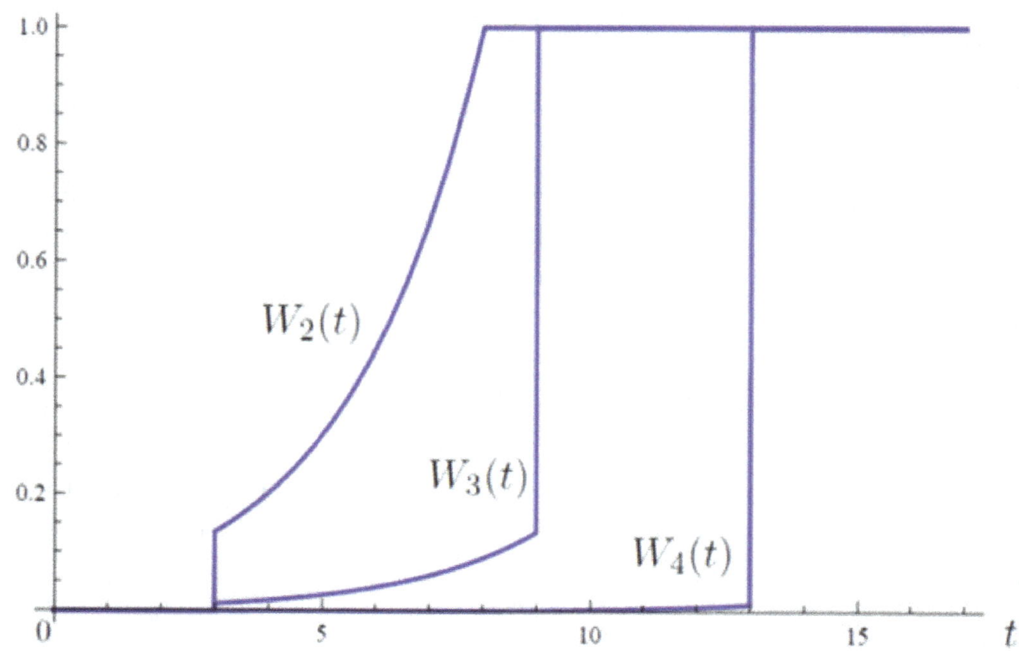

Figure 3. Behavior of the functions $W_k(t)$, $k = 2, 3, 4$.

in particular,

$$W_2(t) = I(t > t_0)\overline{G}(T - t + t_0). \tag{11}$$

Example 3. Let τ has density (Eq. (8)). As initial parameters, we take the following quantities:

$$\lambda = 0.4,\ t_0 = 4,\ T = 8. \tag{12}$$

Graphs of the functions $W_2(t)$ from Eq. (11), $W_3(t)$ and $W_4(t)$ from Eq. (10) with the parameters (Eq. (12)) are depicted in **Figure 4**.

Figures 3 and **4** show that in the case of constant μ_j, the primary delay τ practically does not affect the fourth train and all subsequent ones. This is consistent with the equality $\lim_{k\to\infty} W_k(t) = I(t > t_0 + T)$ which, as it is not difficult to verify, follows from Eq. (10).

Remark 1. It is known that the distribution of sum of the independent random variables is the convolution of their distributions. The convolution of distribution functions F_1 and F_2 is determined by the formula $(F_1{*}F_2)(x) = \int_{-\infty}^{\infty} F_1(x-y)dF_2(y)$, where the integral sign means the improper Riemann-Stieltjes integral. We consider exceptionally piecewise-continuous distribution functions, then the indicated integral exists with the exception of the case when F_1 and F_2 have at least one common discontinuity point (e.g., [12]). The convolution operation is permutable. In the case, when $F_1 = F_2 = \ldots = F_m = F$, we shall use the following notations: $F^{*2} := F{*}F$, $F^{*m} := F{*}F^{*(m-1)}$, $m \geq 2$. By definition, we assume that $F^{*1} := F$. The convolution $(f_1{*}f_2)(x)$ of densities f_1 and f_2 is defined as the improper Riemann integral $\int_{-\infty}^{\infty} f_1(x-y)f_2(y)dy$.

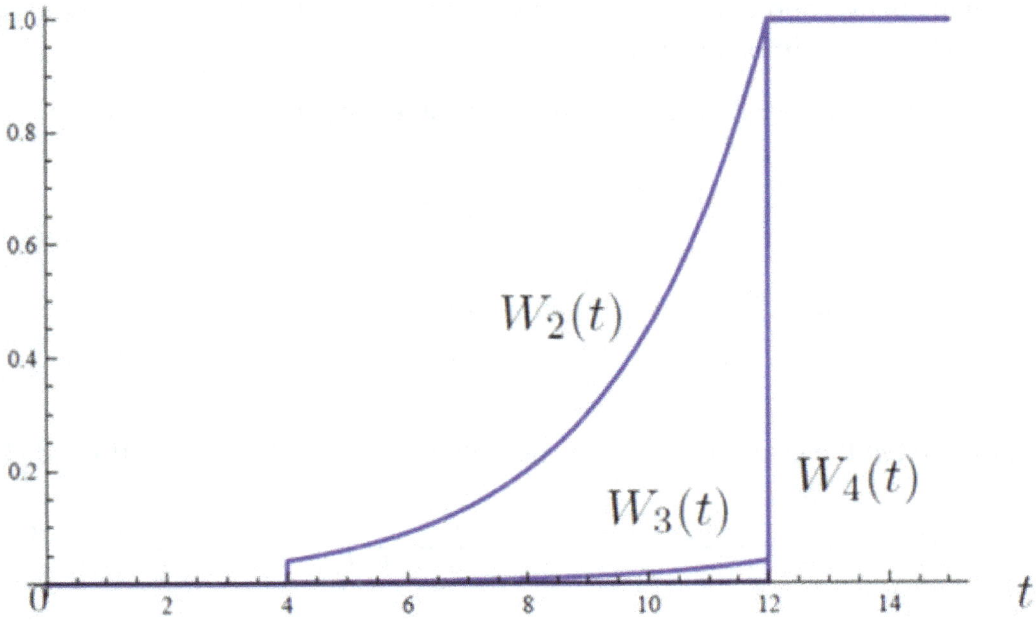

Figure 4. Behavior of the functions $W_k(t)$, $k = 2, 3, 4$.

Corollary 3. *Let μ_j, $2 \leq j \leq n$, be independent identically distributed random variables with a continuous distribution function $\Psi(x)$. Let τ be independent of μ_j, $2 \leq j \leq n$. Then*

$$W_2(t) = I(t > t_0) \int_{-\infty}^{\infty} \overline{G}(z - t + t_0) d\Psi(z), \tag{13}$$

$$W_k(t) = I(t > t_0) \left\{ \Psi(t - t_0) + \int_{-\infty}^{\infty} \left[\int_{t-t_0}^{\infty} \overline{G}(z + u - t + t_0) d\Psi(z) \right] d\Psi^{*(k-2)}(u) \right\},\ 3 \leq k \leq n. \tag{14}$$

Corollary 4. *Let μ_j, $2 \leq j \leq n$, be independent identically distributed random variables with a density function $\psi(x)$. Let τ be independent of all μ_j and has a density function $g(x)$. Then*

$$W_2(t) = I(t > t_0) \int_{-\infty}^{\infty} \left(\int_{z-t+t_0}^{\infty} g(x) dx \right) \psi(z) dz, \tag{15}$$

$$W_k(t) = I(t > t_0) \left\{ \int_{-\infty}^{t-t_0} \psi(z) dz + \int_{-\infty}^{\infty} \left[\int_{t-t_0}^{\infty} \left(\int_{z+u-t+t_0}^{\infty} g(x) dx \right) \psi(z) dz \right] \psi^{*(k-2)}(u) du \right\}, \tag{16}$$

$$3 \leq k \leq n.$$

Remark 2. The integration limit "$-\infty$" can be replaced by 0 in Corollaries 3 and 4 if $\mu_j \geq 0$. On the other hand, we may consider in these corollaries the case when μ_j takes values of different signs. From a practical point of view, such an approach is acceptable if the probability that these random quantities take negative values is small enough. This assumption allows to consider, for example, models in which the random variables μ_j are normally distributed with a variance small enough and to use the property that the class of normal distributions is closed with respect to the convolution operation.

Example 4. Let τ has the density (Eq. (8)), and all μ_j have the same gamma density

$$\psi(t) = I(t > 0)\frac{e^{-t/\beta}t^{\alpha-1}}{\Gamma(\alpha)\beta^{\alpha}}, \tag{17}$$

where $\alpha > 0, \beta > 0, \Gamma(\alpha) = \int_0^{\infty} x^{\alpha-1}e^{-x}dx$ is gamma function. Put

$$\lambda = 0.3,\ t_0 = 5,\ \alpha = 14,\ \beta = 0.5. \tag{18}$$

One can show that in the example under consideration it follows from Eqs. (15) and (16) that

$$W_k(t) = I(t > t_0)\left[1 - \frac{\Gamma(\alpha, (t - t_0 + b)/\beta)}{\Gamma(\alpha)} + ae^{\lambda(t-t_0+b)}\left(\frac{1}{1+\lambda\beta}\right)^{(k-1)\alpha}\frac{\Gamma(\alpha, (1+\lambda\beta)(t - t_0 + b)/\beta)}{\Gamma(\alpha)}\right],$$

where $\Gamma(\alpha, y) = \int_y^{\infty} x^{\alpha-1}e^{-x}dx$ is incomplete gamma function. Graphs of the distribution functions $W_k(t), 2 \le k \le 5,$ with the parameters (Eq. (18)) are depicted in **Figure 5**.

It is not difficult to verify that for $W_k(t)$ from Example 4 the following formula holds:

$$\begin{aligned}
&\lim_{k\to\infty} W_k(t)|_{a=1,b=0} \\
&= \lim_{k\to\infty}\left[I(t > t_0)\left(1 - \frac{\Gamma(\alpha, (t - t_0)/\beta)}{\Gamma(\alpha)} + e^{\lambda(t-t_0)}\left(\frac{1}{1+\lambda\beta}\right)^{(k-1)\alpha}\frac{\Gamma(\alpha, (1+\lambda\beta)(t - t_0)/\beta)}{\Gamma(\alpha)}\right)\right] \\
&= W_{\infty}(t) := I(t > t_0)\left[1 - \frac{\Gamma(\alpha, (t - t_0)/\beta)}{\Gamma(\alpha)}\right].
\end{aligned}$$

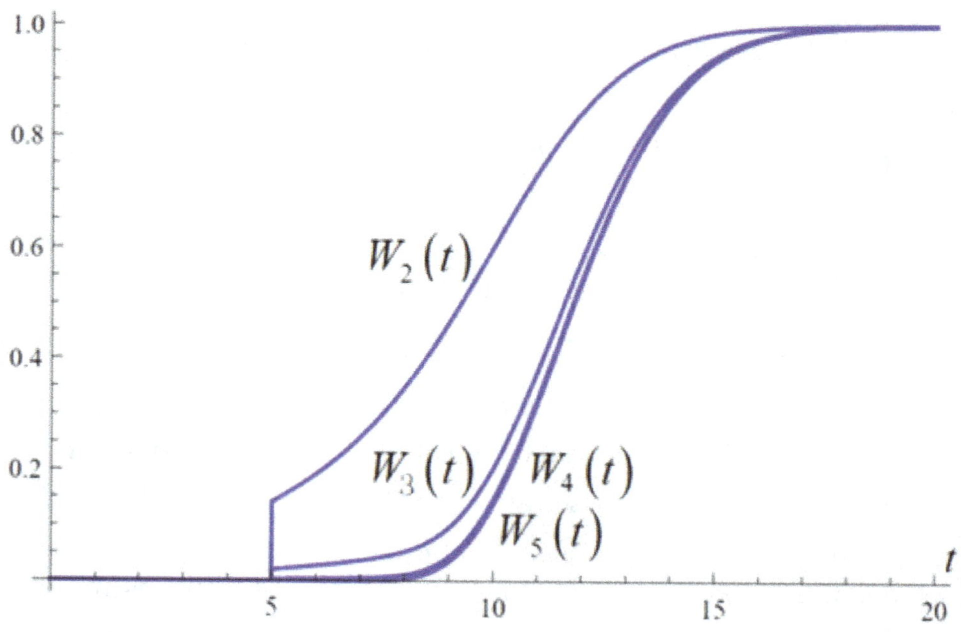

Figure 5. Behavior of the functions $W_k(t)$, $k = 2, 3, 4, 5$.

It can be seen from **Figure 5**, curves $W_4(t), W_5(t)$ and so on are practically merged. Hence, in the case under consideration, one can draw the following conclusion: primary delay τ affects to fifth and all successive trains approximately like on the fourth one.

Remark 3. We define the 0-fold convolution as a generalized function with the following property: the equality $\int_{-\infty}^{\infty} v(t)\psi^{*0}(t)dt = v(0)$ holds for any bounded continuous function $v(t)$. Then, Eq. (16) for $k = 2$ coincides with Eq. (15).

We do not give proofs for the statements of Section 3 because of limitations on the volume. We will make this in another work.

4. Some results on the knock-on delays

Denote by N the random number of knock-on delays (within the framework of the model under consideration).

Lemma 1. *For each fixed integer m,* $1 \le m \le n-1$,

$$P(N \ge m) = P\left(\tau > \sum_{j=2}^{m+1} \mu_j\right). \tag{19}$$

Proof. Easily seen: $\{N=0\} = \left\{t_0 \le \tau + t_0 \le T^{(2)}\right\}, \{N=m\} = \left\{T^{(m+1)} < \tau + mt_0 \le T^{(m+2)} - t_0\right\},$

$m = 1, 2, \ldots, n-2$, $\{N = n-1\} = \left\{\tau + (n-1)t_0 > T^{(n)}\right\}$. This implies that

$$P(N \ge m) = P\left(\tau + mt_0 > T^{(m+1)}\right) = P\left(\tau > \sum_{j=2}^{m+1} \mu_j\right). \qquad \square$$

Here and below, the sign □ denotes the end of the proof.

The corollaries of this lemma are given below. Their proofs are simple and therefore we do not present them.

Corollary 5. *If* $\mu_j = T$ *is a constant value,* $2 \le j \le n$, *then for every fixed integer m,* $1 \le m \le n-1$, *we have the equality* $P(N \ge m) = \overline{G}(mT)$.

Corollary 6. *If* $\mu_j = T$ *is a constant value,* $2 \le j \le n$, *and* τ *is exponentially distributed with parameter* λ, *then for every fixed integer m,* $1 \le m \le n-1$, *the following equality holds,*

$$P(N \ge m) = e^{-\lambda m T}.$$

Corollary 7. If $\mu_2, \ldots, \mu_n$ *are independent identically distributed random variables with a density function* ψ, *then for every fixed integer m,* $1 \le m \le n-1$, *we have the equality*

$$P(N \geq m) = \int_{-\infty}^{\infty} \overline{G}(u)\psi^{*m}(u)du.$$

In what follows, $\tau_1 = \tau$ is the delay duration of the first train, τ_k, $k = 2, \ldots, n$, is the knock-on delay of the k-th train. The problem is to find the distribution functions $G_k(t) = P(\tau_k < t)$, $k = 2, 3, \ldots, n$. Note that the solution of this problem, which we call by the second problem, allows us to find the distribution of the deviations of the real arrival times from the planned ones.

In what follows, we will use the notation $a \vee b$ instead of $\max(a, b)$.

Theorem 3. *The following formula holds:*

$$\tau_k = (\tau_{k-1} - \mu_k) \vee 0, \ 2 \leq k \leq n. \tag{20}$$

Corollary 8. *The following formula holds:*

$$\tau_k = \left(\tau - \sum_{j=2}^{k} \mu_j\right) \vee 0, \ 2 \leq k \leq n. \tag{21}$$

It should be noted that within the framework of our model the deviation of the real arrival time from the planned one for k-th train coincides with τ_k, $1 \leq k \leq n$. **Figure 6** illustrates this statement.

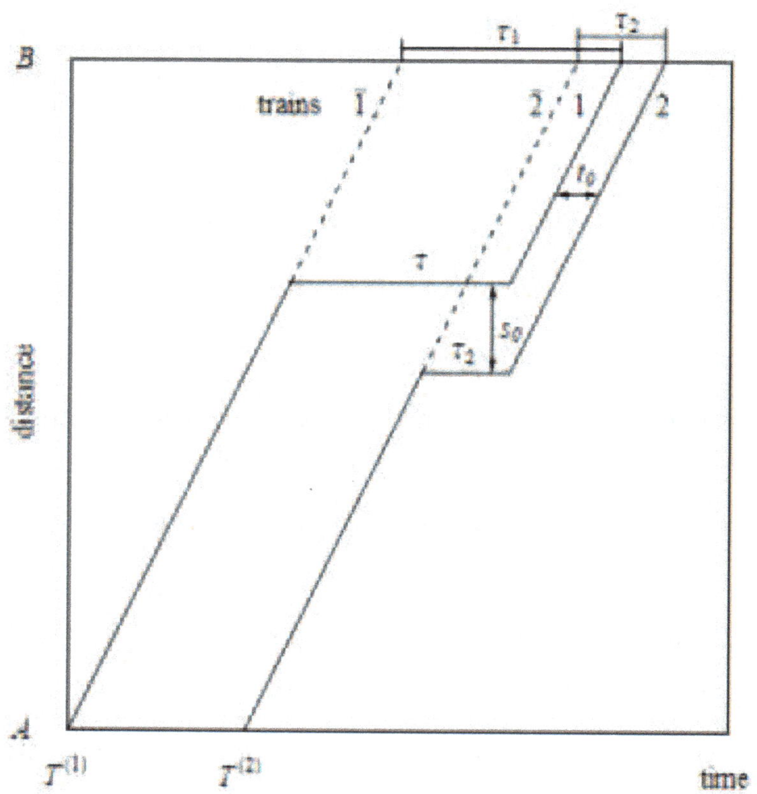

Figure 6. Deviation arrival times from the schedule: delays τ_1 and τ_2.

The dotted lines (lines $\overline{1}$ and $\overline{2}$) represent the scheduled trajectories of trains 1 and 2, solid lines (1 and 2) depict the real trajectories taking into account the delays. It can be seen that the arrival time of the train 1 differs from the schedule at τ and the train 2 on the τ_2.

Denote $\overline{\mu}_k = \sum_{j=2}^{k} \mu_j$, $2 \leq k \leq n$. As it follows from the assumption that the random variables $\mu_2, \ldots, \mu_k$ have the same distribution function $\Psi(t)$. They are mutually independent. The random variable $\overline{\mu}_k$ has the distribution function $\Psi^{*(k-1)}(t)$.

Corollary 9. *The distribution function of τ_k has the following form:*

$$G_k(t) = I(t > 0)\mathrm{P}(\tau - \overline{\mu}_k < t), \ 2 \leq k \leq n. \tag{22}$$

The next Corollaries 10 and 11 follow from Corollary 9 in an obvious way.

Corollary 10. Let *$\mu_j > 0$, $2 \leq j \leq n$ be some constant values. Then*

$$G_k(t) = I(t > 0)G(t + \overline{\mu}_k). \tag{23}$$

Corollary 11. *Let $\mu_j = T > 0$, $2 \leq j \leq n$ be a constant value. Then*

$$G_k(t) = I(t > 0)G(t + (k-1)T). \tag{24}$$

Corollary 12. *Let μ_j, $2 \leq j \leq n$ be independent identically distributed random variables with a continuous distribution function $\Psi(t)$. Let τ be independent of $\mu_2, \ldots, \mu_n$. Then*

$$G_k(t) = I(t > 0) \int_{-\infty}^{\infty} G(t + y) d\Psi^{*(k-1)}(y). \tag{25}$$

Corollary 13. *Let μ_j, $2 \leq j \leq n$ be independent identically distributed random variables with a density function $\psi(t)$. Let τ be independent of μ_j, $2 \leq j \leq n$ and has a density function $g(t)$. Then $G_k(t) = I(t > 0) \int_{-\infty}^{\infty} \left(\int_{-\infty}^{t+y} g(z) dz \right) \psi^{*(k-1)}(y) dy$.*

5. Proof of Theorem 3 and its corollaries

Lemma 2. *The following formula is valid:*

$$\tau_2 = (\tau - \mu_2) \vee 0. \tag{26}$$

Proof. Let $t > 0$ be the time spent by the train on the path length (distance to the place, where an unplanned stop of the train 1 occurred). We show the equality $\tau_2 = 0$ holds under the condition $\tau \leq \mu_2$. The departure time of the train 1 after stopping is $t + \tau$. The time point when train 2 reaches s can be written as $\mu_2 + t_0 + t$. The knock-on delay of train 2 will not occur, i.e., $\tau_2 = 0$, in the case, when the indicated time points are separated by the value $r \geq t_0$,

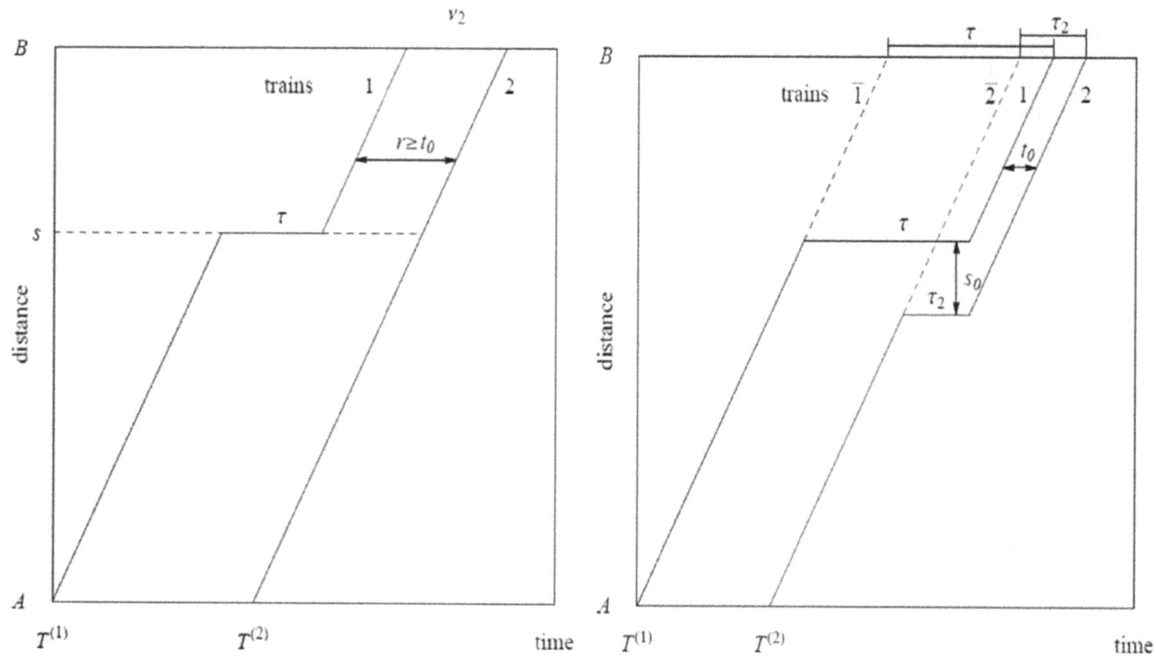

Figure 7. Two traffic scenarios: (a) lack of the knock-on delay; (b) knock-on delay occurs.

i.e., $r = \mu_2 + t_0 + t - (t + \tau) \geq t_0$, or, which is the same thing, $\tau \leq \mu_2$. The considered case is illustrated in **Figure 7a**.

The knock-on delay of the duration $\tau_2 = \tau - \mu_2$ will occur when $\tau > \mu_2$. Indeed, since trains after a random stop depart simultaneously, then the equality $t + \tau = \mu_2 + t_0 + (t - t_0) + \tau_2$ holds, i.e., $\tau = \mu_2 + \tau_2$. The case under consideration is illustrated in **Figure 7b**. Thus, the validity of Eq. (26) is shown. □

Proof of Theorem 3. We shall use the method of mathematical induction. The equality (Eq. (20)) for $k = 2$ is established by Lemma 2. Let Eq. (20) be satisfied. We show that:

$$\tau_{k+1} = \left(\tau_k - \mu_{k+1}\right) \vee 0, 2 \leq k + 1 \leq n. \tag{27}$$

It follows from the inductive hypothesis that $\tau_k = 0$ under the condition $\tau_{k-1} \leq \mu_k$. But if the delay of the k-th train is 0, then the next train does not undergo any delay, that is, $\tau_{k+1} = 0$. The present case is illustrated in **Figure 8**.

In the case, when $\tau_{k-1} > \mu_k$, a knock-on delay of the k-th train occurs and equals to $\tau_k = \tau_{k-1} - \mu_k$ (according to the inductive hypothesis). Further, two cases are possible: either (1) a delay τ_k entails a delay τ_{k+1}, or (2) $\tau_{k+1} = 0$.

Case 1. If the k-th train is delayed, then (k + 1)-th one will be delayed only if $\tau_k > \mu_{k+1}$, and its delay duration is $\tau_{k+1} = \tau_k - \mu_{k+1}$ (this fact follows from the equality of the moments of departure of the k-th and (k + 1)-th trains after an unscheduled stop: $T^{(k)} + t - (k-1)t_0 + \tau_k = T^{(k+1)} + t - kt_0 + \tau_{k+1}$). Case 1 is illustrated in **Figure 9a**.

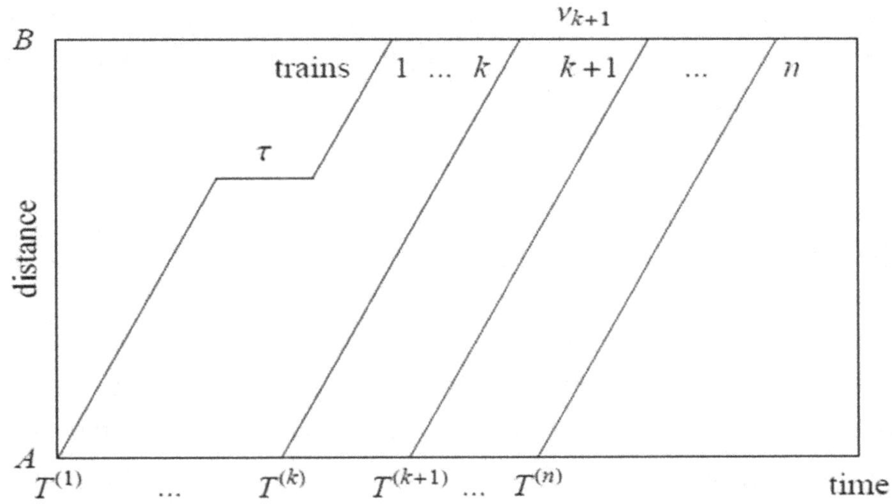

Figure 8. The case, when $\tau_{k-1} \leq \mu_k$: knock-on delays of k-th and consecutive trains are not observed.

Case 2. If the k-th train is delayed, then $(k + 1)$-th one will not be delayed ($\tau_{k+1} = 0$) only if $\tau_k \leq \mu_{k+1}$. Case 2 is illustrated in **Figure 9b**. Note that if the knock-on delay of the k-th train occurs, a conflict of the k-th train with $(k+1)$-th is described similar to the interaction of trains 1 and 2 (see Lemma 2). All described cases lead to Eq. (20). □

Proof of Corollary 8. We indicate that Eq. (21) is similar to Eq. (20). According to the statement of Theorem 3, we have:

$$\tau_2 = (\tau - \mu_2) \vee 0,\ \tau_3 = (\tau_2 - \mu_3) \vee 0,\ \tau_k = (\tau_{k-1} - \mu_k) \vee 0. \tag{28}$$

Using the method of mathematical induction and taking into account that $\overline{\mu}_{k-1} + \mu_k = \overline{\mu}_k$, we obtain Eq. (21) from Eq. (28). □

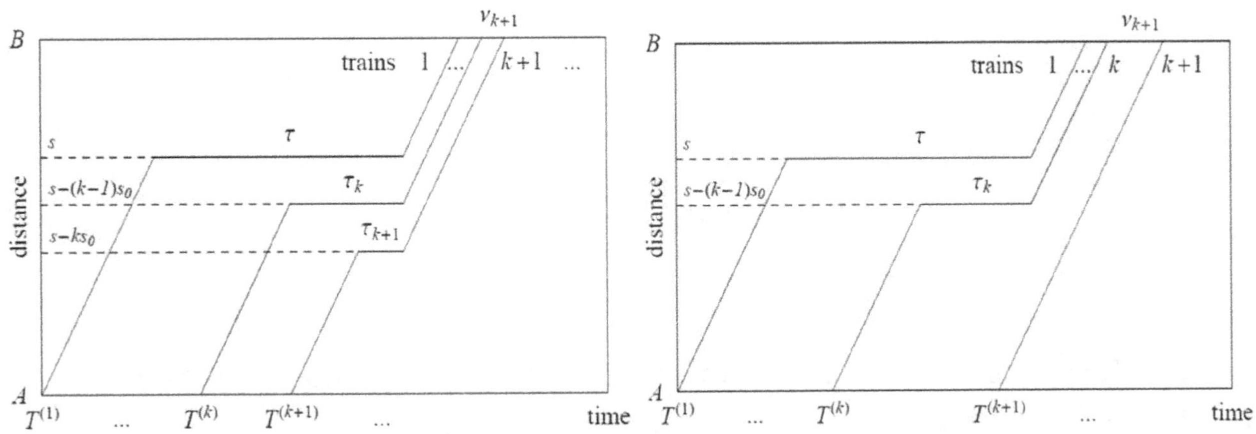

Figure 9. Two traffic scenarios: (a) the case, when $\tau_{k-1} > \mu_k$, $\tau_k > \mu_{k+1}$: all trains up to $(k+1)$-th are detained; (b) the case, when $\tau_{k-1} > \mu_k$, $\tau_k \leq \mu_{k+1}$: all trains from $(k+1)$-th up to n-th are not delayed.

Proof of Corollary 9. It follows from Corollary 8 that $\tau_k = 0$ if $\tau \le \overline{\mu}_k$ (see, e.g., **Figure 8**), and $\tau_k = \tau - \overline{\mu}_k$ if $\tau > \overline{\mu}_k$ (see, e.g., **Figure 9a**). Using the law of total probability, we obtain the following chain of equalities:

$$\begin{aligned} G_k(t) &= \mathrm{P}(\tau_k < t) = I(t>0)\,\mathrm{P}(\tau_k < t) \\ &= I(t>0)\big(\mathrm{P}(\tau_k < t | \tau \le \overline{\mu}_k)\mathrm{P}(\tau \le \overline{\mu}_k) + \mathrm{P}(\tau_k < t | \tau > \overline{\mu}_k)\mathrm{P}(\tau > \overline{\mu}_k)\big) \\ &= I(t>0)\mathrm{P}(\tau - \overline{\mu}_k \le 0) + I(t>0)\mathrm{P}(0 < \tau - \overline{\mu}_k < t) = I(t>0)\mathrm{P}(\tau - \overline{\mu}_k < t). \end{aligned}$$ □

Proof of Corollary 12. Apply the well-known assertion to Eq. (22): if Y_1 and Y_2 are independent random variables, then for any function of two variables $f(\cdot,\cdot)$ and any $c \in R$, the following equality holds: $\mathrm{P}(f(Y_1, Y_2) < c) = \int_{-\infty}^{\infty} \mathrm{P}(f(y, Y_2) < c)dF_1(y)$, where F_1 is the distribution function of Y_1. Consequently, $G_k(t) = I(t>0)\int_{-\infty}^{\infty} \mathrm{P}(\tau - y < t)d\Psi^{*(k-1)}(y)$. This implies Eq. (25). □

Proof of Corollary 13. The assertion follows from Eq. (25). □

Note that the function $G_k(t)$ has a jump at zero which is equal to:

$G_k(0+) = \int_{-\infty}^{\infty} G^+(y)d\Psi^{*(k-1)}(y)$, where $G^+(y) = \lim_{t\to 0+} G(t+y)$.

In the case, when τ and μ_j are absolutely continuous, it follows from Eq. (25) that

$$G_k(t) = I(t>0)\int_{-\infty}^{\infty}\left(\int_{-\infty}^{t+y} g(z)dz\right)\psi^{*(k-1)}(y)dy, \tag{29}$$

where $g(\cdot)$ and $\psi(\cdot)$ are the density functions of τ and μ_1, respectively, $\psi^{*j}(y)$ is the j-fold convolution of the density $\psi(\cdot)$. In this case, we also have

$$g_k(t) := I(t>0)G_k'(t) = I(t>0)\int_{-\infty}^{\infty} g(t+y)\psi^{*(k-1)}(y)dy. \tag{30}$$

If we assume that $\tau \ge 0$, then we deduce from Eq. (29) that

$$G_k(t) = I(t>0)\int_{-t}^{\infty}\left(\int_{0}^{t+y} g(z)dz\right)\psi^{*(k-1)}(y)dy, \tag{31}$$

and we get from Eq. (30),

$$g_k(t) = I(t>0)\int_{-t}^{\infty} g(t+y)\psi^{*(k-1)}(y)dy. \tag{32}$$

6. Corollary of Theorem 2 when the distribution of primary delay is a mixture of exponential and one-point distributions

Consider the cumulative distribution function of the following type:

$$G(x) \equiv \mathrm{P}(\tau < x) = I(x \ge b)\left(1 - ae^{-\lambda(x-b)}\right), \tag{33}$$

where $0 \le a \le 1$, $b \ge 0$, and $\lambda > 0$ are some parameters. Such distribution function is considered, for example, in [8]. It is easy to see that $G(x) = (1-a)G_0(x-b) + aG(x-b;\lambda)$, where $G_0(x)$ is the distribution function of the degenerate distribution concentrated at the point $x = 0$, $G(x;\lambda) = I(x \ge 0)\left[1 - e^{-\lambda x}\right]$.

Let us find out the form of the distribution functions (Eqs. (13) and (14)) in the case of Eq. (33), when the function Ψ is continuous. In what follows, we mean that $n \ge 3$.

Lemma 3. *Let the function G be defined by Eq. (33), and Ψ be continuous. Then*

$$W_2(t) = I(t > t_0)\left(\Psi(t - t_0 + b) + ae^{\lambda(t-t_0+b)} \int_{t-t_0+b}^{\infty} e^{-\lambda z} d\Psi(z)\right), \tag{34}$$

$$\begin{aligned} W_k(t) = I(t > t_0)\Bigg\{ &\Psi(t - t_0) + ae^{\lambda(t-t_0+b)} \Bigg[\int_b^{\infty} e^{-\lambda u} d\Psi^{*(k-2)}(u) \int_{t-t_0}^{\infty} e^{-\lambda z} d\Psi(z) \\ &+ \int_{-\infty}^{b} e^{-\lambda u} \left(\int_{t-t_0-u+b}^{\infty} e^{-\lambda z} d\Psi(z)\right) d\Psi^{*(k-2)}(u)\Bigg] \\ &+ \int_{-\infty}^{b} (\Psi(t - t_0 - u + b) - \Psi(t - t_0)) d\Psi^{*(k-2)}(u)\Bigg\}, \quad k \ge 3. \end{aligned} \tag{35}$$

Proof. According to Eq. (33), one may conclude that function $\overline{G}(x)$ has a unique discontinuity point $x = b$. Hence, the integral $\int_{-\infty}^{\infty} \overline{G}(z - t + t_0) d\Psi(z)$ exists for any continuous distribution function Ψ. Note that if $\Psi(z)$ had a discontinuity point $z = t_1$, then the function $\overline{G}(z - t + t_0)$ would also be discontinuous at the point $z = t_1$ for $t = t_0 + t_1 - b$, and then the considered integral would not exist (see Remark 1). Since

$$\overline{G}(x) = I(x < b) + I(x \ge b) ae^{-\lambda(x-b)}, \tag{36}$$

then

$$\int_{-\infty}^{\infty} \overline{G}(z - t + t_0) d\Psi(z) = \Psi(t - t_0 + b) + ae^{\lambda(t-t_0+b)} \int_{t-t_0+b}^{\infty} e^{-\lambda z} d\Psi(z). \tag{37}$$

In accordance with Eq. (13), the relation (Eq. (34)) is proved.

Let $k \ge 3$. It follows from Eq. (14) that

$$W_k(t) = I(t > t_0)\left[\Psi(t - t_0) + \int_{-\infty}^{\infty} V(u) d\Psi^{*(k-2)}(u)\right], \tag{38}$$

where $V(u) = \int_{t-t_0}^{\infty} \overline{G}(z + u - t + t_0) d\Psi(z)$. Given Eq. (36), it is easy to see that

$$V(u) = V_1(u) + V_2(u), \tag{39}$$

$V_1(u) = ae^{-\lambda(u-t+t_0-b)} \int_{t-t_0}^{\infty} I(z + u - t + t_0 \ge b) e^{-\lambda z} d\Psi(z)$, $V_2(u) = \int_{t-t_0}^{\infty} I(z + u - t + t_0 < b) d\Psi(z)$.

By using equalities

$$\{(u,z): u\geq b, z>t-t_0, z\geq t-t_0-u+b\}=\{(u,z): u\geq b, z>t-t_0\},$$
$$\{(u,z): u<b, z>t-t_0, z\geq t-t_0-u+b\}=\{(u,z): u<b, z\geq t-t_0-u+b\},$$

we receive

$$\int_{-\infty}^{\infty} V_1(u)d\Psi^{*(k-2)}(u)=ae^{\lambda(t-t_0+b)}\left[\int_b^{\infty}\left(\int_{t-t_0}^{\infty} e^{-\lambda(z+u)}d\Psi(z)\right)d\Psi^{*(k-2)}(u)\right.$$
$$\left.+\int_{-\infty}^{b}\left(\int_{t-t_0-u+b}^{\infty} e^{-\lambda(z+u)}d\Psi(z)\right)d\Psi^{*(k-2)}(u)\right]. \tag{40}$$

Since $\{(u,z): u\geq b, z>t-t_0, z<t-t_0-u+b\}=\varnothing$,

$$\{(u,z): u<b, z>t-t_0, z<t-t_0-u+b\}=\{(u,z): u<b, t-t_0<z<t-t_0-u+b\},$$

then

$$\int_{-\infty}^{\infty} V_2(u)d\Psi^{*(k-2)}(u)=\int_{-\infty}^{b}\left(\int_{t-t_0}^{t-t_0-u+b} d\Psi(z)\right)d\Psi^{*(k-2)}(u). \tag{41}$$

It follows from Eqs. (39)–(41) that

$$\int_{-\infty}^{\infty} V(u)d\Psi^{*(k-2)}(u)=ae^{\lambda(t-t_0+b)}\left[\int_b^{\infty} e^{-\lambda u}d\Psi^{*(k-2)}(u)\int_{t-t_0}^{\infty} e^{-\lambda z}d\Psi(z)\right.$$
$$\left.+\int_{-\infty}^{b}\left(\int_{t-t_0-u+b}^{\infty} e^{-\lambda z}d\Psi(z)\right)e^{-\lambda u}d\Psi^{*(k-2)}(u)\right]$$
$$+\int_{-\infty}^{b}(\Psi(t-t_0-u+b)-\Psi(t-t_0))d\Psi^{*(k-2)}(u). \tag{42}$$

The equalities Eq. (38) and Eq. (42) entail Eq. (35). □

Below we give without a proof a corollary of Lemma 3 in the case when μ_j are not random variables, and they are equal to the same constant.

Corollary 14. *Let* $\mu_j=T>0$, $2\leq j\leq n$, *be a constant. Let function G be defined by Eq. (33). Then, for* $2\leq k\leq n$, *the following formula holds:*

$$W_k(t)=I(0\leq b\leq (k-2)T)\left[I(0<t-t_0\leq T)ae^{-\lambda((k-1)T-t+t_0-b)}+I(t-t_0>T)\right]$$
$$+I((k-2)T<b<(k-1)T)\left[I(0<t-t_0\leq (k-1)T-b)ae^{-\lambda((k-1)T-t+t_0-b)}\right.$$
$$\left.+I(t-t_0>(k-1)T-b)\right]+I(b\geq (k-1)T)I(t>t_0). \tag{43}$$

Furthermore,

$$\mathrm{E}\nu_k=I(0\leq b\leq (k-2)T)\left[t_0+T-\frac{a}{\lambda}e^{-\lambda((k-2)T-b)}\left(1-e^{-\lambda T}\right)\right]+I(b\geq (k-1)T)t_0$$
$$+I((k-2)T<b<(k-1)T)\left[t_0+(k-1)T-b+\frac{a}{\lambda}\left(e^{-\lambda((k-1)T-b)}-1\right)\right]. \tag{44}$$

$$\begin{aligned}\mathrm{D}v_k &= I(0\le b\le (k-2)T)\frac{a}{\lambda^2}e^{-\lambda((k-2)T-b)}\left[2\left(1-e^{-\lambda T}\right)-ae^{-\lambda((k-2)T-b)}\left(1-e^{-\lambda T}\right)^2-2\lambda Te^{-\lambda T}\right]\\ &+I((k-2)T<b<(k-1)T)\frac{a}{\lambda^2}e^{-\lambda((k-2)T-b)}\left[2\left(e^{\lambda((k-2)T-b)}-e^{-\lambda T}\right)\right.\\ &\left.-ae^{-\lambda((k-2)T-b)}\left(e^{\lambda((k-2)T-b)}-e^{-\lambda T}\right)^2-2\lambda((k-1)T-b)e^{-\lambda T}\right].\end{aligned} \tag{45}$$

Example 5. Figure 10 depicts the graphs of the functions $W_k(t)$ defined by Eq. (43) with $k=2, 3$ for the following parameters:

$$a=1,\ b=0,\ \lambda=0.26,\ t_0=4,\ T=7. \tag{46}$$

We calculated the values of $\mathrm{E}v_k$ and $\mathrm{D}v_k$ using the formulas (44) and (45) (see **Table 1**).

Remark 4. It can be easily seen that the larger k, $W_k(t)$ from Eq. (43) is closer to $W(t) := I(b\ge 0)I(t>t_0+T)$. This agrees with **Figure 10** and the formulas (44) and (45) due to which we have $\mathrm{E}v_k \to t_0+T, \mathrm{D}v_k \to 0$ as $k\to\infty$, and also with the results of calculations in **Table 1**.

Let the random variable τ be distributed with the density (Eq. (33)) with parameters $a=1$, $b=0$. Now, we find the condition on the parameter T, under which the probability that at least

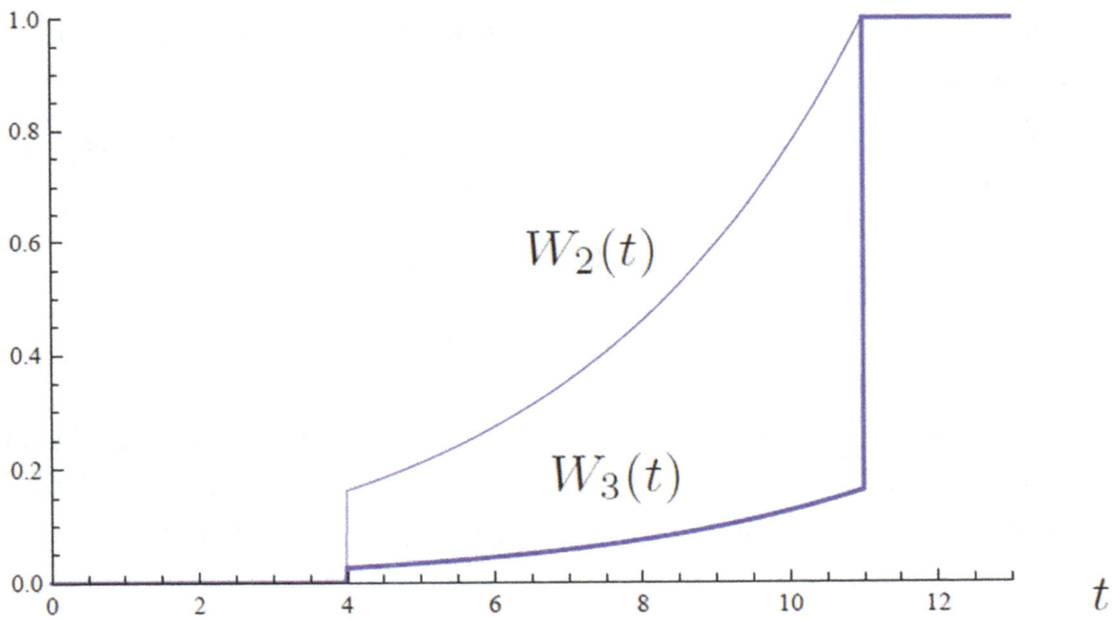

Figure 10. Behavior of the functions $W_2(t)$ and $W_3(t)$.

	$k=2$	$k=3$	$k=5$	$k=8$	$k=10$
$\mathrm{E}v_k$	7.77702	10.47779	10.98629	10.99994	10.99999
$\mathrm{D}v_k$	5.68009	2.33067	0.068156	0.00029	7.63176×10^{-6}

Table 1. The behavior $\mathrm{E}v_k$ and $\mathrm{D}v_k$ with growth of the parameter k.

m of knock-on delays will occur would not exceed a given probability p. Note that the departure headway is equal to $T + t_0$.

According to Corollary 6, it is necessary to solve the inequality $\exp(-\lambda mT) \leq p$. As a result, we obtain the desired condition:

$$T \geq (1/(m\lambda))\ln(1/p) \tag{47}$$

(see also [13]). Denote by $T(m, p, \lambda)$ the minimal T satisfying the inequality (Eq. (47)).

Example 6. Let us fix $\lambda = 0.26$. The behavior of $T(m, p, \lambda)$ as a function of the continuous parameter m with $p = 0.1$ and $p = 0.05$ is shown in **Figure 11a**. Obviously, $T(m, p, \lambda)$ is the decreasing function with respect to the argument p. Exact calculations can be made using the formula:

$$T(m, p, \lambda) = (1/(m\lambda))\ln(1/p). \tag{48}$$

Let $p = 0.1$. The behavior of $T(m, p, \lambda)$ as a function of the continuous parameter m with $\lambda = 0.26$ and $\lambda = 0.15$ is shown in **Figure 11b**. In accordance with Eq. (48), $T(m, p, \lambda)$ is the decreasing function with respect to the argument λ. In the case of exponential density $g(t)$, we have $E\tau = 1/\lambda$. Therefore, the decrease of λ leads to increase in the average of primary delay and the departure headways (if we want to reduce the number of knock-on delays).

We also obtain the corollaries of Lemma 3 in the case when μ_j are distributed according to the gamma-law with the density (Eq. (17)).

Corollary 15. *If primary delay τ has an exponential distribution $g(t) = I(t > 0)\lambda e^{-\lambda t}$ and μ_k, $2 \leq k \leq n$, has the density (Eq. (17)), then the following formulas are true:*

$$G_k(t) = I(t > 0)\left(1 - e^{-\lambda t}(\lambda\beta + 1)^{-(k-1)\alpha}\right), \tag{49}$$

$$g_k(t) = I(t > 0)(\lambda\beta + 1)^{-(k-1)\alpha}\lambda e^{-\lambda t}. \tag{50}$$

Remark 5. The function $g_k(t) = I(t > 0)G_k'(t)$ is not a density, in particular, because of $\int_{-\infty}^{\infty} g_k(t)\,dt = \int_0^{\infty} g_k(t)\,dt = G_k(\infty) - G_k(0+) = 1 - h_k \neq 1$, where h_k is the jump of the function $G_k(t)$ at the origin. At the same time, the function $\tilde{g}_k(t) := \frac{1}{1-h_k} g_k(t)$ is a density.

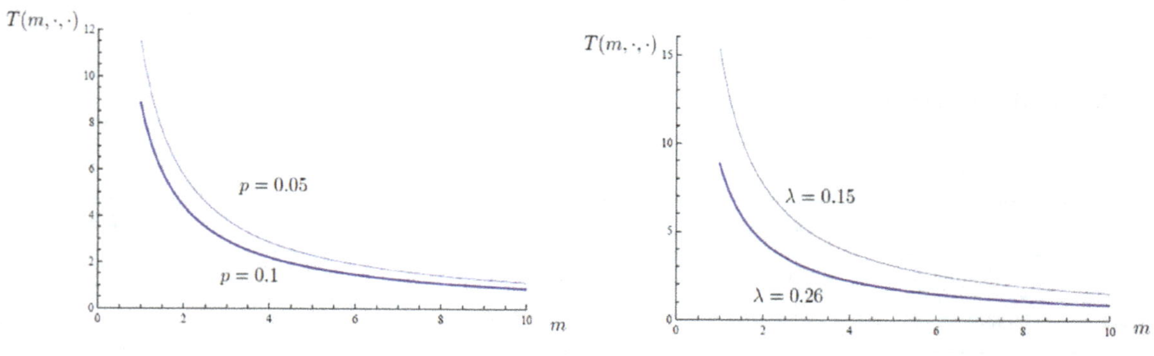

Figure 11. The behavior of the function $T(m, p, \lambda)$: (a) $\lambda = 0.26$, $p = 0.1$, or $p = 0.05$; (b) $p = 0.1$, $\lambda = 0.26$, or $\lambda = 0.15$.

Corollary 15 can be reformulated as follows.

Corollary 15*. *Let primary delay τ is exponentially distributed with a parameter λ, and μ_k, $2 \le k \le n$, have the same gamma distribution with the density (Eq. (17)). Then, τ_k has the distribution function of the form Eq. (33) with $a = (\lambda\beta + 1)^{-(k-1)\alpha}$, $b = 0$, and, consequently,*

$$P(\tau_k = 0) = G_k(0+) = 1 - (\lambda\beta + 1)^{-(k-1)\alpha}.$$

Remark 6. Let $P(\tau_2 = 0) = p$, $0 < p < 1$. Then by Corollary 15*, $P(\tau_k = 0) = 1 - (1 - p)^{k-1}$, $3 \le k \le n$. Hence, $P(\tau_k = 0) \to 1$ as $k \to \infty$.

Example 7. Let $\mu_2, \mu_3, \ldots$ be independent random variables having the same density function (Eq. (17)). We perform three series of experiments and investigate a behavior of distribution of the arrival time deviations τ_k with various combinations of parameters: α, β, k. The results are presented in graphical form in **Figures 12–15** The functions $G_k(t)$ are calculated by formula (49), and the functions $g_k(t)$ by formula (50). Note that product $\alpha\beta$ is the mean of μ_k. Parameter λ is equal to 0.25 and $\alpha\beta = 7$ as it observes in reality.

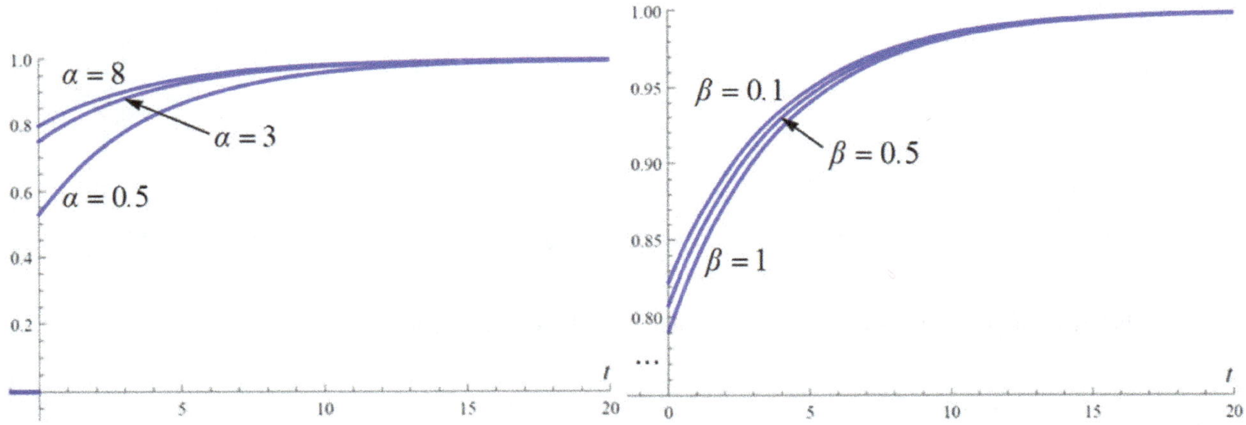

Figure 12. Behavior of distribution $G_2(t)$ when (a) $\alpha = 0.5, 3, 8$ and (b) $\beta = 0.1, 0.5, 1$.

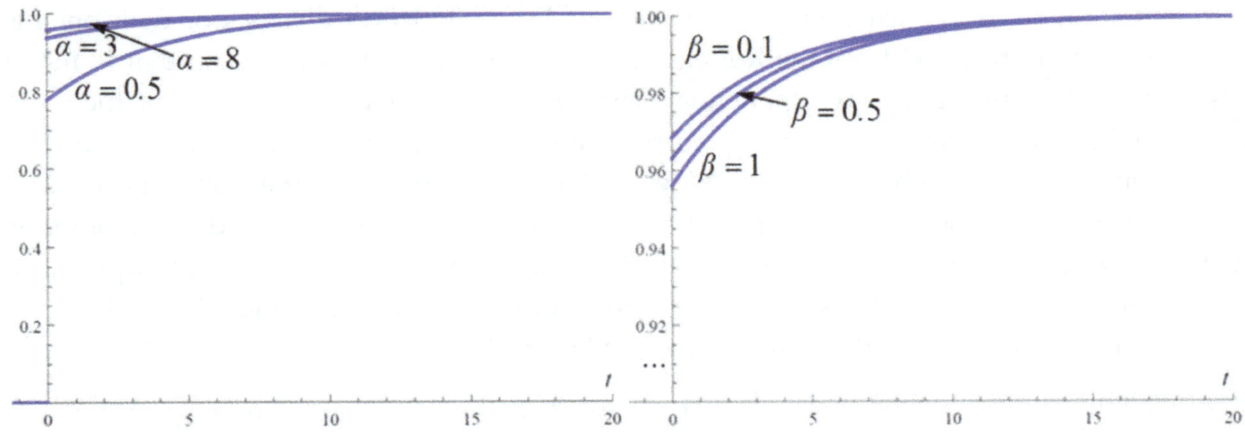

Figure 13. Behavior of distribution $G_3(t)$ when (a) $\alpha = 0.5, 3, 8$ and (b) $\beta = 0.1, 0.5, 1$.

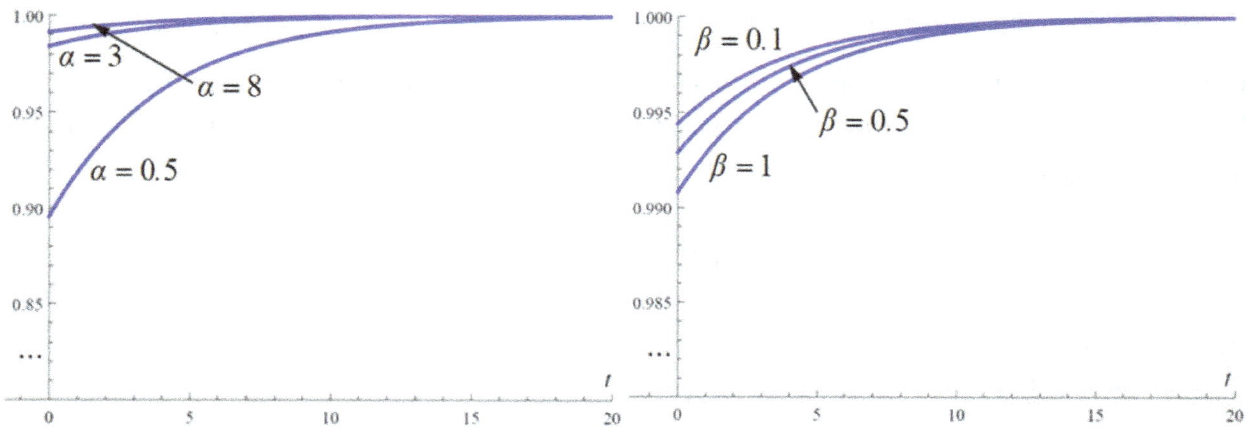

Figure 14. Behavior of distribution $G_4(t)$ when (a) $\alpha = 0.5, 3, 8$ and (b) $\beta = 0.1, 0.5, 1$.

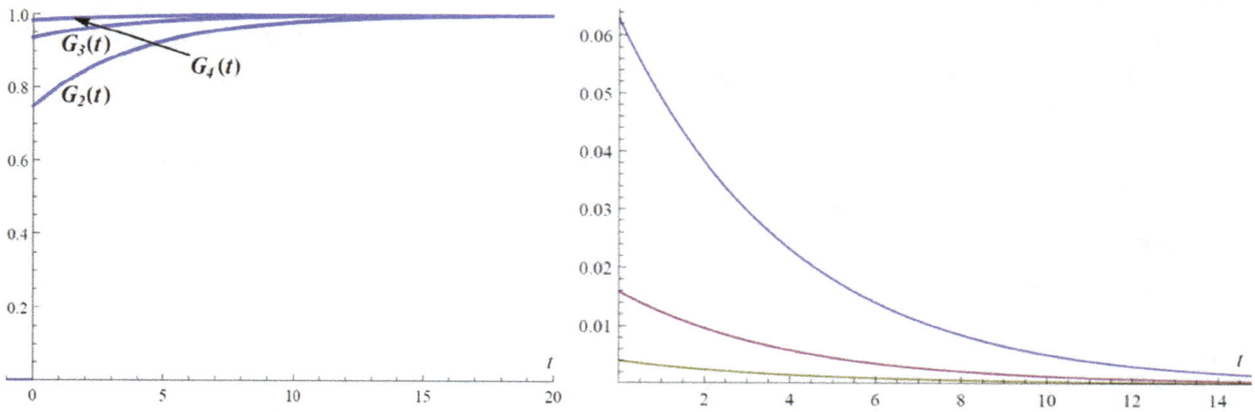

Figure 15. Behavior of distributions $G_k(t)$ (a) and densities $g_k(t)$ (b), $k = 2, 3, 4$ and $\alpha = 3$.

7. Comparison with statistics of real train traffic

Let us consider the following random variable: the deviation of the real moment of arrival at a certain station from the scheduled one. Denote it by ξ. Statistical analysis of data on this random variable, received from the Russian railways, has led to the conclusion that in many cases, they obey the modified exponential law with the distribution function of the form Eq. (33) with $b = 0$. Using data on the suburban trains of the direction "Moscow-Tver" for the period: January, 11–15, February, 1–6, 2016, we obtained a sample from the distribution of ξ of the size $n = 50$ with the sample mean 1.44 and sample variance 2.7. We tested the hypothesis that ξ obeys distribution (Eq. (33)) with $\lambda = 0.35$ and $a = 0.64$. To this end, we applied the Kolmogorov goodness-of-fit test with the significance level $\alpha = 0.05$ and obtained the fit between the hypothesis and the sample data (see **Figure 16**).

Remark 7. It should be noted that in considered example the deviation ξ is nonnegative. But in reality, it can frequently be both positive and negative. Positive values are due to arisen delay. Negative values occur due to the fact that sometimes early arrivals take place.

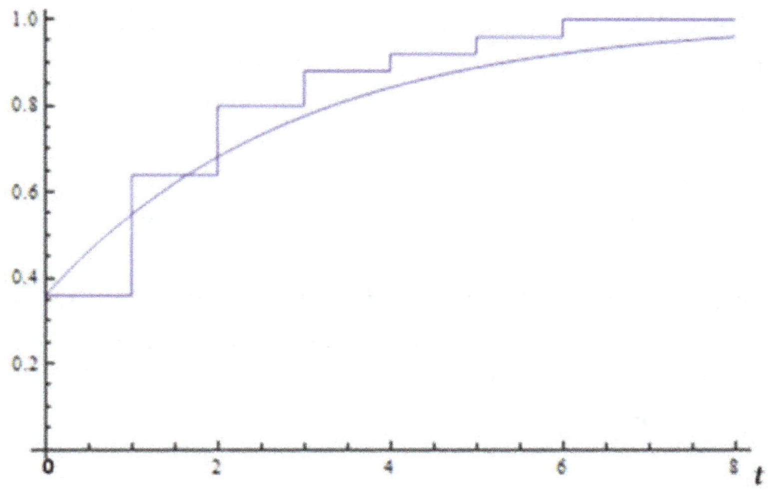

Figure 16. The empirical distribution function and the calculated function of the form Eq. (33).

Remark 8. Although the hypothetical distribution function from **Figure 16** is constructed for deviations without any details about the train number k, it is well correlated with the graph of the function $G_2(t)$ with $\alpha = 0.5$ from **Figure 12**.

This allows us to assume that the distribution of the deviation ξ is mainly determined by the distribution of the delay τ_2.

Remark 9. It was verified that if the length of the random variables μ_j have the same gamma distribution, any variation of the parameters of this distribution (α and β) has a rather small influence on behavior of output distribution (see **Figures 12–15**).

Remark 10. Since the primary delay has a great influence on formation of the output distribution of deviations from the schedule (τ_k), then a knowledge of the primary delay distribution in each particular situation allows to predict the distribution of knock-on delays.

One important practical effect of the considered model is that it enables us to estimate the standard deviation (SD) of the actual arrival delays at the destination station. As an example, we calculated this parameter for the suburban railway line. The data analyzed were collected at the Tver station in the period of January 2016 and February 2016.

Example 8. Due to statistical data, we can consider that τ has the exponential distribution with the parameter $\lambda = 0.25$ (i.e., τ has the distribution function (Eq. (33)) with $\lambda = 0.25$, $a = 1$, $b = 0$), and μ_2 has gamma distribution with the density function (Eq. (17)), where $\alpha = 0.6$, $\beta = 11.7$. Using formulas (49) and (50) with $k = 2$, we have:

$$SD^2 = \int_{-\infty}^{\infty} (t - a_2)^2 d\,G_2(t) = \int_{-\infty}^{\infty} t^2 d\,G_2(t) - a_2^2 = \int_0^{\infty} t^2 g_2(t)\,dt - a_2^2 \approx 10.987.$$

Here $a_2 = \int_{-\infty}^{\infty} t\,d\,G_2(t) = \frac{1}{\lambda}\left(\lambda\beta + 1\right)^{-0.6} \approx 1.763$, $\int_{-\infty}^{\infty} t^2 d\,G_2(t) = \frac{2}{\lambda^2}\left(\lambda\beta + 1\right)^{-0.6} \approx 14.088$,

$$\int_0^{\infty} t^2 g_2(t)\,dt - a_2^2 = \frac{2}{\lambda^2}\left(\lambda\beta + 1\right)^{-0.6} - \frac{1}{\lambda^2}\left(\lambda\beta + 1\right)^{-1.2} \approx 10.987.$$

Thus, theoretical $SD \approx \sqrt{10.987} \approx 3.315$ min. This corresponds with the real statistics which shows the SD amount is 3.32 min for the mentioned station.

8. Conclusions

The mathematical model of train traffic proposed in the chapter allows us to find conditions on initial headways, which provide a smallness of frequency of a large number of delays. In other words, the formulas for the distributions of arrival headways obtained in the chapter enable to optimize the frequency of arriving train delays.

Acknowledgements

The research is funded by the JSC Russian Railways (grant 2016 for development of the scientific school).

Author details

Vladimir Chebotarev[1], Boris Davydov[2] and Kseniya Kablukova[1]*

*Address all correspondence to: kseniya0407@mail.ru

1 CC FEB RAS, Khabarovsk, Russia

2 FESTU, Khabarovsk, Russia

References

[1] Müller-Hannemann M, Schnee M. Efficient timetable information in the presence of delays. In: Ahuja RK, Möhring RH, Zaroliagis CD, editors. Robust and Online Large-Scale Optimization. Springer; 2009;**5868**:249-272

[2] Goverde RMP. A delay propagation algorithm for large-scale railway traffic networks. Transportation Research Part C. 2010;**18**:269-287

[3] Carey M, Kwiecinski A. Stochastic approximation to the effects of headways in knock-on delays of trains. Transportation Research Part B. 1994;**28**:251-267

[4] Carey M, Kwiecinski A. Properties of expected costs and performance measures in stochastic models of scheduled transport. European Journal of Operational Research. 1995; **83**:182-199

[5] Yuan J. Stochastic modeling of train delays and delay propagation in stations [thesis]. The Netherlands: Technische Universiteit Delft; 2006

[6] Meester LE, Muns S. Stochastic delay propagation in railway networks and phase-type distributions. Transportation Research Part B. 2007;**41**:218-230

[7] Berger A, Gebhardt A, Müller-Hannemann M, Ostrowski M. Stochastic delay prediction in large train networks. In: Proceedings of the 11th Workshop on Algorithmic Approaches for Transportation Modelling, Optimization, and Systems (ATMOS'11); 8 September 2011; Saarbrücken, Germany. pp. 100-111

[8] Büker T, Seybold B. Stochastic modelling of delay propagation in large networks. Journal of Rail Transport Planning and Management. 2012;**2**(12):34-50

[9] Yuan J. Statistical analysis of train delays at The Hague HS. In: Hansen IA, editor. Train Delays at Stations and Network Stability (Workshop). Delft, The Netherlands: TRAIL; 2001

[10] Yuan J, Goverde RMP, Hansen IA. Propagation of train delays in stations. In: Allan J, Hill RJ, Brebbia CA, Sciutto G, Sone S, editors. Computers in Railways VIII. WIT Press; 2002. pp. 975-984

[11] Alexandrova NB. Distribution of the train delays duration of due to station malfunctions. In: Proceedings of the Regional Conference "Universities of Siberia and Far East for Transsib". Novosibirsk. 2002. pp. 20-21. (in Russian)

[12] Fikhtengolts GM. Course of Differential and Integral Calculus. 8th ed. Vol. 3. M.: Fizmatlit; 2003. 728p

[13] Davydov B, Dynkin B, Chebotarev V. Optimal rescheduling for the mixed passenger and freight line. In: Proceedings of the 14th International Conference on Railway Engineering Design and Optimization (COMPRAIL 2014), Rome, Italy. 2014. pp. 649-661

6

Probabilistic Analysis of the Influence of Staff Qualification and Information-Psychological Conditions on the Level of Systems Information Security

Igor Goncharov, Nikita Goncharov, Pavel Parinov,
Sergey Kochedykov and Alexander Dushkin

Abstract

Taking into account the criticality of the "human factor," the probabilistic approach for analysis is proposed, including: a model for predicting and assessing the level of systems information security, considering random events, including dependent events; model of information-psychological impact on staff; methodical approach for analyzing an influence of staff qualifications and psychological conditions on the level of system information security. The effectiveness of the application is demonstrated by examples.

Keywords: human factor, information security, information-psychological impact, predicting, assessment

1. Introduction

Information systems are of high importance in organizations, industrial process, banking sector, etc. The "human factor" accounts for approximately 70% of information security breaches. Staff are one of the parts of information system. The influence of the "human factor" on the level of system information security is considered in various articles and standards. In particular, the international standards ISO/IEC 27002 provide recommendations for work with staff at various stages: prior to employment, during employment, termination, and change of employment [14]. The reliability of information system operation and the level of information security depend on different conditions. Wrong actions and inactivity of staff and untimely performance of job duties can lead to violations of integrity, availability, and confidentiality of the information. As a result they influence the level of system information security. The staff of information system have certain characteristics that affect a level of system information security as well as technical and software components. Such characteristics form mental state and psychophysical properties of staff. In addition to attacks on the information system implemented by technical methods, there is also an attack on the staff of the information system. This attack can be carried out by means of information-psychological impact (IPI). In this article, it is proposed to consider mathematical models for predicting and estimating the information security level of information systems, taking into account dependent events and information and psychological impact on staff, methods, and stages of implementing information and psychological impact. The approach to the analysis of staff conditions under the information-psychological impact is considered. A methodical approach is proposed for analyzing the impact of qualification and psychological states of staff on the information security level of the information system. The application of this model is considered.

2. Mathematical models for estimating the level of information security considering the impact of staff qualifications and psychological state

2.1. Model for predicting and assessing the level of systems information security considering dependent events

The boundaries of the conditions for the provision of procedures for modeling secure information systems in terms of compliance with integrity, availability, and confidentiality, and the information circulating in them [9, 12, 25] is estimated by the possibility of realizing their technical characteristics in real devices and conditions [2, 13, 15, 22, 23]. In particular, the ready-made nodes of known information systems are separate technical devices with characteristics corresponding with their passport data. They provide the possibility to choose the topology of the information system within the limits of the compatibility characteristics of the system nodes [2, 4, 13, 15, 22–24]. At the same time, consideration of this approach to modeling allows to choose the priority of providing information security criteria such as integrity, availability, and confidentiality, which are generally interdependent in the construction of an information system and analysis of the possibility of ensuring maximum levels of values of these criteria. It means that depending on the conditions, tasks, which should be solved, and the purpose of building and information system, first of all, it is more important to ensure integrity; second, if availability comes, then it is confidentiality or in another sequence.

This sequence may be due to the complexity of the information system, its configuration, the characteristics of the individual nodes, which are involved in its composition, and external factors that affect the operating conditions. The opinion of experts [12] who make decisions on estimating the values of the parameters of the safety criteria, based on an analysis of the physical characteristics of the information system under consideration, plays an important role in the implementation of this approach. The theorem on the multiplication of the probabilities of dependent events is at the heart of the approach for estimating the parameters of safety criteria [5]. This is due to the dependence of the safety criteria which is described above, estimated by mutual influence in the analysis of the characteristics of the information system. For example, a separate information system node is a complete single device with specific technical characteristics that are individually responsible for the likely conditions for ensuring either integrity or availability or confidentiality. At the same time, by virtue of the technical implementation, this node cannot be ideal from the point of view of safety criteria and cannot provide only either integrity or availability or confidentiality, since the information that must have a certain level of each criterion will circulate in it. And the characteristics of this node will extend to a certain part of the information system, which also estimates the important conditions for ensuring its security [26]. The security of information, in the sense of analyzing the probability of the existence of safety criteria, in the information system can be represented in the diagram of sets shown in **Figure 1**.

If the integrity (I), availability (A), and confidentiality (C) are separate sets, then security (S) is the intersection of these three sets.

It means that it is necessary to ensure both integrity, and availability, and confidentiality to a specific value of the appropriate criterion, estimated for each particular information system in order to ensure security [12, 25]. In its turn, from the point of view of ensuring the probability of the information system security and due to the interdependence described above, integrity, availability, and confidentiality are conditional signs. Then the probability of security should be considered in the following way (Eq. (1)) [12]:

$$p(Sec) = p(I \cap A \cap C). \tag{1}$$

The figure shows a graphical interpretation of the product of the corresponding events I, A, and C for which the following expression is valid (Eq. (2)):

$$p(I \cap A \cap C) = p(I) \cdot p_I(A) \cdot p_{IA}(C). \tag{2}$$

Since the events of ensuring integrity, availability, and confidentiality are dependent, then the probability of producing these events according to the multiplication rule for the probabilities of dependent events, is (Eq. (3)):

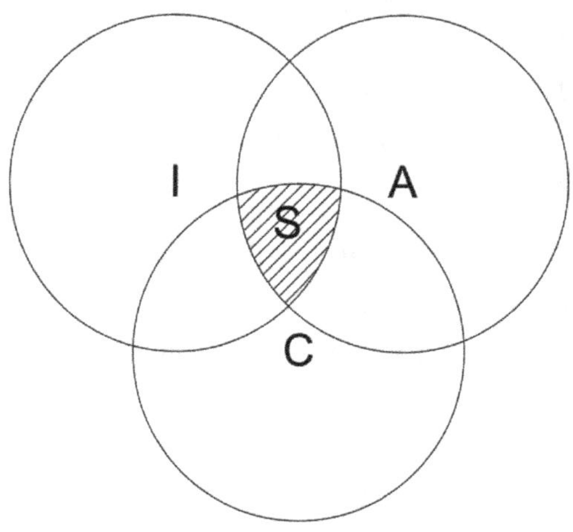

Figure 1. Presentation of integrity, availability, and confidentiality using sets.

$$p(I \cdot A \cdot C) = p(I) \cdot p(A/I) \cdot p(C/I \cdot A). \tag{3}$$

To describe the case, the probability of coexistence of several dependent events is equal to the product of the probabilities of these events, and the probability of each next event in the order of recording is calculated if all the previous ones also take place.

It means that the probability of ensuring both integrity and availability and confidentiality of information is equal to the product of the probability of ensuring integrity to the probability of providing availability if there is ensuring of integrity and the probability of ensuring of confidentiality while integrity and availability are provided.

As it was mentioned before, the priority of the place of writing in the formula of the corresponding probabilities can be estimated by the experts' opinion, taking into account the complexity of their calculation, caused by the need to implement the corresponding values of the safety criteria levels, according to the physical expressions which describe these criteria levels [12].

Thus, the described approach makes it possible to model various information systems based on real physical characteristics that allow to predict and evaluate the levels of safety criteria, taking into account the experts and experts' opinions, and it is actual and necessary in practical implementation nowadays [12]. The information security level of the information system can be estimated according to the calculated values (1).

2.2. Model of information-psychological impact on staff

Along with the impact on the technical and software components of the information system, there are also effects pointed to the staff of the information system (**Figure 2**). They are information-psychological impacts (IPI) [6–11]. They can lead to a change in the characteristics of employees that are the subject of IPI; as a result, the information security level of the information system may change. As a rule, IPI data are usually transmitted through common communication channels.

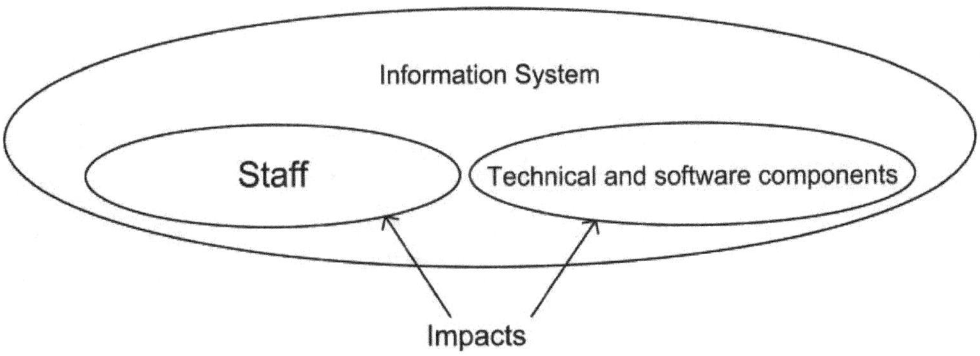

Figure 2. Impacts on the information system.

2.2.1. Stages of implementation of information and psychological impact

It is possible to single out the following stages of IPI implementation [6–11]:

1. The subject determines the goals to be achieved by IPI.
2. The subject determines the object of IPI.
3. The subject collects information about the IPI object and investigates the psychophysical characteristics of the IPI object in order to detect subject matters of the IPI object and their characteristics (the subject is understood to be a component of the IPI object that determines its possible characteristics; one characteristic may belong to several IPI objects).
4. The subject chooses the most appropriate means of influencing the IPI object and the communication channel, based on the data of points 1–3. Each of the means affects the relevant objects of influence and their characteristics.
5. The subject forms a message for the IPI object.
6. The subject implements an impact on the IPI object, with the aim of achieving a sustainable change in characteristics. To do it, the generated message is coded using the selected IPI tools and sent via the selected communication channel to the object.
7. The IPI object decodes the received message.
8. The decoded message affects the characteristics of the IPI object; as a result, they change, and there is some possibility of appearing/disappearing new characteristics.

In **Figure 3**, the scheme of IPI is shown.

2.2.2. Formal IPI model

The formal model of the IPI process is proposed [7, 8, 11]. For the IPI object *Obj* there is a set of subject matters Sub_i, and a set of characteristics (Eq. (4)) is defined for each of them:

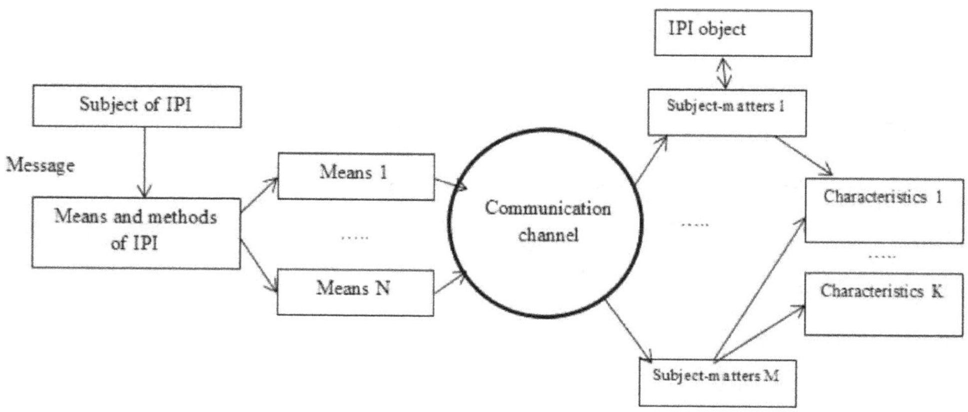

Figure 3. Scheme of IPI.

$$Obj = \{Sub_1, \ldots, Sub_n\}, \quad Sub_i = \{Char_1, \ldots, Char_k\}. \tag{4}$$

Each object can have several characteristics. Dependence of objects and their characteristics is estimated in the matrix of properties. In the columns, the subject-matters of the IPI object are indicated; in the lines, characteristics are indicated; at the intersection, their correspondence is denoted (Eq. (5)):

$$Obj = \begin{pmatrix} Sub_1(Char_1) & Sub_2(Char_1) & \ldots & Sub_n(Char_1) \\ Sub_1(Char_2) & 1 & \ldots & Sub_n(Char_2) \\ \ldots & \ldots & \ldots & \ldots \\ 1 & Sub_2(Char_k) & \ldots & Sub_n(Char_k) \end{pmatrix} \tag{5}$$

The subject has many means of impact that do not always correspond with articles of the IPI object; this is proposed that the set of means of the subject's impact is defined as $S = \{S_1, \ldots, S_n\}$.

Each of the means of impact S_j is applied for the purpose of changing the property $Sub_i(Char_m)$ and affects different objects and their characteristics in different ways. The result of such a change will be denoted $Ef_{i,m,j}$, which can be equal to zero or be negative (that means it has the opposite effect to the aims of IPI) and can be positive (it means it can have an effect corresponding with the goals of the IPI).

Realization of IPI for m-characteristics (Eq. (6)):

$$(Sub_i(Char_m), S_j) = Ef_{i,m,j}. \tag{6}$$

For the case when the IPI object possesses articles with characteristics, and the subject has means of impact, this is proposed to obtain the matrix of efficiency of IPI; in the columns, the articles of the IPI object are indicated; in the lines, characteristics are indicated; at the intersection, their correspondence is denoted (Eq. (7)):

$$S = \begin{pmatrix} Ef_{1,1,1} & Ef_{1,1,2} & \cdots & Ef_{1,1,l} \\ Ef_{1,2,1} & Ef_{1,2,2} & \cdots & Ef_{1,2,l} \\ \cdots & \cdots & \cdots & \cdots \\ Ef_{n,1k1} & Ef_{n,k2} & \cdots & Ef_{n,k,l} \end{pmatrix}. \tag{7}$$

The sum of all impacts on the m-characteristic is described by Eq. (8):

$$\sum_{j=1}^{l} (Sub_i(Char_m), S_j) = Ef_{i,m}, \tag{8}$$

where the efficiency is provided when the matrixes are added in stages, which means that several IPI tools can affect one characteristic. The formal model of IPI implementation can be written in the following form (Eq. (9)):

$$(Obj,S)=\sum_{j=1}^{l}\begin{pmatrix}1 & \dots & Sub_n(Char_1)\\ \dots & \dots & \dots\\ Sub_1(Char_k) & \dots & Sub_n(Char_k)\\ 1 & \dots & 1\\ \dots & \dots & \dots\\ 1 & \dots & 1\end{pmatrix}\cdot S_j,\quad Obj=\begin{pmatrix}1 & \dots & Ef_{n,1}\\ \dots & \dots & \dots\\ Ef_{1,k} & \dots & Sub_n(Char_k)\\ 1 & \dots & Sub_n(Char_{k+1})\\ \dots & \dots & \dots\\ Ef_{1,f} & \dots & 1\end{pmatrix}. \tag{9}$$

Operation « · » has the following properties:

1. $Sub_i(Char_m)\cdot S_j=Ef_{i,m,j}$.

2. $Sub_i(Char_m)\cdot S_j=Ef_{i,m,j}=1$—the property $Sub_i(Char_m)$ has disappeared.

3. $Sub_i(Char_m)\cdot S_j=Ef_{i,m,j}=0$—the property $Sub_i(Char_m)$ has not changed.

$$1\cdot S_j=\begin{cases}1,\text{ if the property continues being absent,}\\ Sub_s(Char_f),\text{ if the property has appeared.}\end{cases} \tag{10}$$

The result of the malefactor's attack on the IPI object is a matrix of properties, which will take the changed form (Eq. (11)):

$$Obj=\begin{pmatrix}1 & Ef_{1,2,l} & \dots & 1\\ Ef_{2,1,j} & 1 & \dots & Sub_n(Char_2)\\ \dots & \dots & \dots & \dots\\ 1 & Sub_2(Char_k) & \dots & Sub_n(Char_k)\end{pmatrix}. \tag{11}$$

Some of the properties resulting from IPI may remain unchanged; others are replaced by $Ef_{i,m,j}$.

2.2.3. Mathematical model of information-psychological impact

The change in the property which undergoes IPI can be described by equation or model [11, 16–19] (Eq. (12)):

$$K=f(H,P), \tag{12}$$

where P is the characteristics of the IPI object, H is the characteristics of the IPI (means of impact), and K is the response (the level of change). As the characteristic of IPI, we will use H as the effectiveness of implementing the means of influencing the property $Ef_{i,m,j}$. Thus, the equation takes the form (Eq. (13)):

$$K=f((Obj,S),P). \tag{13}$$

Eq. (12) makes it possible to evaluate the change in the properties and the response of an object to IPI. In our case, staff are considered as the IPI object. Eq. (13) of the change in the property and the human reaction to the effects is given in articles [11, 16–19] and has the form (Eq. (14)):

$$R\frac{d^2Y}{dt^2}+\frac{2F\sqrt{RA}}{QF_0}\cdot\frac{dY}{dt}+\frac{A}{Q^2}Y=X, \tag{14}$$

where F is the frustration; F_0 means some value of the level of frustration, considered normal or threshold; A is the aggression; Q means the time parameter; R is the stiffness; X means the effectiveness of information-psychological impact; and Y is the reaction level. These parameters are measured in conditional scores. They can be estimated using psychological tests and an expert method. This is proposed to use Eq. (11) to estimate the change in the property of the IPI object as a result of the action. The transfer equation for Eq. (14) has the form (Eq. (15)):

$$W(p)=\frac{Q^2F_0}{RQ^2p^2F_0+2QF\sqrt{RA}p+AF_0}. \tag{15}$$

2.3. Methodological approach for analyzing the impact of staff qualifications and psychological conditions on the level of systems information security

Employees' qualifications, mental state, and psychophysical properties can act as their characteristics.

2.3.1. Staff qualification assessment

This is a proposed estimate of staff's qualification in an expert way:

- k = 0 if the staff of the information system are idle in the case of vulnerabilities, technical malfunctions are idle in the technical and software components of the information system [27, 28].
- k < 1 if the staff of the information system fail to remove vulnerabilities, technical malfunctions fail in the technical and software components of the information system in time [27, 28].
- k = 1 if the staff of the information system eliminate vulnerabilities, technical malfunctions eliminate in the technical and software components of the information system in time [27, 28].
- k > 1 if the staff of the information system independently detect and fix vulnerabilities (temporary solutions, before the release of the update from the manufacturer) in the technical and software components of the information system, technical malfunctions are prevented [27, 28].

The limiting minimum value for the staff's qualification k is 0, because staff does not create vulnerabilities and technical malfunctions in the technical and software components of the information system. The maximum value for the staff's qualification k is 3; in this case the security service includes a large number of highly skilled employees who can increase labor productivity working together.

The estimation should be carried out separately for each component because maintenance of various components of the information system is implemented in different ways. This is proposed to define the malfunction as various malfunctions in the operation of the information system components that require staff intervention to eliminate them. This is proposed to understand vulnerability as a defect of information system that can violate its integrity, availability, and confidentiality and cause a malfunction.

2.3.2. Impact of staff's psychological conditions on labor activity

During the work activity, the staff of the information system may be in different psychological conditions. The effectiveness of the staff depends on what psychological state they are in. The following states can be distinguished as [3, 20, 29]:

- Optimum working condition ensures the greatest efficiency of activity. It is characterized by the presence of a conscious goal of activity, high concentration of attention, aggravation of memory, and activation of thinking. The electroencephalogram shows that in this state, the brain rhythms mainly lie in the beta range.
- The state of tense activity arises in the course of work in abnormal situations. Mental tension develops directly in proportion to the difficulty of the task. Easy tasks are solved with minimal effort; complex and new actions require a higher degree of mental pressure. Mental tension is a physiological reaction of the organism, mobilizing its resources to perform more difficult tasks. Mental tension stimulates the physical and mental processes of the human body, which increases its adaptive abilities. The tension reaction develops in a responsible environment, as well as when people perform complex production tasks, if they change the stereotype of actions and habitat and if they are under the influence of extreme conditions. Under the influence of mental stress, vital body functions such as metabolism, circulation, and respiration change. If in the behavior of a person, there is some general concentration, the actions become clearer, the speed of motor reactions increases, and physical performance improves. At the same time, perception becomes aggravated, the process of thinking is accelerated, memory is improved, and concentration of attention is increased.

It should be remembered that the dependence of the efficiency of labor activity (working capacity) of employees on the level of tension of its functional systems is parabolic. It was found out that mental stress has a positive effect on the result of labor up to a certain limit. Exceeding the critical level of activation leads to a decrease in the results of labor up to a complete loss of efficiency.

- Fatigue is a functional state of a person, temporarily occurring under the influence of prolonged or intensive work, accompanied by a decrease in its effectiveness. Fatigue is caused by the depletion of body resources during prolonged or excessive activity and is characterized by a decrease in motivation to work, a violation of attention and memory.

At the physiological level, the appearance of a protective inhibition of the central nervous system is noted. Fatigue may eventually go into the exhaustion, which requires a longer rehabilitation to get it over.

- Stress is a state of increased and prolonged pressure associated with the inability to adapt to the requirements of the habitat. This condition is caused by the long-term impact of environmental factors, exceeding the possibilities of the organism adaptation. It is characterized by mental stress, a sense of frustration, anxiety, and worry, and in the last stage, indifference and apathy appear. At the physiological level, there is a depletion of adrenal hormone stores, muscle tension, and a two-phase activation of the autonomic nervous system.

Figure 4 shows the possible dynamics of staff states. The transition between states can occur both as a result of labor activity and under the influence of information-psychological impact.

The effectiveness of staff for different psychological conditions is a quantity with no dimension and can be estimated in the following way:

- For an optimal working condition, the efficiency of labor activity will be estimated as (Eq. (16)) [3, 20, 29]

$$E = k, \tag{16}$$

where k is the qualification of the staff.

- For the state of intense activity, the efficiency of labor activity will be estimated as (Eq. (17)) [3, 20, 29]

$$E = -at^2 + bt + k, \tag{17}$$

where k is the qualification of the staff, a [1/h^2] and b [1/h] are parameters that estimate the rate of staff fatigue, and t [h] is time.

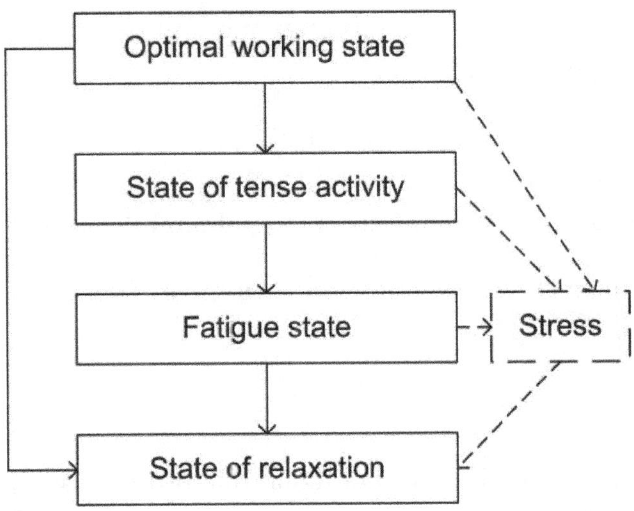

Figure 4. Possible dynamics of staff transitions in the course of labor activity.

- For the state of fatigue, the efficiency of labor activity will be estimated as (Eq. (18)) [3, 20, 29]

$$E = k - bt, \tag{18}$$

where k is the qualification of the staff, b [1/h] is the parameter that estimates the rate of staff fatigue, and t [h] is time.

- For the state of fatigue, the efficiency of labor activity will be estimated as (Eq. (19)) [3, 20, 29]

$$E = k - sbt, \tag{19}$$

where k is the qualification of the staff, b [1/h] is the parameter that estimates the rate of staff fatigue, s is the reaction to information-psychological impact, and t [h] is time.

- The efficiency of labor activity is equal to zero for the state of relaxation.

2.3.3. Mathematical model of information system operation

The information system consists of various technical and software components; each of them can have vulnerabilities and fail due to a technical malfunction. Vulnerabilities and technical faults pose a threat to the confidentiality, integrity, and availability of information. This is proposed to represent the information system as a set of queuing systems [26–28]; each of them simulates the dynamics of vulnerabilities and technical faults that threaten the confidentiality, integrity, and availability of information. The input of the described system receives a non-stationary Poisson stream of requests (vulnerabilities and faults). This model is presented in **Figure 5**, where $\lambda(t)$ is the speed of detection of vulnerabilities or faults that threaten the confidentiality, integrity, or availability of information; E is the effectiveness of staff to ensure the confidentiality, integrity, or availability of information; and T_a is the average time to eliminate the vulnerability or malfunction.

The average speed of elimination of vulnerabilities and faults of the information system will be described in the following way (Eq. (20)):

$$\mu = E\mu_a, \tag{20}$$

where μ_a is the average speed of elimination of the vulnerability and malfunction.

The assessment of μ_a will be estimated in the following way (Eq. (21)):

$$\mu_a = \frac{1}{T_a}. \tag{21}$$

The average number of vulnerabilities and faults in the information system will be the sum of the average number of vulnerabilities and faults that threaten the confidentiality, integrity, and availability of information (Eq. (22)):

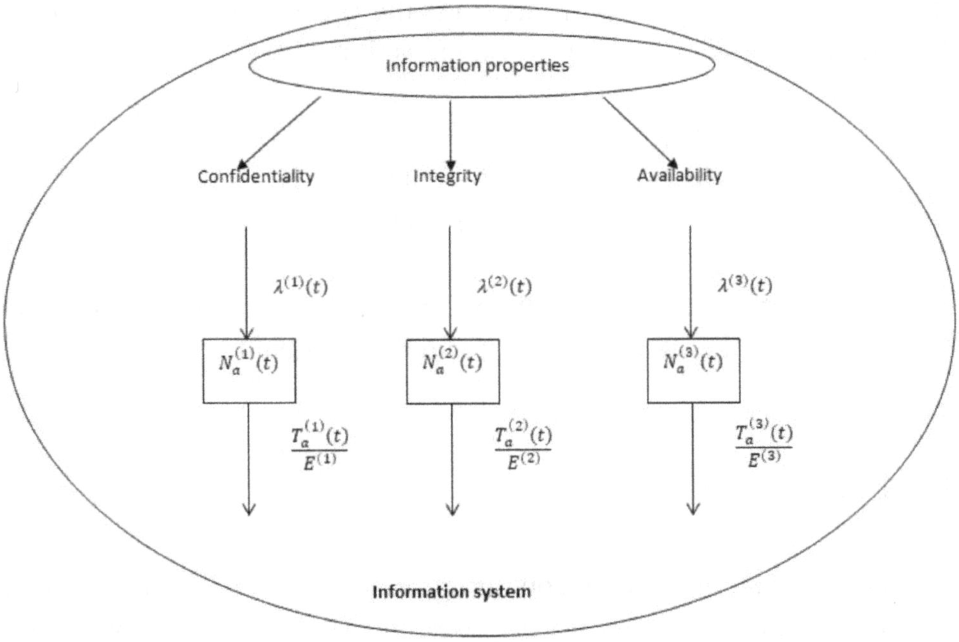

Figure 5. Model of changes in the state of security of the information system, taking into account the staff activities.

$$N_a(t) = \sum_{m=1}^{3} N_a^{(m)}, N_a^{(m)}(t) = \frac{T_a^m e^{-t}}{E^{(m)}} \left(\lambda^m(t) + \int_0^t \lambda^m(\tau) e^{\tau} d\tau \right). \tag{22}$$

When $E^{(m)}$ is equal to zero of vulnerability and faults, the average number of vulnerabilities and faults will be estimated as (Eq. (23))

$$N_a^{(m)}(t) = \int_0^t \lambda^m(\tau) e^{\tau} d\tau. \tag{23}$$

There is a probability of a number of vulnerabilities and faults (Eq. (24)):

$$P_n(t) = \frac{[N_a(t)]^n}{n!} e^{-N_a(t)}. \tag{24}$$

Thus, the probability of the absence of vulnerabilities and faults is (Eq. (25))

$$P_0(t) = e^{-N_a(t)}. \tag{25}$$

In general, based on the proposed models, it is proposed to estimate the security of the information system $P_{IS}(t)$, taking into account the impact of staff qualifications and psychological conditions, as (Eq. (26))

$$P_{IS}(t) = P_{Sec}(t) + P_0(t)(1 - P_{Sec}(t)), \tag{26}$$

where $P_{Sec}(t)$ is estimated from Eq. (3).

To analyze the influence of the human factor on the properties of each component of the investigated information system, one can consider, as (Eq. (27)) [1]:

$$\begin{aligned} P_I(t) &= P_I(t) + P_{0I}(t)(1 - P_I(t)), \\ P_A(t) &= P_A(t) + P_{0A}(t)(1 - P_A(t)), \\ P_C(t) &= P_C(t) + P_{0C}(t)(1 - P_C(t)), \end{aligned} \tag{27}$$

where $P_{0I}(t)$, $P_{0A}(t)$, and $P_{0C}(t)$ are the likelihood of the absence of vulnerabilities and faults in the component providing integrity, availability, and confidentiality.

3. Example of using models

Let us consider an information system, consisting of an X router and a file server under the management of the operating system Y. Users who are allowed to have an access connect to the router through a Wi-Fi connection and get an access to files according to the permitting access system.

In this information system, confidentiality, integrity, and availability are provided by means of a router and a server running the operating system Y.

It is possible to infringe the security of the information system by violating the performance of one of the components which are responsible for confidentiality, integrity, and availability.

As the experience of practical studies [12] has shown for 802.11 wireless networks in calculating the probability values of safety criteria, it is advisable to take noise immunity coding into account for the estimation of integrity. But it is necessary to take modulation efficiency and bandwidth usage technology into account for the estimation of availability, and it is important to take cryptographic strength of encryption into account for the estimation of confidentiality. Then the expression for the probability of ensuring the security of information takes the form (Eq. (28)):

$$p(\,Sec\,) = p(I) \cdot p(A/I) \cdot p(C/IA), \tag{28}$$

where

$$p(I) = p(coding\_immunity), \tag{29}$$

$$p(A) = p(Effect\_of\_modular\_techno\log ical\_use\_of\_frequencies), \tag{30}$$

$$p(C) = p(cryptographic\_strenght\_of\_encryption), \tag{31}$$

With a more detailed representation of the parameters (Eq. (32)):

$$p(I) = p(r, R), \tag{32}$$

$$p(A) = p\big(S, SNR, V_m, p_{er}, parametr\_t\big), \tag{33}$$

$$p(C) = p\left(N, p_{vulnerability}, com\right), \tag{34}$$

where R is coding rate, r is relative redundancy of coding, S is spectral efficiency, SNR is a signal-to-noise ratio, V_m is modulation rate, C is the real throughput, p_{er} is the probability of a bit error, *parametr_t* is a parameter that estimates the effectiveness of the selected technology for the use of the frequency band, N is the number of possible combinations with the selected encryption (coding), $p_{vulnerability}$ is probability of the protocol's vulnerability, and *com* is password complexity. This makes it possible to choose the most flexible algorithm for modeling an information system with the required level of security [9].

Thus, perhaps there are five more options for writing and using the applied expression for multiplying dependent probabilities. Perhaps, because of the complexity of accounting for modeling the network with a great number of parameters in the above expressions, experts believe that in the proposed formula for calculating security, the probability of availability should be put on the first place, the second one should be given to the conditional probability of confidentiality, and then the conditional probability of integrity comes.

If it is possible to ensure security while ensuring integrity and confidentiality considering integrity and availability in the context of integrity and confidentiality, the expression for the probability of network security will take the following form (Eq. (35)):

$$p(Sec) = p(I) \cdot p(C/I) \cdot p(A/IC), \tag{35}$$

and so on.

Different variants of writing these expressions are fair to use then; it is more advantageous to calculate safety when taking into account the corresponding described conditions. For different networks, the probabilities of security criteria will be described by different physical expressions and different number of parameters in these physical expressions [5, 12].

For different information systems at different stages of the technological process that they implement, it may be expedient to differentiate the priority of providing information security criteria (integrity, availability, confidentiality), including the exclusion of some of them. For example, in information retrieval systems that provide users with a legislative basis or a database of threats, it is primarily necessary to ensure the integrity and availability of information, while ensuring confidentiality is not required, since information is publicly available.

Obtaining probability values is a separate research area and requires a separate assessment technique [12]. Values of the probability of ensuring integrity, availability, and confidentiality for various information systems are given in **Table 1**. These values are obtained on the basis of practical experience [21].

Table 2 shows the average time to resolve vulnerabilities and faults for components of various information systems.

Table 3 provides statistics on the intensity of vulnerability and fault detection for components of various information systems.

Let Obj_A (object IPI) be the staff possessing such things as Sub_1 which is the relation to any facts, events, phenomena, and members of a society [11]; Sub_2 is a mental state [11]; Sub_3 is the physiological state of the staff [11]. Things such as Sub_1, Sub_2, and Sub_3 have intersecting sets of properties (concentration, fatigue, understanding, emotionality, etc.).

Using Eq. (14), this is proposed to estimate the reaction to the information-psychological impact. Depending on the characteristics of the staff, the reaction can be both sustainable (staff can do their duties; their effectiveness is defined as Eq. (19)) and unstable (staff is incapable). In the case of an unstable reaction, the graph of the reaction level of the staff is periodic; in the case of a stable reaction, the graph of the reaction level of the staff will not be periodic. **Figure 6** presents examples of the dependence of the level of staff reaction on the information and psychological impact.

Let the staff in question have the following characteristics, obtained from the results of the psychological tests of Eysenck: $F = 16$, $F_0 = 10$, $A = 4$, and $R = 9$. Doing so, this is proposed to assume that the staff, being under the IPI, will not purposefully violate the technical and software components of the information system.

Using Eq. (26), this is proposed to estimate the probability of the security of the information system. **Figure 7a–c** shows the probability of the security of the information system, depending on the coefficient of staff work and their state, for the first, second, and third cases.

Probability $p(Sec)$	Availability	Confidentiality	Integrity
For Case 1	0.85	0.88	0.86
For Case 2	0.74	0.85	0.9
For Case 3	0.91	0.82	0.64

Table 1. Probabilities of ensuring integrity, availability, and confidentiality.

T_a	Availability	Confidentiality	Integrity
For Case 1	0.019	0.016	0.023
For Case 2	0.04	0.021	0.3
For Case 3	0.01	0.001	0.03

Table 2. Average time and speed of vulnerability and malfunction elimination.

λ	Availability	Confidentiality	Integrity
For Case 1	0.00366	0.002	0.0077
For Case 2	0.001	0.0047	0.01781
For Case 3	0.00146	0.0023	0.00724

Table 3. Statistics of the intensity of vulnerability and fault detection for components of the information system.

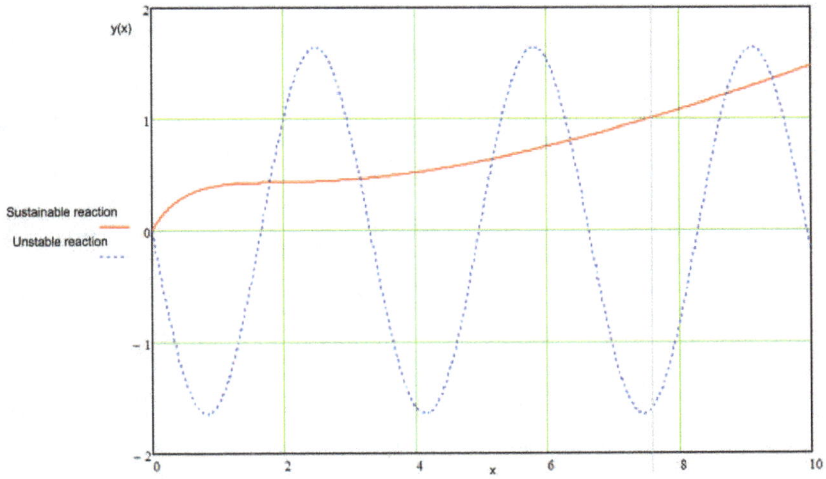

Figure 6. The level of the subject's reaction to information and psychological impact.

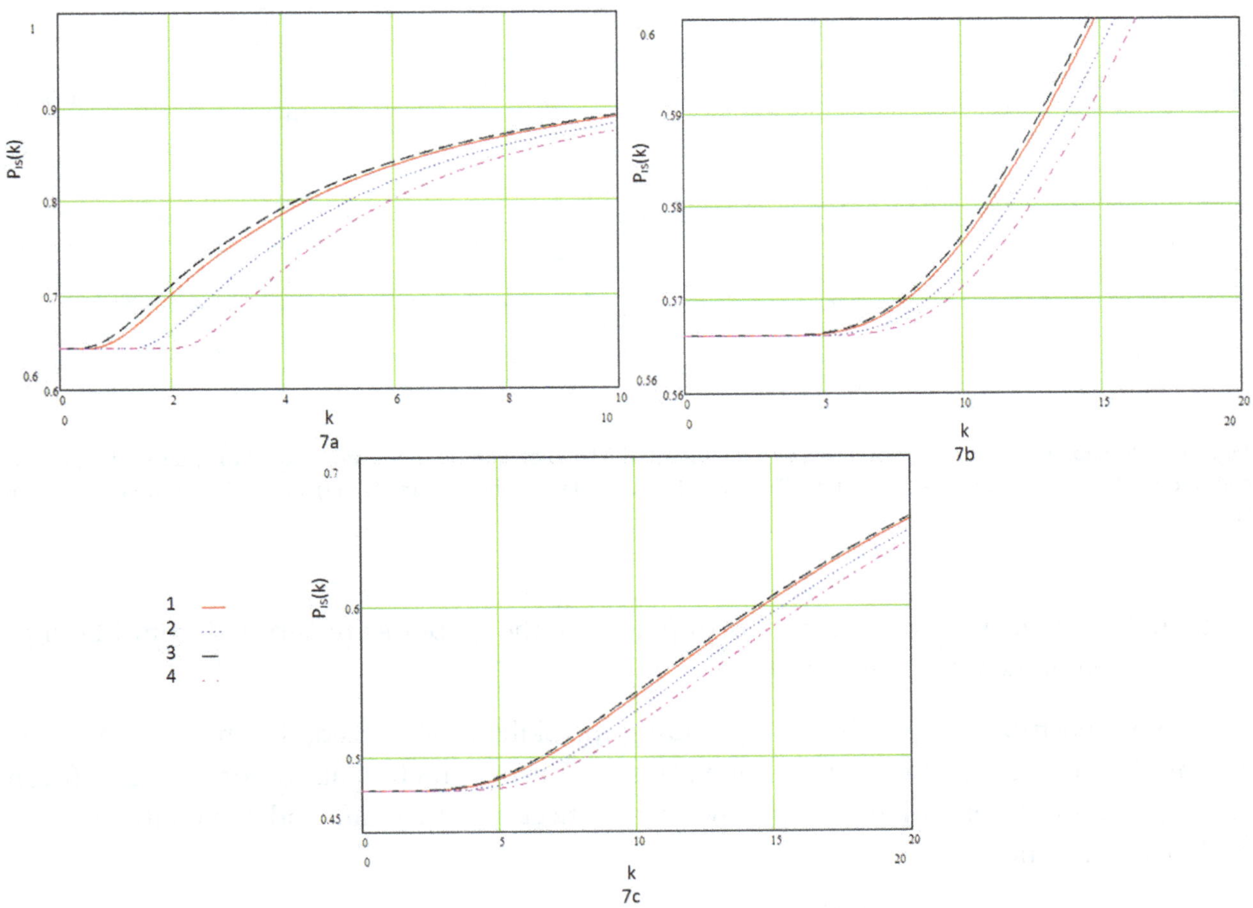

Figure 7. Probability of the information system security, depending on the employees' workload and their condition [(1) optimal condition, (2) fatigue status, (3) state of stressful activity, (4) stressful condition (impact on staff)].

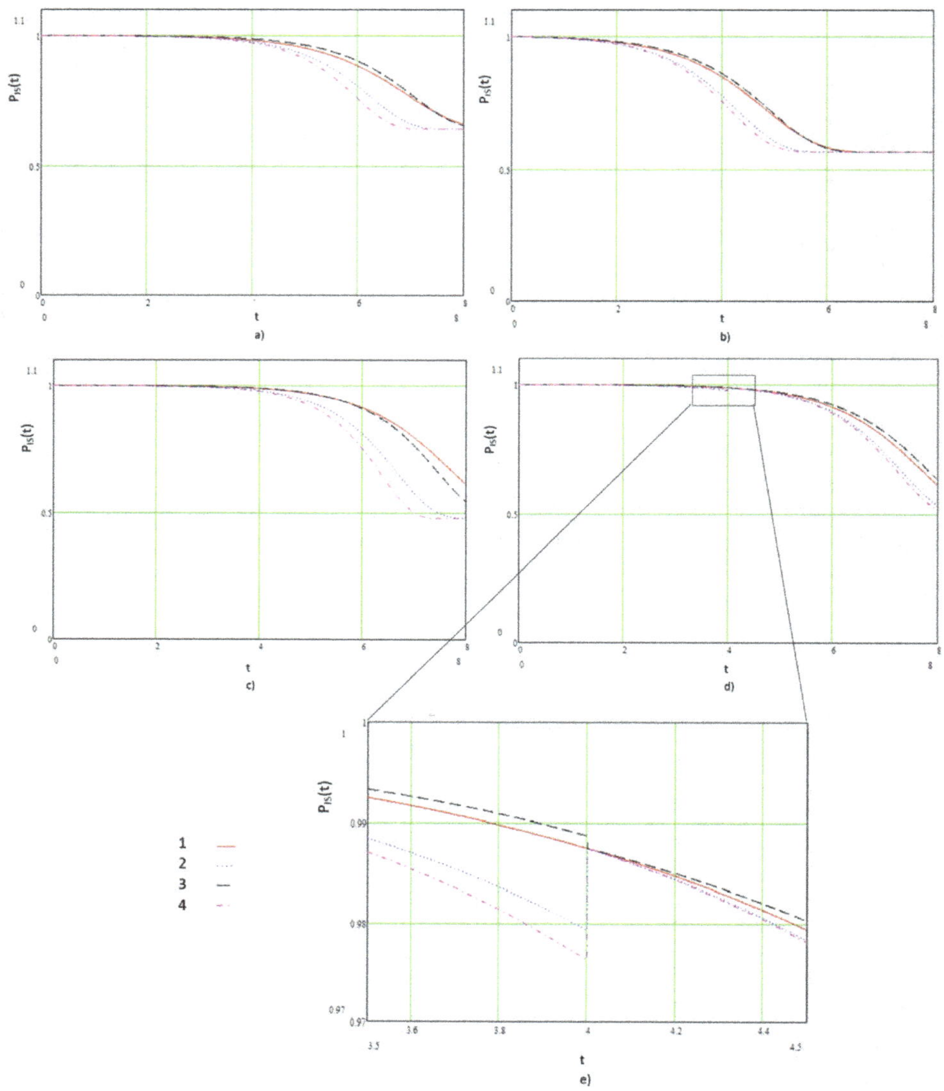

Figure 8. Probability of the information system security at the level of staff qualification is equal to 1, depending on the condition of the staff [(1) optimal condition, (2) state of fatigue, (3) state of tense activity, (4) stressful condition (impact on staff)].

At first, the results of IPI on staff are not apparent, so the graphics are depicted from 1 hour of the operation of the information system.

It can be seen from the graph that upgrading the skills of staff leads to an increase in the probability of security of the information system. Thus, the high qualification of the staff can compensate the information and psychological effects on the staff and their fatigue from prolonged activities.

Figure 8a–c shows the probability of security of the information system for the first, second, and third cases, respectively, if a staff qualification level is equal to one, depending on the condition of staff. **Figure 8d** shows the probability of the security of the information system for

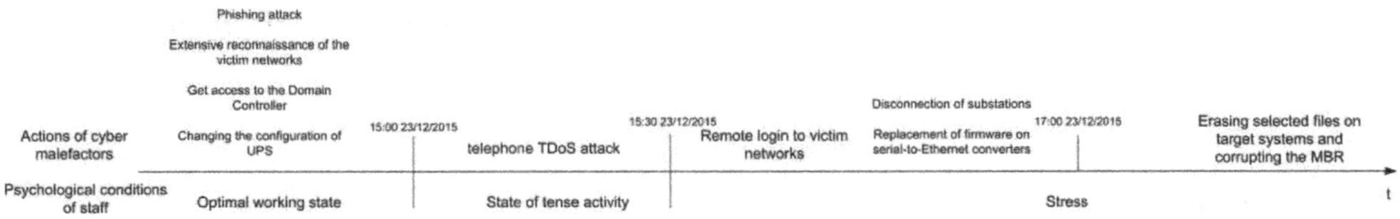

Figure 9. Time diagram of exemplary actions of cyber malefactors and psychological conditions of staff.

the third case, taking into account the recess for recovery. However, the time for the restoration process itself was not taken into account. **Figure 8e** shows an enlarged transition fragment after recovery for **Figure 8d**. A time interval equal to the average working day was taken for consideration.

At the initial stage of operation with a stressed state, the probability of ensuring the security of the information system is higher than at the optimal state, but this is a temporary effect; as it can be seen from **Figure 8a** with prolonged operation in the stressed state, the probability of the information system safety is lower than at the optimal state. With an optimal state, the probability of ensuring the security of the information system is higher than if staff are in a state of fatigue or under the influence of IPI in a stressful state. **Figure 8d** shows that if the staff use the break to restore their original characteristics, the probability of the information system safety increases.

For example, on December 23, 2015 [1], Ukrainian power companies experienced unscheduled power outages impacting a large number of customers in Ukraine. The outages experienced on December 23, 2015, were caused by external cyber attackers. After extensive reconnaissance of the victim networks, the telephone tdos attack was conducted on staff. As a result staff did not notice that substations disconnected in time. Exemplary actions of cyber malefactors and psychological conditions of staff are shown in the time diagram of **Figure 9**.

The received results coincide with the data obtained in the course of practical activity by interviewing the staff and owners of information systems, so it confirms the effectiveness of the proposed model for estimating the level of systems information security based on probabilistic analysis of the impact of their staff qualifications and psychological state.

Thus, to ensure the security of the information system, it is essential to take into account the abilities of staff. It is necessary to take into account the qualification of staff, which can change the probability of security of the information system characterized by technical and functional construction according to Eq. (1), from values $p(Sec)$ to 1, to monitor the condition of the staff, keeping them in in an optimal working condition with breaks.

4. Conclusions

The proposed method allows to use the probabilistic assessment of the system information security, taking into account the technical characteristics of the components of the information system, the qualifications of the staff, the mental state of the employees, and their psychophysical characteristics. Their permanent use in system life cycle helps to increase information security and decrease a potential danger of "human factor."

Author details

Igor Goncharov[1]*, Nikita Goncharov[1], Pavel Parinov[1], Sergey Kochedykov[2] and Alexander Dushkin[2]

*Address all correspondence to: goncharov@infobez.org

1 JSC "NGO" Infosecurity, Voronezh, Russia

2 Voronezh Institute of the Federal Penitentiary Service of Russia, Voronezh, Russia

References

[1] Cyber-Attack Against Ukrainian Critical Infrastructure. Available form: https://ics-cert.us-cert.gov/alerts/IR-ALERT-H-16-056-01

[2] Deev V. Methods of modulation and coding in modern communication systems. SPb: Science. 2007. 207 p

[3] Fress P, Piaget J. Experimental Psychology (Ed.-Comp.) Moscow: Progress; 1975. pp. 120-125

[4] Feer K. Methods of Modulation and Spreading the Spectrum. Moscow: Radio and Communication; 2000. 518 p

[5] Gnedenko B. Course of the theory of Probability. Moscow: Editorial URSS; 2007. 448 p

[6] Goncharov I, Demyanenko N, Khachumov A, Nozdrachev S. Analysis of the possibilities and systematization of technical means characterizing the construction of a channel for information and psychological impact. In: Proceedings of the Russian Scientific and Technical Conference. Voronezh: Publishing house VSU; 2009. p. 168-174

[7] Goncharov I, Demyanenko N, Mishina Y. Formalization of the Process of Information-psychological Influence. Vestnik VGU, System Analysis and Information Technologies. Voronezh: Publishing house VSU; 2012;**2**(36):41

[8] Goncharov I, Demyanenko N, Mishina Y. Possibility of modeling the process of information-psychological impact with the help of neural networks. In: XIII International Scientific-methodical Conference "Informatics: Problems, Methodology, Technologies". Voronezh: Publishing house VSU; 2013

[9] Goncharov I, Gerasimenko V, Vorobyova E, Dmitriev Y. Technical Means of Ensuring Information Security. Voronezh: VSTU; 2004

[10] Goncharov I, Mishina Y. Description of the approach to the representation of the states of objects and subjects of the process of information-psychological impact with the help of

wavelet transform. In: International Scientific-practical Conference "Technique and safety of the objects of the penal system-2013". Voronezh Institute of the Federal Penitentiary Service of Russia; Voronezh. 2013

[11] Goncharov I, Parinov P. Models of Information-psychological Impact. Vestnik VSU System Analysis and Information Technologies. 2017;**3**(65):71c

[12] Goncharov N. Justification of the approach to assessing the security of information in modern wireless networks. In: Sirota A, Goncharov I. Scientific Publication "Fundamental Problems of System Security" Materials of the III School-seminar of Young Scientists, May 26–28, 2016, Part 1 – Yelets: YSU them. I.A. Bunin. 2016. p. 87-100

[13] IEEE Standard Association. Available from: http://odysseus.ieee.org/

[14] ISO/IEC 27002 Information technology — Security techniques —Code of practice for information security management. Available from: http://www.iso27001security.com/html/27002.html

[15] Korolev A. Codes and Devices of Noisome Encoding. Minsk. Mn. 2002. 286 p

[16] Kudinov A, Chusova E. The research of loss stability of the level of psychical reaction of a human with the power of informational influence on him, bulletin of PFUR. Series mathematics. Information Sciences Physics. 2014;**2**:259-262

[17] Lieberman Y, Lieberman M. Experience of investigating the effectiveness of memoarherapy. Ekaterinburg: USPU; 2013. p. 192-200

[18] Lieberman Y, Matveva T. Training as a process of managing the level of knowledge and skills. Ekaterinburg: Economics of Education; 2006. p. 192-199

[19] Lieberman Y, Metelkov V. Mathematical model of the level of a person's psychic reaction and its investigation. Successes of Modern Natural Science. 2004. p. 10-14

[20] Naenko N, Ovchinnikova O. Problems of Engineering Psychology. Moscow: Nauka; 1969

[21] National Vulnerability Database [Electronic resource]. National Institute of Standards and Technology. Available from: http://nvd.nist.gov

[22] Prokis J. Digital Communication (trans. with English). Moscow: Radio and Communication; 2000

[23] Roshan P, Liery D. Fundamentals of Building Wireless LANs of the Standard 802.11 (Per. with English) Moscow: Williams Publishing House; 2004

[24] Sklyar B. Digital Communication: Theoretical Foundations and Practical Applications. Moscow: Vilams; 2016. 1104 p

[25] Standards for Security Categorization of Federal Information and Information Systems. Available from: https://www.nist.gov/publications/standards-security-categorization-federal-information-and-information-systems

[26] Venttsel E, Ovcharov L. Chapter 10: Markov processes: Streams of events: Theory of queuing. In: Theory of Probability. Moscow: "Science" (The Main Publishing House of the Physics and Mathematics Literature); 1969. 368 p

[27] Vyalykh A. Dynamics of vulnerabilities in modern secure information systems. In: Vyalykh A, Flacky S. Bulletin of the Voronezh State University. Series: System Analysis and Information Technology. Voronezh: Publishing house VSU; 2011;**2**:59-63

[28] Vyalykh A. Assessment of the vulnerability of modern information systems. In: Vyalykh A, Vyalykh S. Informatics: Problems, Methodology, Technologies: Mater. XI International. Sci. Method. Conf., Voronezh, February 10–11, 2011. Vol. T.1. Voronezh: CPI VSU; 2011. pp. 168-172

[29] Yerkes RM, Dodson JD. The relationship of strength of stimulus to rapidity of habit-formation. Journal of Comparative Neurology and Psychology. 2014;**18**(5). Available from: http://onlinelibrary.wiley.com/doi/10.1002/cne.920180503/pdf

7

Analysis of Terrorist Attack Scenarios and Measures for Countering Terrorist Threats

Dmitry O. Reznikov, Nikolay A. Makhutov and
Rasim S. Akhmetkhanov

Abstract

The chapter will present the classification of the types of modern terrorism and describe scenarios and probabilistic models of ordinary, technological, and the so-called intelligent terrorism that are distinguished by their triggering events, propagation modes, damaging factors, probabilities, and consequences. A comparative assessment of these three types of terrorism is presented. Dynamic three-sided models allow assessing the situation from standpoints of terrorists and law enforcement agencies, the administration of the complex engineering system, and analyzing actions and counteractions of various sides involved. A new comprehensive approach to ensuring complex engineering system security is described. This approach is focused not only on the development of protection barriers and safeguards against predetermined list of design-basis scenarios of terrorist attacks but also on increasing the system's resilience toward beyond design-basis attack scenarios.

Keywords: complex engineering system, terrorist attack, risk assessment, protection barrier, resilience

1. Introduction

Complex engineering systems (CESs), such as nuclear and thermal power stations; hydro engineering facilities; chemical, metallurgical, and oil refinery plants; etc., are critical in terms of population life support and ensuring sustainable economic development. The functioning of complex engineering systems is connected with storing, processing, and transportation of huge amounts of energy and hazardous materials. The unauthorized release of energy and hazardous material at a CES may cause disastrous consequences and trigger cascading failures in interrelated infrastructures. This makes complex engineering systems attractive targets for terrorists and requires special attention in countering terrorist threats [1–8].

Complex engineering systems are characterized by a complex structure, complicated behavior, and interaction between their components, which determine the ability of systems to redistribute loads and to resist cascading failures occurring after local failure of their individual components. Owing to the high level of uncertainty concerning the governing parameters of CESs, environmental conditions, and external impacts, the estimation of the complex engineering system performance should be probabilistic. Their evolution should be described by multivariate scenario trees [9–11].

Through the efforts of specialists from many countries, an extensive bank of knowledge has been developed for analyzing accidents and catastrophes at complex engineering systems, studying scenarios by which they might be initiated, and reducing the vulnerability of CESs with regard to natural and man-made disasters [12]. This bank of knowledge should be used as widely as possible to ensure security against the impacts of terrorism. This approach to analyzing terrorism-related threats presupposes that emergency situations triggered by terrorist attacks develop according to laws analogous to the development of emergency situations caused by natural or industrial disasters. Therefore, they may be analyzed by methods and models used to address classical problems in risk and safety theory [13–16].

The threat of terrorist attacks must be included in the system of studies of possible scenarios of how emergency situations might develop. In particular, event trees used in risk analysis at critically important infrastructure sites must be augmented with scenarios taking into account the possibilities of terrorist attacks that substantially change the scenarios themselves as the structure of primary initiating factors in emergency situations. They also lead to the initiation of cascading processes in the development of accidents and catastrophes with the most serious losses to the population, economic objects, and other vital resources. A classification and probabilistic models of basic scenarios of terrorist attacks were developed (**Figure** 1).

The need to include in the range of problems being considered the analysis of terrorism risks and terrorist mechanisms for initiating extreme situations requires developing and adapting existing models and methods for studying catastrophes with the aim of taking into account the special characteristics of their initiation with the help of unauthorized and terrorist actions that could be taken to attack at the most vulnerable and significant targets critically important for the national security infrastructure.

As it is imperative that terrorist risks and terrorist mechanisms of triggering emergencies be included into the framework of traditional risk assessment, the existing models and methods for analysis of accidents at CES should be modified, and new ones have to be developed in order to take into account specific properties of emergency initiation by terrorist impacts which can be targeted at the most vulnerable facilities of critical infrastructures. Most of the

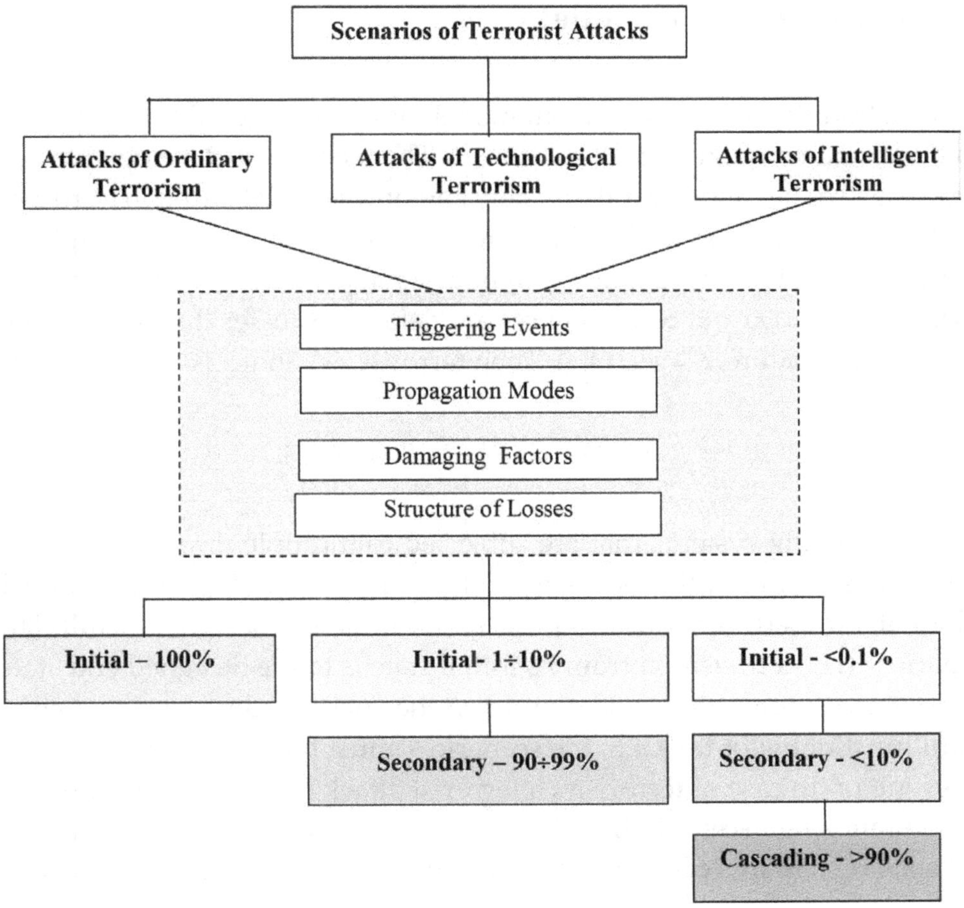

Figure 1. Basic scenarios of terrorist attacks.

components of complex engineering systems were however constructed in conformity with national and international regulations and norms for design, construction, and maintenance without direct consideration of terrorist threats [17, 18]. In this context, two major security-related problems arise:

1. Ensuring protection of the existent CES against terrorist attacks
2. Designing and constructing of a new CES with special protection barriers against terrorist attacks

To cope with these fundamental problems, it is necessary that a special analysis of methods and scenarios of terrorist acts be carried out and a study into how the existing and new protection barriers respond to terrorist attacks be conducted.

Conventional safety analysis for CES is to be focused on the question: What is the way for an accident scenario to be realized in the given system?

When addressing security problems for complex engineering systems, one should also consider the situation from the terrorist's standpoint. Hence, the modified question for security analysis should be: What is to be done for the given scenario to be realized at a CES?

2. General risk assessment model

According to the traditional risk assessment model, risk is considered to be a function of threat T, vulnerability V, and consequences C: $R = f(T, V, C)$. The model was developed to assess risks of technological catastrophes and natural disasters and now is widely used in terrorist risk assessments. Here threat is defined as probability of terrorist attack on a certain complex engineering system, $T = P(A)$; vulnerability is estimated as conditional probability of a system's failure given the attack occurs, $V = P(F \mid A)$ and consequences are defined as losses that occur as a result of the attack and the system failure, $C = E(U \mid A, F)$. Then terrorist risk index is determined by Eq. (1):

$$R = P(A) \cdot P(F \mid A) \cdot E(U \mid A, F). \tag{1}$$

For complex engineering systems that are subjected to multiple threats and multiple failure scenarios, risk assessment implies assessment of a scenario tree (**Figure** 2). This is being done using graph models called scenario trees [6, 7, 9]. The system is designed to fulfill the so-called success scenario S_0 (i.e., a transition from its initial state IS to the designed end state ES_0). Since any failure scenario S_* presents a deviation from the success scenario S_0 that corresponds to the successful functioning of the CES, the scenario S_* must have a disturbance point at which an extreme event, or, in case of terrorism, a terrorist attack (A_k), occurs (**Figure** 2). Each attack gives rise to a branch of a scenario tree that has a corresponding set of scenarios S_i that ends with an end state (ES_i). In this case, one can get a similar risk index using matrix expression:

$$R = \underbrace{\{P(A_1); P(A_2);\ldots; P(A_n)\}}_{Threat\ T} \times \underbrace{\begin{bmatrix} P[ES_1 \mid A_1]\, P[ES_2 \mid A_1] \cdots P[ES_m \mid A_1] \\ P[ES_1 \mid A_2]\, P[ES_2 \mid A_2] \cdots P[ES_m \mid A_2] \\ \cdots \\ P[ES_1 \mid A_n]\, P[ES_2 \mid A_n] \cdots P[ES_m \mid A_n] \end{bmatrix}}_{Vulnerability\ \mathbf{V}} \times \underbrace{\begin{Bmatrix} U_{ES_1} \\ U_{ES_2} \\ \cdots \\ U_{ES_m} \end{Bmatrix}}_{Consequences\ \mathbf{C}} \tag{2}$$

Eqs. (1) and (2) give first-order indicators of terrorist risk. They also determine three main ways of risk reduction: Reduction of terrorist threat is in the sphere of responsibility of law enforcement and intelligence communities, while reduction of vulnerability and consequences are the domains of engineering community and emergency management agencies, respectively.

In terrorist risk assessment framework, the main challenge is to estimate the probability of a terrorist attack. Some specialists believe that probabilistic measure is not adequate for the terrorist risk assessment since terrorist attack is not a stochastic event but a deliberate action based on the assessment made by terrorists regarding their skills and capabilities and the system's vulnerabilities.

Assignment of probabilities to the terrorist attack is a task which has a substantial human and behavioral dimension. The main problem is to describe the intentions of terrorists, their preferences, system of values (i.e., utility function), and decision rule. This allows one to assess the probability of different attack scenarios. The probability of each attack scenario is a function of the scenario's successful realization and their preferences regarding the expected consequences of that scenario.

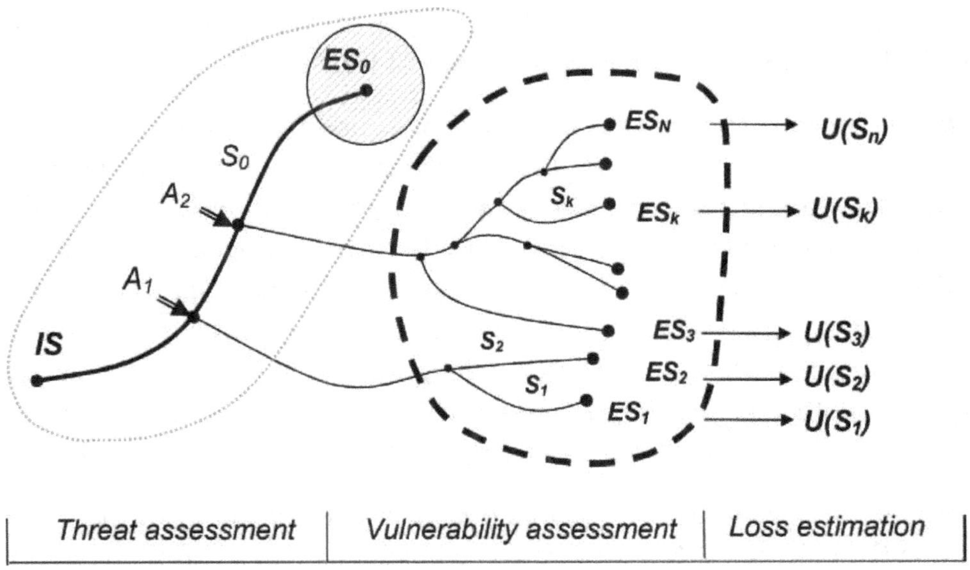

Figure 2. General risk assessment framework.

Unfortunately, Eqs. (1) and (2) could only be considered first-order indicators of the terrorist risk. The problem is that these equations do not allow one to account for a number of specific features of terrorism.

3. Specific features of terrorist threats

When assessing security-related problems for complex engineering systems, one should take into account the following characteristics of the terrorist threat [17, 19, 20].

High level of uncertainty: In modeling terrorist scenarios, we encounter a higher level of uncertainty. In addition to the uncertain factors inherent in threats of a natural or man-made nature, terrorist threats entail new factors of uncertainty resulting from the complexity of evaluating terrorists' system of values and behavioral logic as well as their organizational-technical potential and the resources at their disposal.

High level of dynamism: Terrorist attack scenarios and impact factors are more dynamic by nature than scenarios and impact factors for natural and man-made disasters to which the system is subject. A change in the spectrum and intensity of terrorism-related extreme effects on the system is significantly more rapid than in the case of natural or man-made threat. This is due to the terrorists' capacity for constantly expanding their arsenal of mechanisms for initiating emergency situations using modern means of attack, reacting to changes in protection barriers, and learning lessons from mistakes made during previous attacks on the system similar to it.

The capability of terrorists to choose attack scenarios deliberately: This refers to terrorists' deliberate selection of attack scenarios (places, times, and types of actions), taking into account the system vulnerability parameters and the losses expected if an attack is successfully carried out. That is, terrorists are capable of analyzing the vulnerability matrix and structure of losses for various types of actions against the CES and selecting the attack scenario that maximizes the harm to society (taking into account secondary and cascading losses). Here, in addition to probability analysis, it is also necessary to apply the tools of game theory, which makes it possible to take into account the intentional actions of terrorists.

Complex nature of the terrorist threat: The presence of a terrorist organization in a region may give rise to the possibility of a broad spectrum of attack scenarios. Thus, to counter terrorist threats and terrorist mechanisms for initiating emergency situations to an even greater degree than for natural and man-made risks, a systemic approach is needed for ensuring security and developing an optimal strategy for counterterrorism force and resource deployment. Inasmuch as concentrating resources on protecting one system element (or protecting a target from one scenario of terrorist action) could prove useless because, after evaluating the situation, the terrorists could redirect the attack against another element of the system or switch to a different attack scenario. In this case, counterterrorism efforts will fail to reduce risk and increase the system's level of protection.

Presence of two-way linkages between the terrorist threat and system vulnerability: The structure of linkages among the risk factors for the given *CES* in case of natural or manmade catastrophes is presented in **Figure** 3**a**. One differentiating feature of a terrorist risk assessment is the presence of two-way linkages (feedbacks) between the terrorist threat and (a) vulnerability of the system to the threat and (b) the magnitude of expected losses if the threat is successfully realized (see **Figure** 3**b**). This characteristic of terrorism must be examined in detail. In particular, reducing the vulnerability of a given system makes it possible to reduce substantially the level of the terrorist threat it faces.

In terrorist risk assessment framework, the main challenge is to estimate the probability of a terrorist attack. Some specialists believe that probabilistic measure is not adequate for the terrorist risk assessment since terrorist attack is not a stochastic event but a deliberate action based on the assessment made by terrorists regarding their skills and capabilities and the system's vulnerabilities.

Assignment of probabilities to the terrorist attack is a task which has a substantial human and behavioral dimension. The main problem is to describe the intentions of terrorists, their preferences, system of values (i.e., utility function), and decision rule. This allows assessing probability of different attack scenarios.

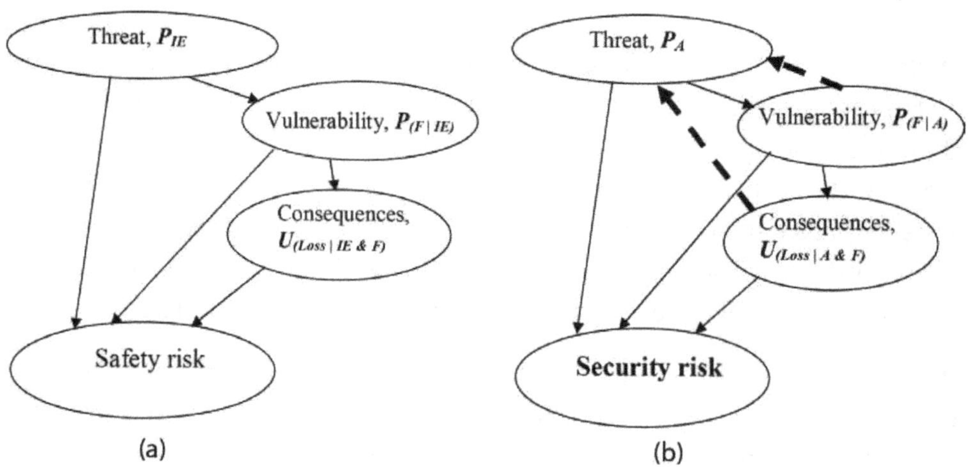

Figure 3. (a) System of linkages among risk factors for natural or man-made hazards (safety context). (b) System of linkages among risk factors for terrorist threat (security context).

Terrorists' capacity for self-learning: Because terrorists are capable of analyzing the results of previous attacks and drawing conclusions from them, their experience in "successful" and "unsuccessful" attacks can have a noticeable effect on the selection of a scenario for the next attack. Attack scenarios that proved their effective in the past are most likely to be repeated by terrorists in the future, while scenarios that ended unsuccessfully will most likely to be less attractive to terrorists and consequently are less likely to be repeated. Therefore, in assessing the chances that various attack scenarios will be realized, statistical self-learning models are more effective than traditional frequency methods.

In solving the above problem of security analysis, it is necessary to assess the resources the terrorists possess. In security analysis, by resources we mean a broad set of factors that determine the potential of a terrorist organization. These include:

- Material resources: technical means, equipment, and "human material" that can be used for terrorist attack
- Nonmaterial resources: experience and skills of terrorists, their knowledge, and access to the CES internal procedures

To answer the question of security analysis, experts should consider the quality of equipment the terrorists have, their skills and knowledge of *CES*, and their ability to take advantage of the existing vulnerabilities (and even create new ones) in order to organize the attack.

The ability of terrorists to select the most vulnerable and critical elements of *CES*, choose the time and place of an attack, adapt to changes of safety barriers and defense strategies, and learn lessons from previous attacks requires that the game theory approaches be included into probabilistic risk assessment models. That means that (a) traditional scenario trees used in safety risk assessment, which include only chance nodes, have to be supplemented by decision nodes that describe rational deliberate actions and counteractions of terrorists and counterterrorists; (b) models for terrorist risk assessment should be multi-sided and describe the situation from the perspective of terrorists and counterterrorist forces [11]; (c) these models should be dynamic and allow one to update actions and counteractions of various sides involved at different time steps.

4. Three types of terrorist attack scenarios

Scenarios of terrorist attacks can be divided into three types, scenarios of ordinary, technological, and intelligent terrorism, that differ in resources used by terrorists to carry out the attacks and structure of losses inflicted by the attacks (**Figure** 1) [17–19].

Scenarios of ordinary terrorism imply organization of explosions, fires, and assassinations of officials, public figures, and people at large in order to intimidate people and destabilize political situation in the country or region. Scenarios of ordinary terrorism are not considered in this paper since these scenarios are not focused on complex engineering systems. We are going to deal with two other types of terrorist attack scenarios that are directly related to *CES*.

4.1. Scenarios of technological terrorism

Scenarios of technological terrorism (*STT*) imply powerful unauthorized impacts at complex engineering system capable of:

- Breaking through the *CES* protection system
- Initiating secondary catastrophic processes due to hazardous substances (*W*), energy (*E*), and information (*I*) stored or processed at the *CES*
- Escalation of the accident outside the *CES* boundaries with substantially increased secondary and cascade losses

Technological terrorism is based on taking advantage of the existing vulnerability of the system. To perform an attack of technological terrorism, it is necessary to preliminarily:

- Analyze the *CES* structure and vulnerability, i.e., to reveal potential sources of secondary catastrophic processes (stocks of *W,E,I*), the weak points in the *CES* protection systems, and to devise the most efficient attack scenarios.
- Identify the *CES* key elements and links whose failure would disrupt the system.
- Calculate the strength of the initial impacts that might break through the *CES* protection barriers.
- Assess the *CES* scenario tree and determine the end states ES_* capable of initiating major secondary catastrophic processes outside the *CES.*

Scenarios of technological terrorism do not require that the attacking party have any insider information and can inflict point impacts imperceptible by the *CES* monitoring systems;

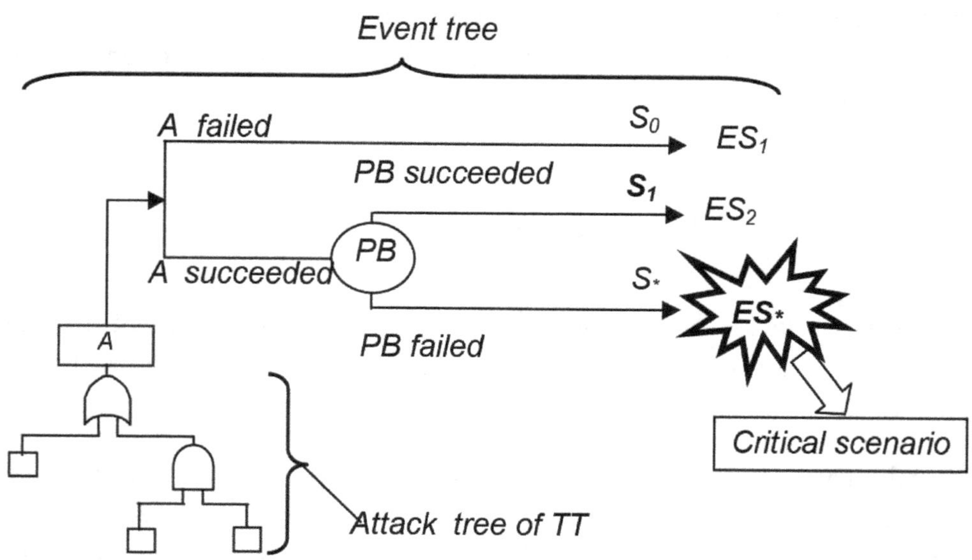

Figure 4. The scenario tree for technological terrorism.

therefore, they have to prepare a powerful action capable of breaking through the *CES* protection barriers [20]. It is necessary for the terrorist to select the method for the attack resulting in the CES end state that would initiate the accident propagation outside the *CES* boundaries.

The selection of the attack scenario is made through a hybrid scenario tree that in case of *TT* could be quite simple. It incorporates several attack trees describing the abilities and resources of terrorists and the event tree describing the *CES* vulnerability (**Figure** 4).

4.2. Scenarios of intelligent (or highly sophisticated, insiders') terrorism

Intelligent terrorism (*IT*) is a deliberate unauthorized interference into the process of designing, building, and/or operating the CES aimed at increasing its existing vulnerabilities and creating new ones in the system so that these input vulnerabilities, insider's knowledge of the system, and access to its elements are used for future realization of most disastrous scenarios of a terrorist attack.

IT implies:

- A comprehensive vulnerability assessment of a system under design, construction, or operation with respect to various scenarios of terrorist impacts and identification of the most effective way of realization of the initiating impact upon the system
- Insertion of latent changes into the system at the stage of its being designed, built, or operated, in order to give rise to new vulnerabilities in the CES
- Disconnection or disruption of the *CES* monitoring and protection systems
- Triggering cascading failures in the system and the environment

As a rule, scenarios *of IT* require that a member of a terrorist group penetrate into the staff of the organization that is designing, building, or operating the *CES*. The terrorist must possess insider's information on the *CES* and be able to perform well-camouflaged actions in order to weaken protection systems and create latent defects undetectable by the existing monitoring systems.

Consequently intelligent terrorism implicates detailed knowledge of the *CES* structure and working principles. It also implies awareness of its existing and potential vulnerabilities, possible end states, possible scenarios of accident propagation, and initial impacts that can trigger them. Additionally, *IT* can anticipate distortion of the success scenario, formulate false targets, and generate new disastrous scenarios.

Attacks of intelligent terrorism can be carried out at any stage of the *CES*'s life cycle:

- At the stage of design, some latent defects can be intentionally introduced into the system.
- At the stage of construction, additional vulnerabilities can be input into the *CES* through intentional violations of the technological processes.
- At the stage of operation, some maintenance procedures that are critical for the *CES*'s safety can be intentionally violated.

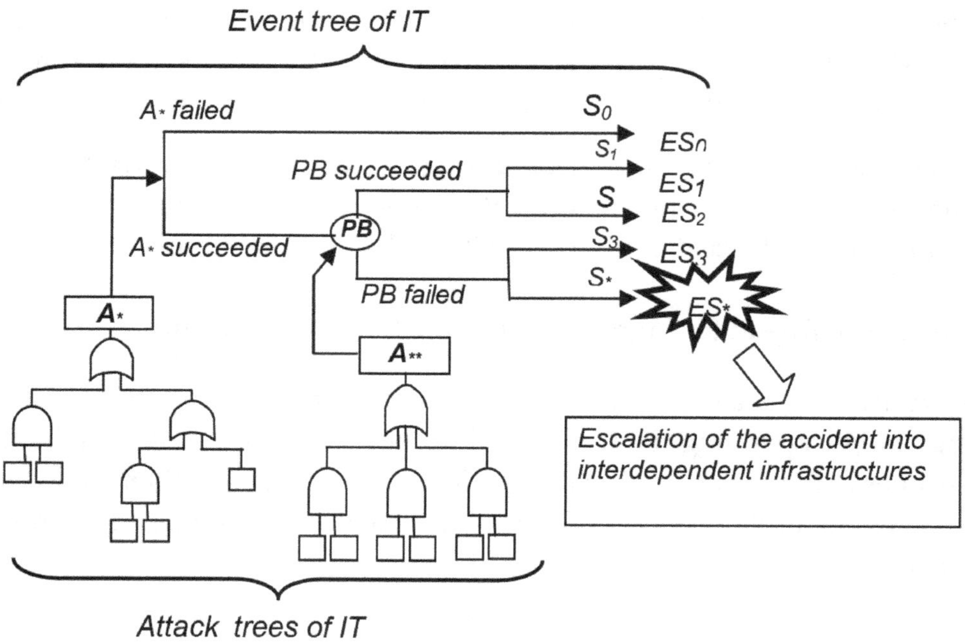

Figure 5. The scenario tree for intelligent terrorism.

Intelligent terrorism implies maximal level of the terrorist competence (comprehensive knowledge of the *CES* and its control, operation, and protection barriers), which enables it to select the most disastrous accident scenarios and find the most effective way of their initiation, disconnection, or disruption of the *CES* monitoring systems in order to prevent prompt response to failures. The assessment of the attack scenarios is made through a hybrid scenario tree that in case of *IT* could be more complicated (**Figure** 5). It incorporates several attack trees describing the abilities and resources of terrorists and the decision tree describing the system's vulnerability.

5. Development of dynamic multi-sided models for analyzing scenarios of terrorist attacks and developing counterterrorist measures

In view of the specific features of terrorist threats addressed in p.3 and the analysis of the scenarios of terrorist attacks on CESs presented in p.4 of this chapter, an integrated (three-sided) terrorist risk model based on the approaches developed in Bayesian networks and game theory has been developed [8, 21–23]. The schematic representation of the model is given in **Figure 6**. Each of the three graphs represents an influence diagram from the perspective of the following players: terrorist group, administration of industrial facility subjected to terrorist threat, and municipal authorities. These three diagrams are separated to keep the decisions made by different parties separate. Oval nodes represent random variables or events with their possible realizations and probabilities assigned. Rectangular nodes represent decisions and are characterized by possible options. The arrows represent probabilistic dependences between the events, state of variables or decision variables.

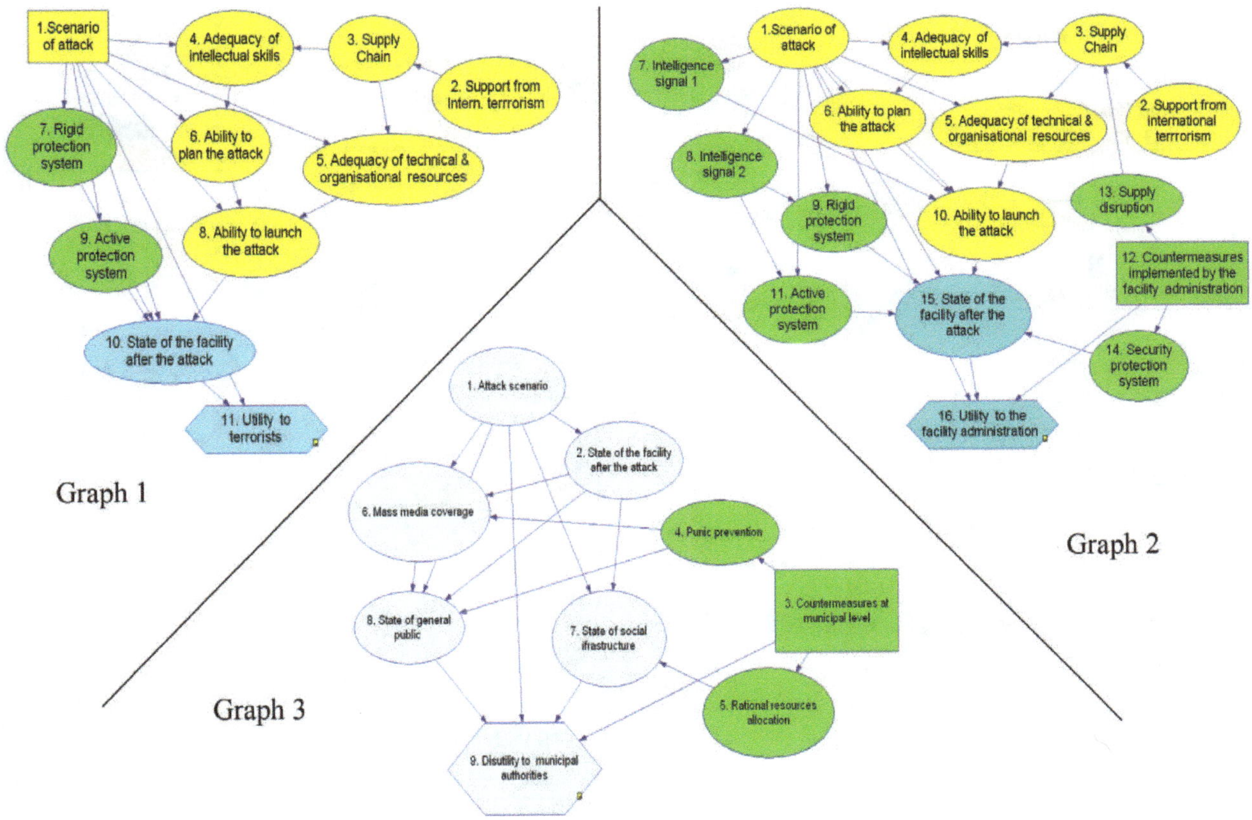

Figure 6. Multi-sided terrorist risk assessment model.

The model is based on the assumption that all the players act in such a way as to minimize their maximum losses. This strategy is governed by so-called minimax criterion: Counterterrorist players don't know which attack scenario the terrorist group will select, that is why they should choose the defense strategy that results in the lowest possible worst-case expected losses.

Graph 1 (**Figure** 7) represents an influence diagram from the perspective of terrorists. It allows one to assess (a) the probabilities that the specified attack scenario will result in damage and (b) the expected utility of terrorist of different attack scenarios[1].

$$EU(s_i) = \sum_{j=0}^{m} \left[Ut(s_i; v_j) \times P\left(V = v_j \mid S = s_i\right) \right] \; (i = 1, 2, \ldots, n), \tag{3}$$

where $Ut(s_i; v_j)$ is an element of utility matrix.

$$\begin{bmatrix} W(s_1; v_0) - Z(s_1) & W(s_1; v_1) - Z(s_1) & \cdots & W(s_1; v_m) - Z(s_1) \\ W(s_2; v_0) - Z(s_2) & W(s_2; v_1) - Z(s_2) & \cdots & W(s_2; v_m) - Z(s_2) \\ \cdots & & & \\ W(s_n; v_0) - Z(s_n) & W(s_n; v_1) - Z(s_n) & \cdots & W(s_n; v_m) - Z(s_n) \end{bmatrix} \tag{4}$$

[1]Figures on the diagram are conditional and are presented for the illustrative purpose.

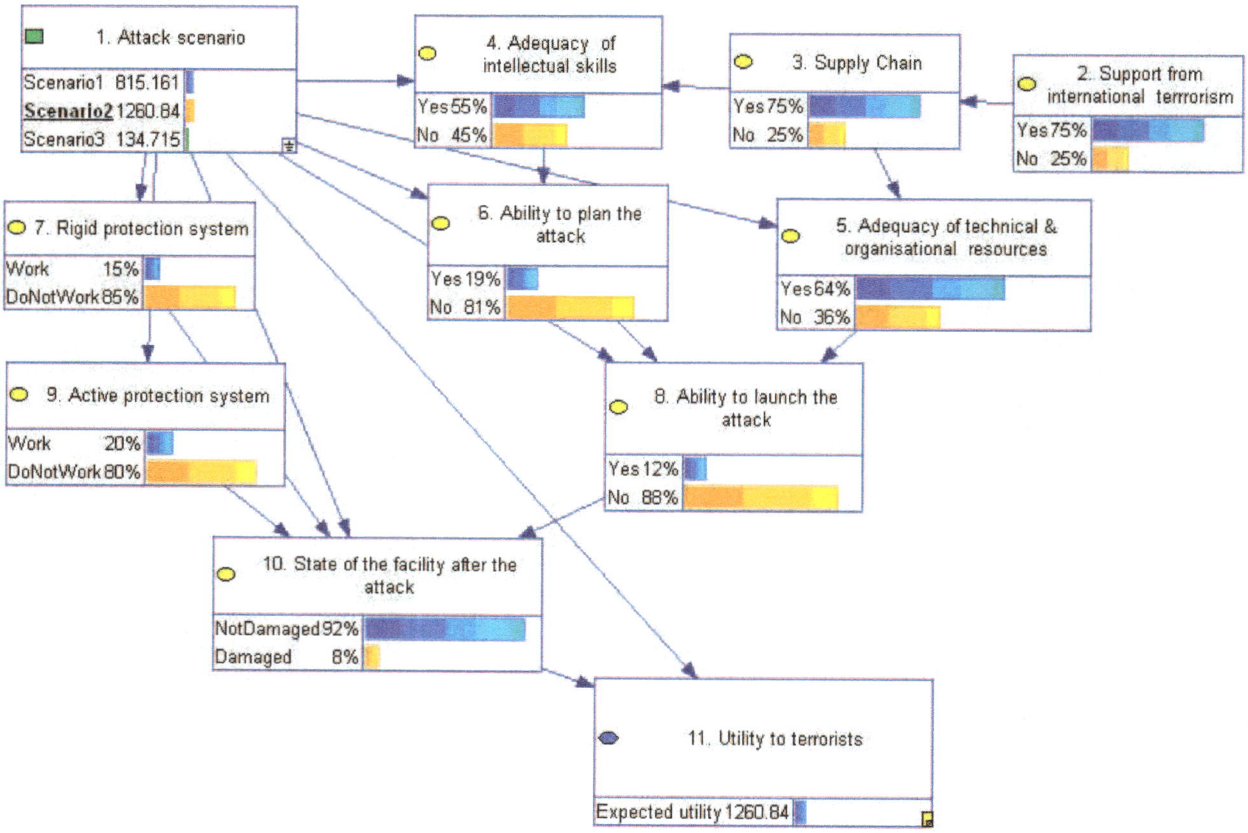

Figure 7. An illustrative example of the influence diagram from the perspective of terrorist group.

s_i is attack scenario; v_j is damage factor of the facility inflicted by the attack ($j = 0, 1, \ldots n$: $j = 0$ corresponds to a not damaged system, while $j = n$ corresponds to completely destroyed system); $P(V = v_j | S = s_i)$ is conditional probability of inflicting damage factor j to the facility provided that attack scenario i was carried out; $W(s_i; v_j)$ is the outcome in case of attack scenario i and damage state j; $Z(s_i)$ are the costs of implementing attack scenario i.

Calculation of expected utility values for different attack scenarios allows one to estimate probabilities of these scenarios (Eq. (5)) [8, 11]:

$$P_t(S = s_i) = \frac{EU_t(s_i)}{\sum_{k=1}^{n} EU_t(s_k)} \quad (i = 1, 2, \ldots n). \tag{5}$$

Eq. (5) assumes that (a) different attack scenarios are mutually exclusive and (b) the decision taken by terrorists is rational (i.e., they chose attack scenarios that maximize the expected utility). The results obtained in Graph 1 are then used as inputs to Graphs 2 and 3. The results of Graph 2 are then used in Graph 3.

Graph 2 (**Figure 8**) represents an influence diagram from the perspective of administration of industrial facility subjected to terrorist threat. It allows one to assess expected disutilities related to various countermeasures made by the administration of the facility involved. The probabilities $P_t(S = s_i)$ (Eq. (5)) are used in Graph 2 as state probabilities of the chance node 1.

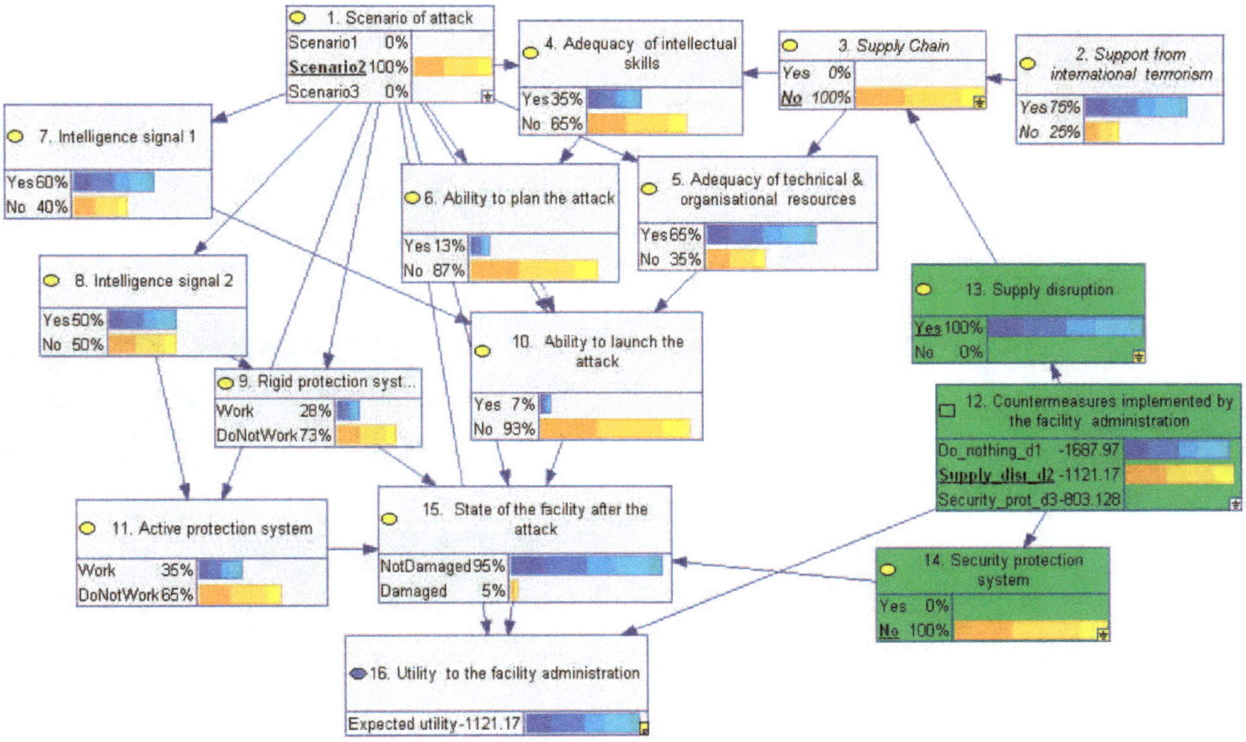

Figure 8. An illustrative example of the influence diagram from the perspective of CES's administration.

The graph permits estimation of expected disutilities to facility administration in case of various countermeasures adopted by the facility administration, to rank countermeasures.

Graph 3 (**Figure 9**) represents an influence diagram from the perspective of local community authorities. Graph 2 and Graph 3 permit assessment of risk reduction benefits of different countermeasures and their costs.

The structure of the influence diagrams and probabilistic dependences between the variables should be developed by the joint efforts of specialists representing a broad spectrum of disciplines (these include specialists in terrorist threat assessment, reliability theory, social sciences, loss estimation), each providing insights in their relevant area of expertise. The model permits identification of effects of different factors and parameter values on the likelihood of success of different attack scenarios and on the expected utilities to different sides involved.

The model described above can be used in dynamic fashion via discrete time steps. At each step, each player updates his beliefs, objectives, and decisions based on his previous step. Each of the players is uncertain about the other's actions and state of knowledge. To address the dynamics of security problem, one needs to model moves and countermoves of all three sides involved, changes in the structure of terrorist organizations and systems of protection, and lessons learned by all parties from previous attacks.

At each consecutive time period, all three parties make decisions regarding their actions in the upcoming time period based on the information accumulated so far (Blocks I_{t_k} and $I_{t_{k+1}}$, **Figure 10**). Estimations of probabilities of various attack scenarios and countermeasures adopted by facility administration and community authorities obtained at time step t_k could be treated as prior

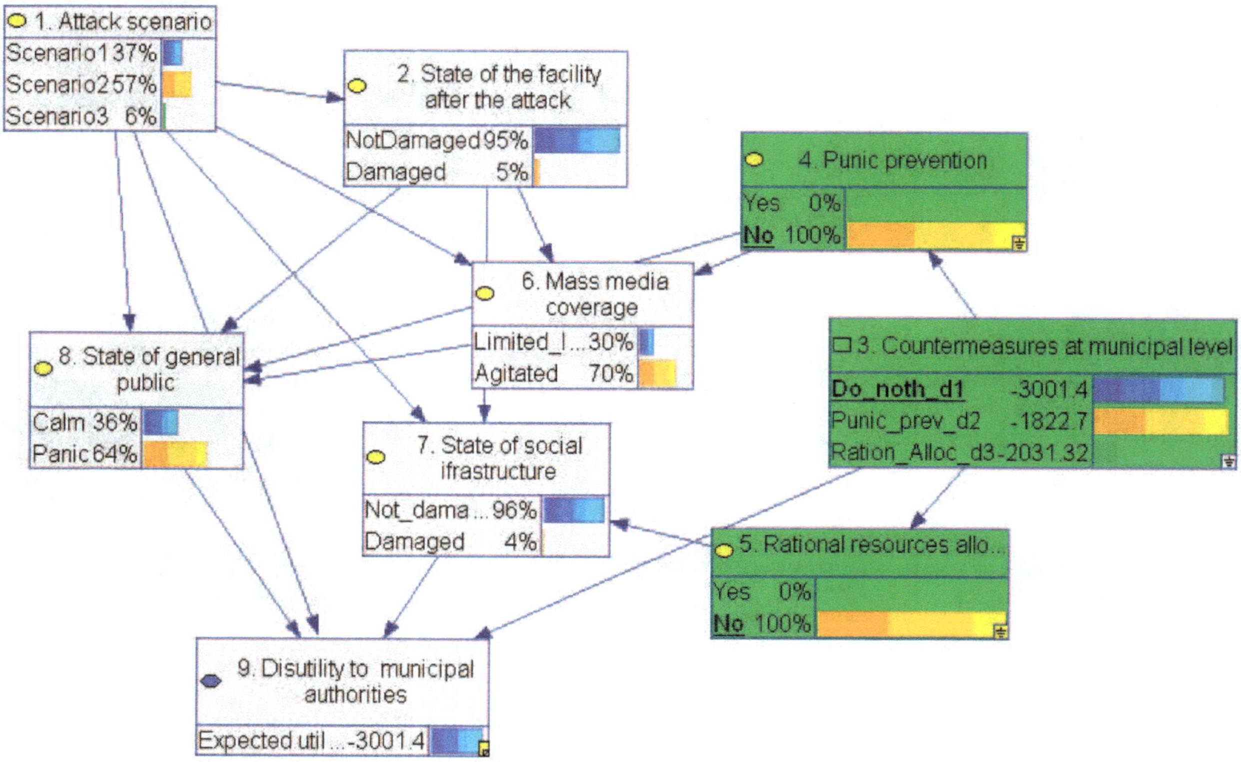

Figure 9. An illustrative example of the influence diagram from the perspective of community authorities.

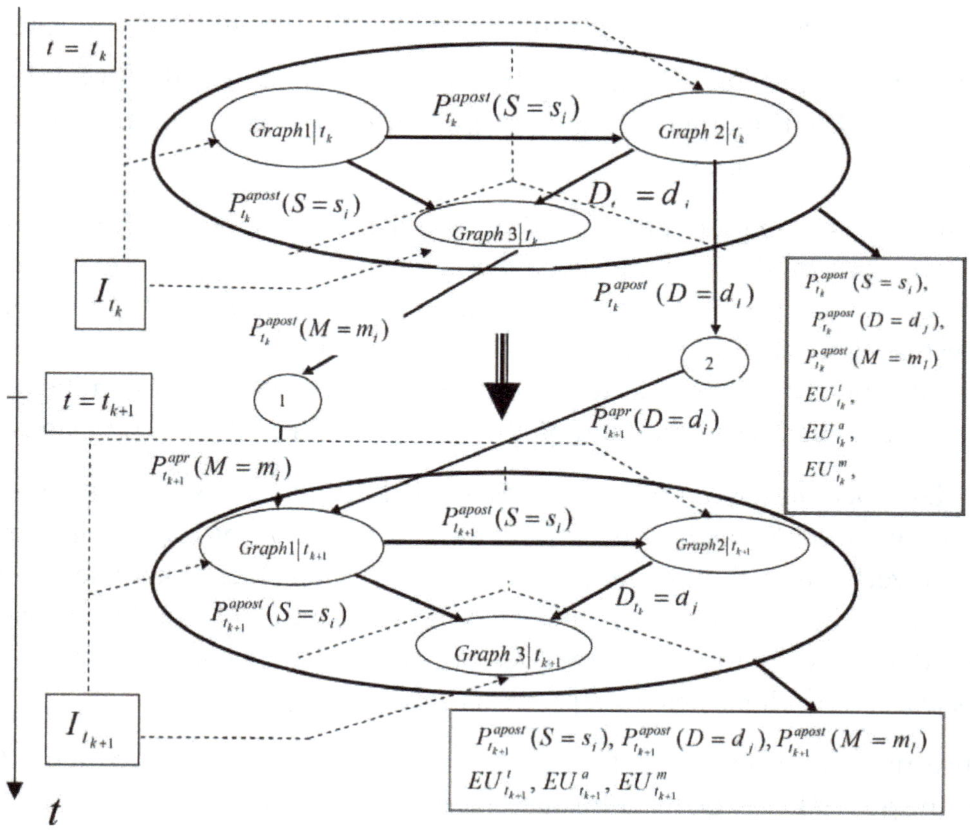

Figure 10. Dynamic multi-sided terrorist assessment model.

estimates for the time period t_{k+1}. Terrorist may take into account countermeasures of counterterrorist forces by including the respective chance nodes into Graph 1 at time step t_{k+1} and estimate probabilities of countermeasures adopted by facility administration d_j and municipal authorities m_l using Eq.(6) similar to Eq.(5):

$$P_a(D = d_j) = \frac{EU_a(d_j)}{\sum_{g=1}^{3} EU_a(d_g)}, k = 1, 2, 3; P_m(M = m_l) = \frac{EU_m(m_l)}{\sum_{f=1}^{3} EU_m(m_f)}, l = 1, 2, 3 \quad (6)$$

6. Measures for countering terrorist threats

6.1. Measures aimed at increasing protection of a CES from terrorist threats

The complexity of modern engineering systems and their interdependence with other systems make them vulnerable to attacks of technological and intelligent terrorism. This complexity stems largely from the vast functional and spatial dependencies and nonlinear interactions between the components of CES as well as from interdependencies that exist among the CESs which enable failures to cascade within one system and pass from one system to another.

Different historical, economic, political, social, as well as cultural traditions have formed different approaches to ensuring safety of complex engineering systems. Contemporary CESs, i.e., power, transport, and telecommunication networks, are becoming transboundary. Their significant spatial extension makes their functioning dependent on many factors and events in different parts of the world. The ensuring of CES's security is a complex interdisciplinary problem. It is impossible to solve this problem without joining efforts of experts in different fields and taking into account technical, social, psychological, and cultural-historical aspects.

Analysis of major disasters at CES in different countries shows that high-risk engineering systems in many cases are being designed and constructed according to traditional design codes and norms that are based on common and quite simple linear "sequential" risk assessment models and employ traditional design, diagnostics, and protection methods and procedures. This is being done in the assumption that a bounded set of credible design-basis impacts and subsequent failure scenarios could be determined for the CES, thus allowing one to create a system of protection barriers and safeguards that could secure the CES from the identified impacts with required substantially and high probability. This bounded set of impacts referred to as design-basis impacts includes normal operation events as well as abnormal events (component failures, human errors, extreme environmental loads, attacks of technological terrorism on CES) that are expected to occur or might occur at least once during the lifetime of the CES.

The currently available approach to ensuring security of complex engineering systems is based on the so-called protection approach that provides for the development of a set of protection barriers against the list of terrorist attack scenarios that were identified in advance. Within this approach, attacks of technological terrorism should be included into the list of design-basis events. To protect CESs from these scenarios of terrorist attacks, the following types of protection barriers should be developed (see **Figure 11**):

- Rigid protection barrier (protection barrier that requires a powerful impact to be broken)
- Functional protection barrier (protection barrier that in case of an accident could take on certain system's functions for a limited time or could prevent an accident from progressing further)
- Natural protection barrier (involves the use of passive natural phenomena and processes aimed at limiting the scales of the accident)
- Security guards

Circles "1," "2," and "3" stand for separate types of protection barriers. Areas of intersection ('1-2," "2-3," "1-3," and "1-2-3") – correspond to combination of correspondent types of protection barriers. Security guard barrier "4" is organized to ensure protection of all of the above mentioned barriers ("1," "2," "3," "1-2," "2-3," "1-3," and "1-2-3").

Application of this protection approach allows one to reduce risks of design-basis scenarios of technological terrorism (compare FN curves 1 and 2; **Figure 12**). However, it should be noted that this protection-based approach does not allow one to reduce risk of unforeseen "low-probability-high-consequence" scenarios of intelligent terrorism that could not be included into the list of design-basis events.

In currently applied protection-based approach, a number of low-probability impacts of extreme intensity are neglected as being practically incredible. Other impacts (such as attacks of intelligent terrorism) are not identified and, consequently, not analyzed. Such impacts are classified as beyond design-basis impacts. Thus, the issue of protection of CES from beyond design-basis impacts has not been addressed in a proper manner. These impacts however can cause large-scale disasters of extreme severity and induce tremendous property losses and a great number of victims.

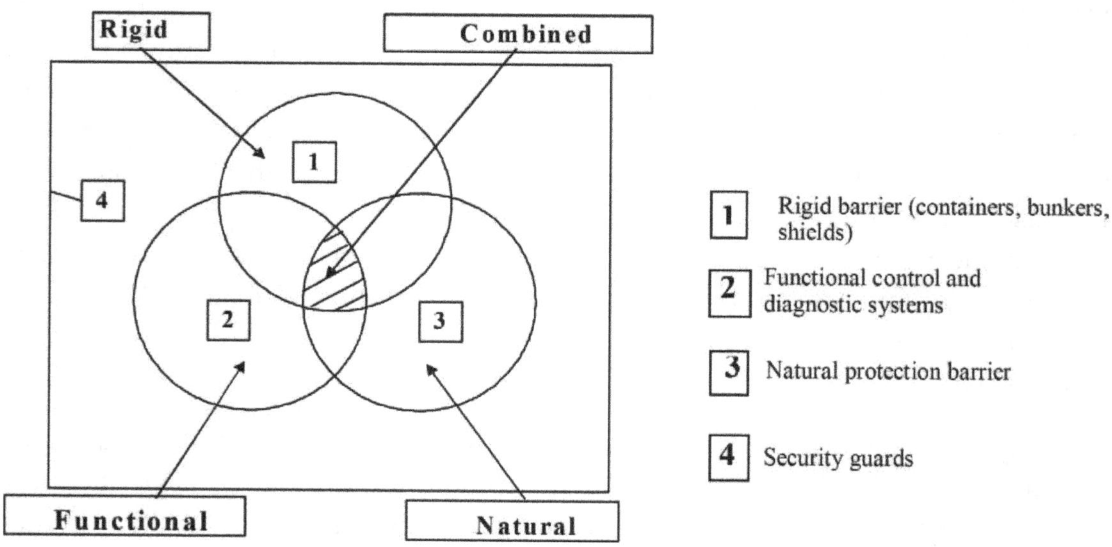

Figure 11. Types of protection barriers.

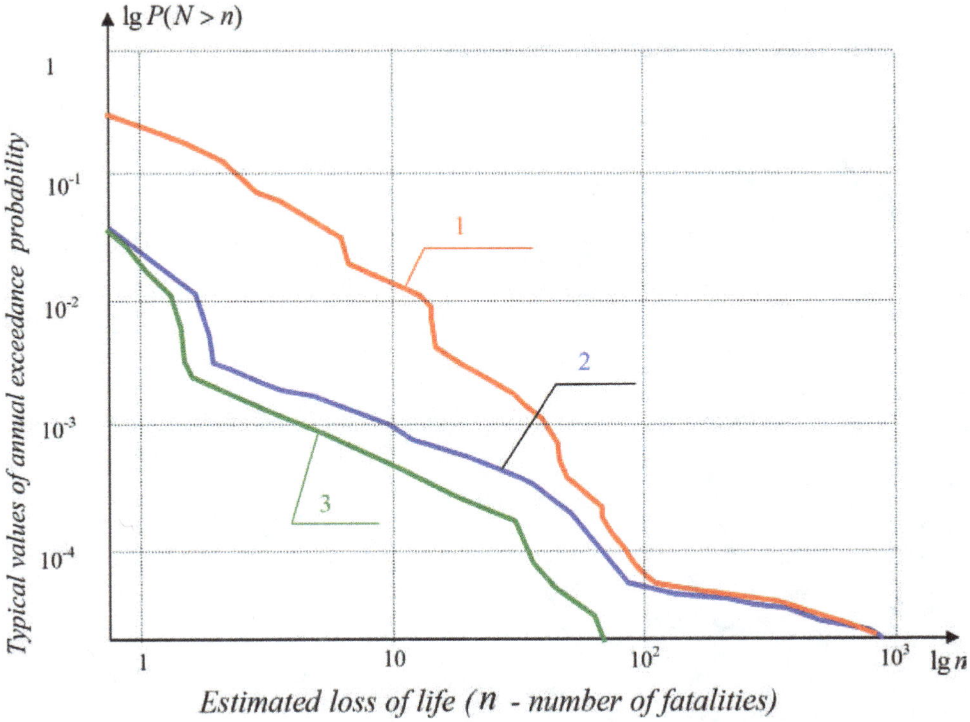

Figure 12. FN curves before and after realization of protection and resilience measures. (1) FN, curve before realization of any measure; (2) FN, curve after realization of protection measures; (3) FN, curve after realization of protection and resilience measures.

6.2. Measures focused on ensuring CES's resilience to beyond design-basis events

Complex engineering systems are becoming global networks. The currently available methodologies of risk assessment and reliability engineering were developed for technological systems with fixed boundaries and well-specified hazards for which exists statistical and/or actuarial data on accident initiation events, component failure rates, and accidents' consequences which allow one to quantify and verify models taking into account uncertainties deriving from both natural variations of the system parameters (and performance conditions) and from lack of knowledge of the system itself.

The protection-based approach is focused on developing safety barriers for countering the identified scenarios of terrorist attacks that were included in the list of design-basis events. This approach however has the weakness of neglecting the possibility of beyond design-basis events. To overcome this weakness, a new comprehensive strategy is needed. This strategy should not only include measures aimed at development protection barriers against design-basis attacks of technological terrorism but also development of special measures aimed at increasing the system's resilience to future yet-to-be-determined scenarios of attacks of intelligent terrorism (**Figure 13**) [24, 25].

The current accident models and risk assessment techniques such as fault and event tree analysis are not adequate to account for the complexity of modern engineering systems. Due to rapid technological and societal developments of the recent decades, modern engineering systems are becoming steadily more complex. It means that (a) in safety assessments for CES, there are too many details to be considered, and (b) some modes of CES's operation may

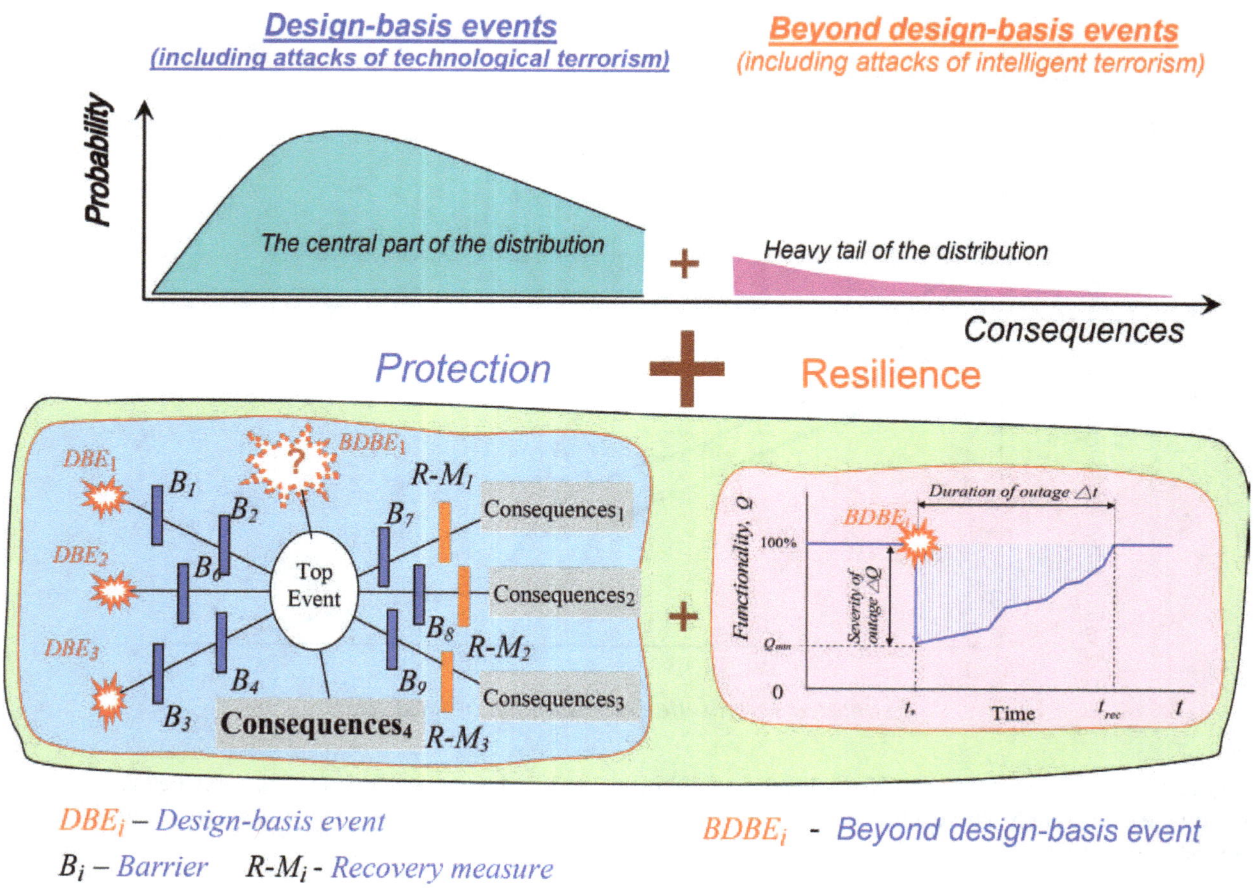

Figure 13. A new comprehensive approach to ensuring CES's security based on implementation of protection measures and measures for improving resilience of CES.

be incompletely known due to complex nonlinear interactions between components of CES, due to tight couplings among different systems, and because CES and its environment may change faster than they can be described. As a result, it is impossible to describe the performance of CESs in every detail. In other words for complex engineering systems, it is practically impossible to define a bounded set of design-basis impacts that are expected to occur or might occur at least once during the lifetime of the CES.

This problem can be solved by including the concept of resilience in the processes of designing and ensuring the safety and security of CESs [26, 27]. The proposed approach should not be considered as a substitute but rather a supplement to the traditional one. Adopting this view creates a need to move beyond traditional "threat-vulnerability-consequence" models that are limited to analyzing design-basis events and deal with beyond design-basis impacts and impact combinations. This comprehensive approach will be based on such concepts as resilience to provide more adequate explanations of accidents as well as identify ways to reduce risks caused by beyond design-basis impacts.

In other words, the new security paradigm for complex engineering systems should focus the efforts not only on development of protection barriers and safeguards against design-basis accidents but also on increasing the CES's resilience toward beyond design-basis impacts (**Figure 13**).

The CES's resilience is the capacity of the system potentially exposed to hazards to adapt, by resisting or changing in order to reach and maintain an acceptable level of functioning. This is determined by the degree to which the CES is capable of organizing itself to increase its capacity, of learning from past disasters for better future protection, and to improve risk reduction measures.

Figure 14 presents the so-called resilience profile of the system: a powerful beyond design-basis event (BDBE) occurs at the time moment t_* resulting in a slump of the system's performance characteristics Q which recovers at the time moment t_{rec}. A ratio of the square F_e of the figure BDEF that is located under the chart of the CES's performance characteristics in the period between the time moment t_*, when the beyond design-basis event occurs, and the moment t_{rec} when the system returns to its normal operation level and the square F_n of the rectangular ADEF can be considered as a quantitative measure of the system's resilience [26, 28]:

$$Res = \frac{F_e}{F_n} = \frac{\int_{t_*}^{t_c} Q(t)dt}{(t_{rec} - t_*) \cdot Q_n} \times 100\% \tag{7}$$

Two groups of measures aimed at increasing the CES resilience can be identified:

- Measures focused on reducing the severity of outage ΔQ (**Figure 15a**)
- Measures focused on the reducing the duration of the outage Δt (**Figure 15b**)

As previously stated, due to the complexity of modern engineering systems and their potentially large-scale catastrophes, in order to ensure security of such systems, one needs to move beyond traditional design-basis risk management framework. The new paradigm needs to be focused on increasing CES's resilience (**Figure 13**). That means that if the beyond design-basis accidents are to be considered, the scope of the analysis should be widened. Security-related efforts should be focused not only on the development of protection barriers and safeguards from predetermined (postulated) set of design-basis attacks of technological terrorism but

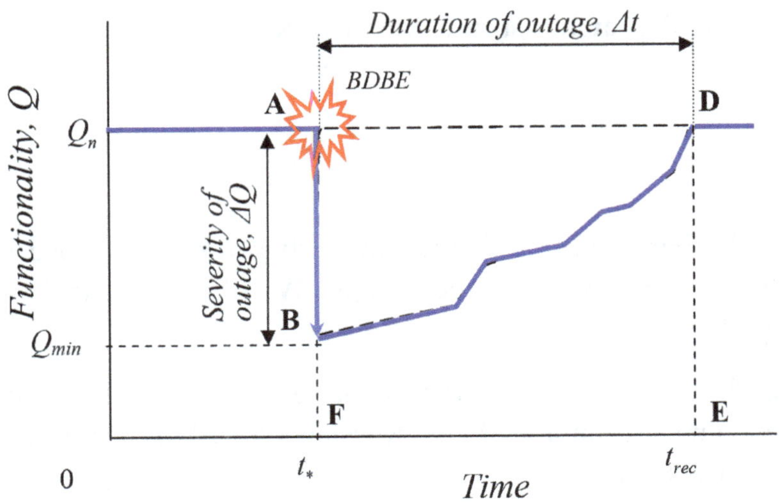

Figure 14. Resilience profile of CES.

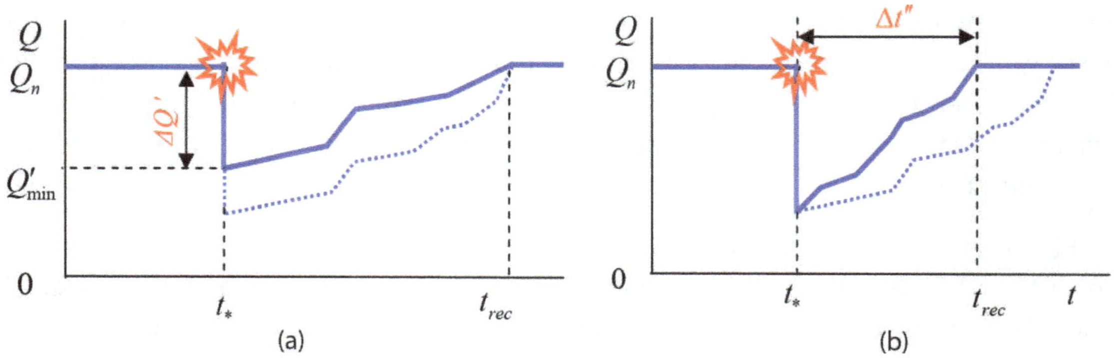

Figure 15. Measures to increase CES resilience. (a) Reduction of the outage severity. (b) Reduction of the outage duration.

also on additional set of measures aimed at increasing complex engineering system resilience that would prevent catastrophic failure and long-term dysfunctioning of CESs in case of beyond design-basis attacks. Application of such comprehensive (protection and resilience focused) approach allows one to reduce risks of beyond design-basis scenarios of intelligent terrorism (compare FN curves 2 and 3; **Figure 12**).

Acknowledgements

This work was financially supported by the Russian Foundation for Basic Research (grant no. 16-29-09575).

Author details

Dmitry O. Reznikov*, Nikolay A. Makhutov and Rasim S. Akhmetkhanov

*Address all correspondence to: mibsts@mail.ru

Mechanical Engineering Research Institute, Moscow, Russia

References

[1] Schweitzer G, Sharber C, editors. Countering Urban Terrorism in Russia and the United States: U.S.-Russian Workshop Proceedings. Washington: The National Academies Press; 2006. p. 241

[2] Schweitzer G, editor. Countering Terrorism: Biological Agents, Transportation Networks, and Energy Systems: Summary of U.S.-Russian Workshop. Washington: The National Academies Press; 2009. p. 239

[3] Terrorism: Reducing Vulnerabilities and Improving Responses: U.S.-Russian Workshop Proceedings. Washington: The Academies Press; 2004. p. 239

[4] Frolov K, Baecher G, editors. Protection of Civilian Infrastructure from Acts of Terrorism. Dordrecht: Springer; 2006. p. 244

[5] Makhutov N, Baecher G, editors. Comparative Analysis of Technological and Intelligent Terrorism Impacts on Complex Technical Systems. Amsterdam: IOS Press BV; 2012. p. 194

[6] Kaplan S. Applying the general theory of quantitative risk assessment (QRAC) to terrorism risk. In: Haimes Y, Moser D, editors. Risk-Based Decision-Making in Water Resources X: Proceedings of the Conference. Reston: ASCE Publications; 2002. pp. 77-81

[7] Garrick B, Hall J, et al. Confronting the risk of terrorism: Making the right decisions. Reliability Engineering and Safety Systems. 2004;**86**:129-1768

[8] Pate-Cornell E. Probabilistic modeling of terrorist threats: A systems analysis approach to setting priorities among counter-measures. Military Operations Research. 2002;**7**:5-23

[9] Berman A, Nikolaychuk O, Yurin A. Intellectual data system for analyzing failures. Journal of Machinery Manufacture and Reliability. 2012;**41**(4):337-343

[10] Makhutov N, Reznikov D. Assessment and regulation of risks related to operation of complex technical systems. Problems of Safety in Emergency Situations. 2012;**5**:3-9 (in Russian)

[11] Makhutov N, Reznikov D, Zatsarinny V. Two types of failure scenarios in complex technical systems. Problems of Safety in Emergency Situations. 2014;**2**:28-41 (in Russian)

[12] Frolov K, Makhutov N, editors. Multi-Volume Addition Safety of Russia. Legal, Social, Economic, Scientific and Engineering Aspects. Znanie publ. Vol. 1-54; 1997-2018 (in Russian)

[13] Akhmetkhanov R. Stability of social system under terrorist impacts. In: Makhutov N, Baecher G, editors. Comparative Analysis of Technological and Intelligent Terrorism Impacts on Complex Technical Systems. Amsterdam: IOS Press; 2012. pp. 157-166

[14] Akhmetkhanov R. Risk management in natural and societal systems: Taking into account terrorist threats. In: Frolov K, Baecher G, editors. Protection of Civilian Infrastructure from Acts of Terrorism. Dordrecht: Springer; 2006. pp. 7-20

[15] Makhutov N, Akhmetkhanov R, Dubinin E, Kuksova V. Problems of rationing of terrorist risks to critical facilities, taking into account the risks increase of regular functioning. Problems of Safety in Emergency Situations. 2017;**2**:30-44 (in Russian)

[16] Makhutov N, Reznikov D. Characteristics of technological terrorism scenarios and impact factors. In: Countering Terrorism: Biological Agents, Transportation Networks, and Energy Systems: Summary of a U.S.–Russian Workshop. Washington: The National Academy of Sciences Press. 2009. pp. 53-70

[17] Reznikov D. Technological and intelligent terrorism: Specific features and assessment approaches. In: Makhutov N, Baecher G, editors. Comparative Analysis of Technological and Intelligent Terrorism Impacts on Complex Technical Systems. Amsterdam: IOS Press; 2012. pp. 45-60

[18] Makhutov N, Reznikov D, Petrov V. Engineering infrastructures: Problems of safety and security. In: European Perspective on Security Research and Safety Aspects. Berlin/Heidelberg: Springer; 2010. pp. 93-106

[19] Makhutov N, Reznikov D, Khaziakhmetov R. Basic scenarios of terrorist attacks at hydropower engineering facilities. In: Escuder-Bueno et al., editors. Risk Analysis, Dam Safety, Dam Security and Critical Infrastructure Management. London: Taylor & Francis Group; 2012. pp. 389-394

[20] Makhutov N, Reznikov D, Dubinin E, Kuksova V. Assessment of terrorist risk and making decision regarding the expedience of creation of protection systems against terrorist impacts. Problems of Safety in Emergency Situations. 2007;**1**:88-105 (in Russian)

[21] Woo G. Quantitative terrorism risk assessment. The Journal of Risk Finance. 2003;**4**(1): 15-24

[22] Makhutov N, Reznikov D. Application of Bayesian networks for assessment of terrorist risk and identification of optimal counterterrorist strategy. Problems of Safety in Emergency Situations. 2007;**1**:89-104 (in Russian)

[23] Makhutov N, Akimov V, Akhmetkhanov R, et al. Safety of Russia: Human Factor in Problems of Safety. Moscow: Znanie Publishing; 2008. p. 687 (in Russian)

[24] Critical Thinking: Moving from Infrastructure Protection to Infrastructure Resilience. Critical infrastructure protection program. Discussion Paper Series. George Masson University [Internet]. 2007. p. 109. Available from: http://citeseerx.ist.psu.edu/viewdoc/download;jsessionid=C359CF09E0E785A43C91C0A1871A9B4E?doi=10.1.1.169.9384&rep=rep1&type=pdf [Accessed: January 04, 2018]

[25] Makhutov N, Reznikov D, Petrov V. Specific futures of ensuring critical infrastructures safety. Safety in Technosphere. 2014;**3**(1):3-14

[26] Hollnagel E, Woods D, Leveson N, editors. Resilience Engineering: Concepts and Precepts. Farnham: Ashgate; 2007. p. 410

[27] Hollnagel E, Paries J, Woods D, Wreathall J, editors. Resilience Engineering in Practice: A Guidebook. Farnham: Ashgate; 2011. p. 362

[28] Cimellaro G, Reinhorn A, Bruneau M. Quantification of seismic resilience. In: Proceedings of the 8-th U.S. National Conference on Earthquake Engineering, USA [Internet]. 2006. Available from: http://citeseerx.ist.psu.edu/viewdoc/download?doi=10.1.1.538.8029&rep=rep1&type=pdf. [Accessed: January 04, 2018]

8

Probabilistic Analysis of Transportation Systems for Oil and Natural Gas

Yuriy V. Lisin, Nikolay A. Makhutov,
Vladimir A. Nadein and Dmitriy A. Neganov

Abstract

In this chapter, the need of probabilistic modeling for design, construction, and operation of oil and gas pipelines is justified. Such modeling should use information and databases on deterministic and statistical dependencies related to deformation, damage accumulation, failure, fracture accidents, and catastrophes. The probabilistic design equations and their parameters for the characteristics of strength, durability, fracture toughness, risks of accidents, and manmade catastrophes are given. The economic efficiency of pipeline management based on controlling probabilistic characteristics through conducting diagnostic, repair-and-renewal operations while ensuring the acceptable levels of reliability and safety parameters is substantiated. The results of studies in the field of statistics and probabilities of emergency situations during manufacturing, construction, and operation conducted by Russian and foreign specialists are presented.

Keywords: oil and gas transportation, pipeline transport, main pipelines system, pipe steel, pipeline strength, yield strength

1. Introduction

Oil, gas and chemical complex (OGCC) is one of the system and fund forming in our country. It includes tens of thousands of oil and gas production facilities, over 500,000 km of field and main pipelines for transportation of liquid and gaseous hydrocarbons, thousands of large oil and gas storage facilities, and hundreds of major oil and gas refineries for fuel and chemical products for civil and military use.

These figures indicate the exceptional importance of the integrated safety and security of the national oil, gas, and chemical complex, which constitute a significant part of the national and international safety problems. The scientific analysis of these problems, and the solution of fundamental, practical and economically significant tasks in the field of safety are becoming more relevant as the scope and geography of OGCC expands in Russia.

In the second half of the twentieth century and the beginning of the twenty-first century, environmental and economic damage, accidents, and injuries at the facilities of the OGCC (including objects of the main pipeline systems (MPS)) became the subject of active interaction between state authorities, sectorial scientists, and design, technological, construction, and operating organizations. The leading roles in this interaction belong to the Security Council, Rostekhnadzor, the Russian Academy of Sciences, the research centers of the largest companies (Transneft, Gazprom, Rosneft), and the leading universities in the country.

In the traditional and advanced safety developments for OGCC and MPS facilities, the priority will be under scientifically grounded combination of research, rationale, regulation, and expertise, as well as improvement of strength, durability, and safety of the technologies in the light of the emerging spectrum of threats and risks in the context of diversifying economy.

The solution of these problems mainly lies in deterministic, statistical, and probabilistic methods of modeling, calculations, tests, and justification of performance of OGCC and MPS facilities.

Therefore, the major focus is on the probabilistic, statistical, and deterministic analysis of strength and durability of the main pipelines for oil and gas transportation.

2. Basic design dependencies

In the second half of the twentieth century and the beginning of the twenty-first century in Russia and abroad, branched pipeline systems for hydrocarbons transportation, including the main and field oil and gas product pipelines have been constructed. At present, one of the world's largest pipeline systems operates in Russia (**Table 1**) with a total length of more than 500,000 km.

Design, construction, and operation of pipelines for many decades were based [1–5] mainly on the strength standards. These standards (in the form of state standards (GOST), industry standards (OST), building norms and rules (SNiP), guidelines (RD), technical regulations (TR), federal rules and regulations (FNiP), methodological recommendations (MR)) were based on:

- Classical strength theories (I) maximum normal stresses σ_{max} (II) maximum deformations e_{max} (III) maximum tangential stresses τ_{max} (IV) maximum forming energy V_{max}
- Analysis of designed operational nominal stresses $\sigma_n^э$ by methods of material resistance and the theory of rods, plates, and shells

No.	Type	Purpose	Length (ths. km)
1	Main pipelines	Gas pipelines	180.2
		Oil pipelines	55.3
		Product pipelines include: Ammonia pipelines NGL pipelines	22.2 1.4 4.3
		Total	**257.8**
2	Field pipelines	General purpose	250.0
Total			**507.2**

Table 1. Types, purposes, and length of pipeline systems.

- Use of the calculations of allowable stresses $[\sigma]$ or limiting resistances R_u
- Basic characteristics of the mechanical properties of pipe steels that determine the resistance to plastic deformation, failure, and loss of stability

Generally, the conditions of pipeline's strength, at present, can be described (**Figure 1**) [1–3] by the functional relation:

$$\begin{gathered}\sigma^{s}_{n\max} = F_{\sigma}\{p, N, M_{u}, M_{k}, \delta, D, E, R_{u}, \mu\} \leq [\sigma] = \frac{\sigma_{on}}{n_{\sigma}}, \\ \sigma_{on} \leq F_{o}\{\sigma_{\max}, \varepsilon_{\max}, \tau_{\max}, V_{\max}\} = F_{\sigma}\{\sigma_{\text{т}}, \sigma_{в}, \sigma_{y}\}\end{gathered} \tag{1}$$

where $\sigma^{s}_{n\max}$—maximum designed stress for the most dangerous operating conditions (taking into account internal and external pressure p, axial forces N, bending M_{u}, and torque M_{κ} in a critical section and a critical point); σ_{on}—critical (ultimate) stress, determined from the test

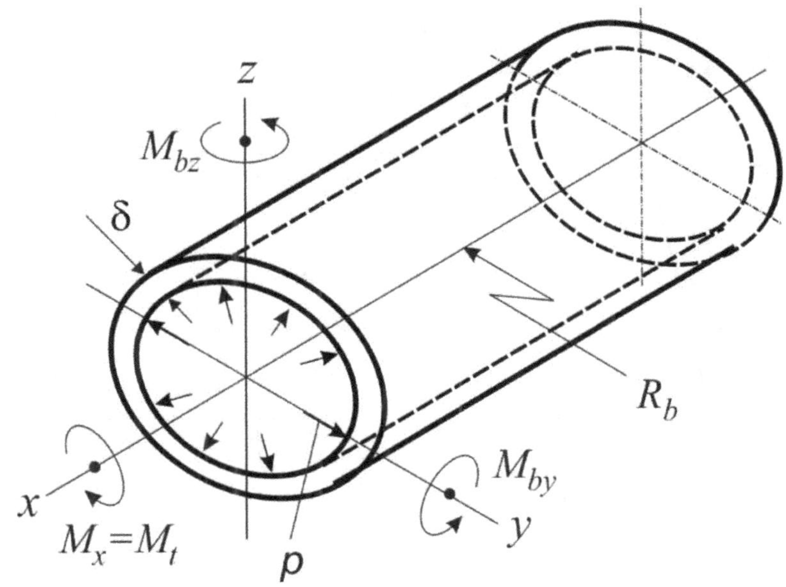

Figure 1. Scheme of operational loading of the pipeline.

data of standard specimens on strain (compression) at the stages of the beginning of fluidity (yield point σ_y), reaching the ultimate strength (ultimate stress limit σ_u) or the beginning of buckling (critical stress σ_y); $N = N_x$—longitudinal force along the x-axis; $M_{вy}, M_{вt}, M_{uz}$—bending moments around the y- and z-axes; $M_t = M_x$—torque around the x axis; n_σ—margin of safety ($n_\sigma \geq 1$); δ—pipeline wall thickness; D—diameter of the pipeline (external, internal, or mean); E—modulus of longitudinal elasticity; μ—Poisson's ratio; and $R_в$—bend radius of the pipeline axis.

All the calculated parameters of Eq. (1) can be considered in deterministic, statistical, and probabilistic formulation, taking into account the complication of operational conditions and the improvement of engineering methods of mathematical modeling, physical experimentation, and normative calculations.

The calculation of stresses $\sigma^s_{n\max}$ as a function F_σ in Eq. (1) is the initial independent goal of solving boundary value problems—analysis of nominal stress-strain states under complex operational and exploitational loading regimes at all stages of the life cycle of pipes and pipelines.

In expression (1), based on the static tension diagram of a standard sample (**Figure 2**) in the conditional coordinates «$\sigma - e$» (without taking into account the reduction in the cross-sectional area and increasing the sample length), as critical stress σ_{on} is used [3–5]

- In the yield zone: the yield strength σ_y as the ultimate resistance to elastic deformation—the limit of proportionality σ_p, the yield strength σ_y at the yield plateau, the conditional yield strength corresponding to the achievement of a given plastic deformation, for example, 0.2% ($\sigma_{0,2}$), or a specified elastoplastic deformation, for example, 0.5 ($\sigma_{0,5}$) or 1% ($\sigma_{1,0}$)

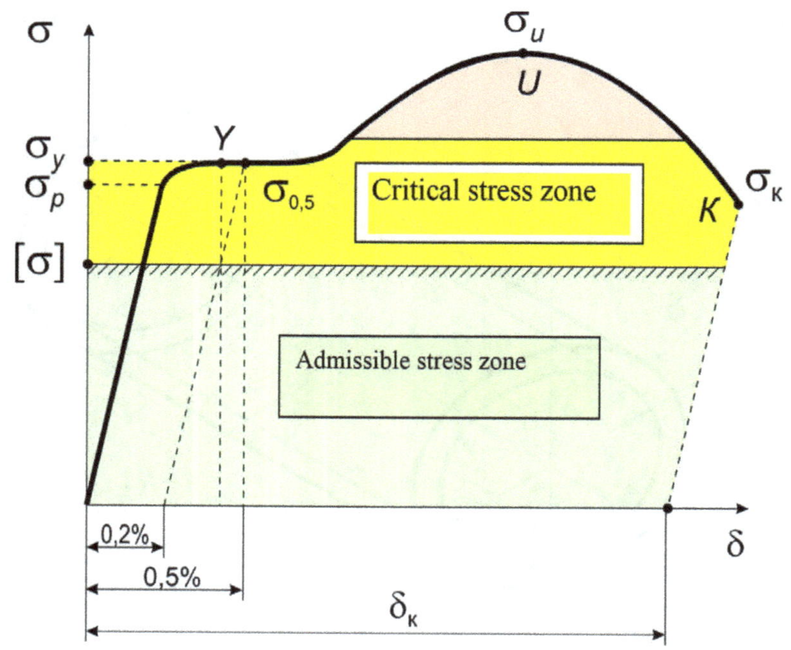

Figure 2. Static tension diagram of a standard sample.

- In the ultimate stress zone: ultimate strength-ultimate resistance σ_u as the maximum engineering stress at the stage of uniformity loss of plastic deformations and neck formation under tension

The calculated plastic (e_p = 0.2%) and elastoplastic deformations (e = 0.5% and e = 1%) for modern tube steels are substantially smaller than the relative elongation δ_κ in case of failure. In this connection, for tube steels $\sigma_p \leq \sigma_y \leq \sigma_{0,2} \leq \sigma_{0,5} \leq \sigma_{1,0}$.

Introduction to calculation (1) stresses σ_{on} in the form of the above characteristics makes it possible to exclude the appearance of mechanical properties of three dangerous limit states:

- Beginning of fluidity and formation of plastic deformations (σ_p, σ_y, $\sigma_{0,2}$, $\sigma_{0,5}$, $\sigma_{1,0}$).
- Failure after reaching the ultimate strength (σ_u).
- Total loss of stability after reaching critical stresses.

This required the use of three safety margins n_u:

- Yield strength n_y.
- Tensile strength n_u.
- Critical stress σ_c under loss of stability n_c

Hence, in accordance with Eq. (1), the allowable stress $[\sigma]$ must be minimal:

$$[\sigma] = \min\left\{\frac{\sigma_y}{n_y}, \frac{\sigma_u, \sigma_c}{n_u, n_c}\right\} \tag{2}$$

Since for the first two limiting states $\sigma_y \leq \sigma_u$ for tube steels hardening in the elastoplastic range, then safety margins are $n_y \leq n_u$.

According to the third limiting state, there are two possible cases:

If $\sigma_c \leq \sigma_в$ then $n_c \leq n_u$.

If $\sigma_c \leq \sigma_y$ then $n_c \leq n_y$.

When calculating pipeline's strength in limiting states in accordance with national standards and when the design resistances R_y (inadmissibility of plastic deformation development) and R_u (inadmissibility of destruction) are used, then

$$[\sigma] = \min\left\{\frac{R_y \cdot m\psi}{n \cdot K_2 \cdot K_н}, \frac{R_u \cdot m\psi}{n \cdot K_1 \cdot K_н}\right\} \tag{3}$$

where m—condition load effect factor; $K_н$—design safety factor; n—load safety factor; K_1, K_2—material resistance factor; and ψ—factor for biaxial stress states.

№.	Factor	Symbol	Value
1.	Condition load effect factor	m	0.6–0.9
2.	Load reliability factor	K_1	1.1–1.5
3.	Material resistance factor	K_2	1.34–1.55
4.	Design safety factor	$K_н$	1.0–1.05

Table 2. Calculated normative values of factors.

From Eqs. (2) and (3), it follows that margins n_y and n_u in the calculations for the allowed stresses are related to the factors m, K1, K2, and Kн, in Eq. (3) for calculations on the limiting states at $R_y = \sigma_y$ and $R_u = \sigma_u$:

$$n_y = \frac{K_2 \cdot K_н}{\psi m}, \quad n_u = \frac{n \cdot K_1 \cdot K_н}{\psi m} \tag{4}$$

In essence, the safety margins n_y, n_u and stability n_c according to Eqs. (2)–(4) reflect the role of statistical and probabilistic uncertainties, inaccuracies, ignorance, and responsibility of pipeline systems.

Based on strength and stability calculations under Eq. (1) with addition of Eqs. (2) and (3) for the pipeline with given p, N, M_θ, M_t, R_θ, and D, the wall thickness δ is chosen to be greater than the minimum ratio of yield strength σ_y and strength σ_u to margins n_y and n_u with subsequent binding of stability with σ_c and n_c.

Equation (2) defines the area of allowable stresses for deterministic normative calculations of pipeline strength (**Figure 1**).

The values of the factors in the calculations according to the norms [2] are given in **Table 2**.

3. Trends in improving methods of rationing, calculation, and management of mechanical properties of pipe steels

In the evolution (τ) of pipeline transport in Russia and abroad, three trends have been and are currently dominant (**Figure 3**) in view of Eqs. (1)–(4) in deterministic formulation [3–7]:

- Increase of the diameter of pipelines D (from 250–300 to 1200–1400 mm) and pressures p (from 2.0–2.5 to 14.0–16.0 MPa)
- Increase of mechanical properties of pipe steels (yield strength σ_y) (from 200–250 to 600–800 MPa) and strength σ_u (from 400–450 to 700–900 MPa)
- Decrease in safety margins n_y (from 1.8–3.2 to 1.2–1.5) and n_u (from 2.4–3.5 to 1.6–1.8)

At the first stages (1930–1960) of the development of pipeline systems, carbon (with a carbon content of 0.22–0.35%), unalloyed steels with larger of the abovementioned margins n_y and n_u,

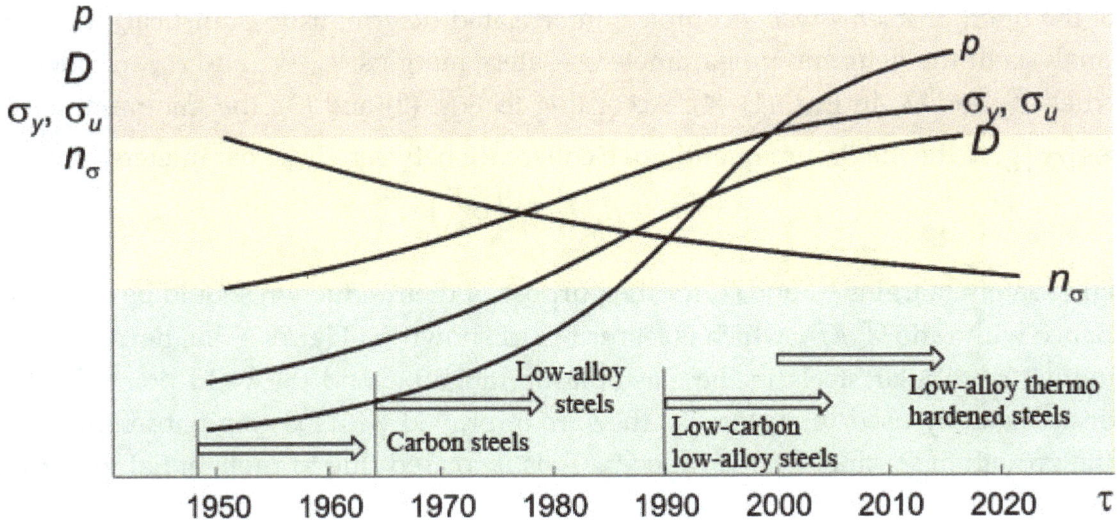

Figure 3. Basic determinate variations in design parameters of pipelines.

and lesser p, D, σ_y, and σ_u were predominantly used. Under these conditions, when determining the thickness of the pipe wall δ, the margins n_y and yield strength σ_y proved to be key factors, because they gave smaller permissible stresses $[\sigma]$ under Eqs. (2) and (3).

The idea that increasing the pipe steels yield strength σ_y is crucial in those years led to the desire of metal scientists, technologists, and designers to reduce the material consumption of pipelines by increasing the yield strength σ_y by all available methods and means (alloying steels, thermomechanical processing of sheets and pipes while reducing margins n_y). The same approach was typical for the development of general engineering, energetics, oil and gas chemistry, transport, and construction.

In the process of accelerated development of pipeline systems, low-alloy steels, low-carbon low-alloy steels, and low-alloy thermo-hardened steels have been consistently used since the 1960s.

This aspiration not supported by the necessary scientific justifications led to:

- Significant problems with increased damageability of objects such as pressure vessels and pipelines with high parameters of pressure P and temperature t in thermal power engineering, bearing structures of civil and industrial buildings
- Extended brittle fractures and loss of stability of the main pipelines

From the generalized statistical analysis of damage and destruction of various objects (including those working under increased pressure), it follows that engineering materials, design, and technological solutions associated with increase of σ_y and decrease of n_y are insufficient to prevent large-scale emergency and sometimes catastrophic situations. It became clear that the existing engineering practice of calculation focused on the designation of independent margins n_y and n_u and the basic characteristics of strength σ_y and σ_u is entailed with the danger of a real and reliable operation of pipeline systems.

One of the main problems was a complex, interrelated deterministic, statistical, and probabilistic analysis of the determining parameters—safety margins n_{σ}, n_y, and n_u and mechanical properties σ_y and σ_u in Eqs. (1)–(4). According to Eqs. (2) and (3), the minimum allowable stresses $[\sigma]$ give the maximum quantitative coherence between these parameters:

$$n_y = n_u \{ \sigma_y / \sigma_u \} \tag{5}$$

Managing safety margins n_y and n_u for the purpose of their reduction should be carried out in accordance with ratio σ_y/σ_u, which is featuring, as shown on **Figure 1**, the hardening degree (or module) of tubular steels in the elastoplastic range beyond the yield point σ_y. For the majority of actually used pipe steels as they are improved with existing hardening methods, with the growth of σ_y and σ_u, the ratio σ_y/σ_u is increased due to preferential growth of σ_y (**Figure 4**).

In the nomenclature and types of the previously used tube carbon steels (**Figures 1** and **2**) with reduced yield strength σ_y (less than 300 MPa) and a ratio σ_y/σ_u (less than 0.6), the traditional calculations of the yield strength σ_y with margins n_y were of primary importance. With a further increase in the yield strength σ_y and decrease in the safety margin n_y, the calculations for the ultimate strength σ_u with margins $n_в$ have become determinative, in accordance with Eq. (5).

However, in this case, the problem of increasing the danger of stability loss under $\sigma^s = \sigma^c$ and an uncontrolled dangerous transition to large plastic deformations according to Eq. (2) remains, in fact, not explicitly reflected in Eq. (5), due to a reduction in the degree of hardening of steels with a simultaneous increase of σ_y and the ratio σ_y/σ_u. Such conclusion in the framework of modern concepts of strength calculations [1, 3–6] required a gradual transition from calculations in stresses σ to calculations in deformations e. This transition already received not only its scientific justification [6–8] but also its practical implementation in norms

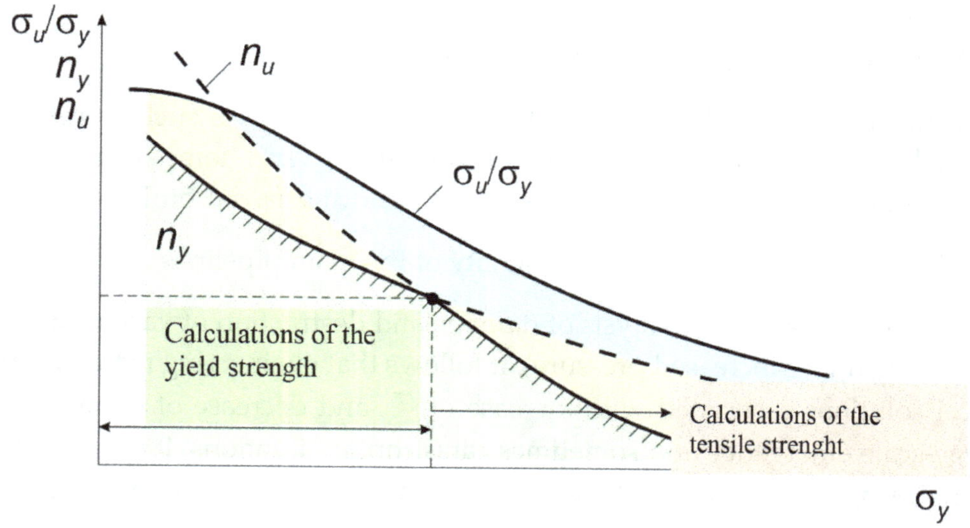

Figure 4. Coherence between strength margins and mechanical properties of pipe steels.

and substantiation of strength of the vessels and pipelines in nuclear reactors [8–10] and space and missile systems [11].

4. Modern problems of justifying the strength of pipeline systems

Four strategic tasks are being solved by methods of deterministic, statistical, and probabilistic modeling and calculation nowadays in Russia:

- Design and construction of new pipelines for liquid and gaseous hydrocarbons transportation (including marine and harsh climatic conditions of Siberia, the North Sea and the Arctic Sea)
- Extension of operation of existing pipelines within the limits of modern regulatory requirements for strength and durability
- Resolving the issues of complex technical diagnostics, repair, and restoration works in the damage areas beyond the norms of permissible defects for the prolongation of safe exploitation within the assigned terms
- Decommissioning in cases of significant exhaustion and formation of dangerous critical and un-repairable defects

The solution of these tasks must meet the modern requirements of:

- The federal legislation on justification and ensuring industrial safety by risk criteria
- Industry norms and rules for justifying strength, durability, and reliability

The tasks of justifying and ensuring industrial safety of pipeline systems in accordance with the criteria of strength, resource, and risks in compliance with the Federal Law No. 116-FZ "On Industrial Safety of Hazardous Production Facilities" are resolved with the coordinating and decisive role of Rostekhnadzor with the participation of the Russian Academy of Sciences, leading oil and gas companies as Transneft, Gazprom, Rosneft, the Russian Union of Oil and Gas Constructors and leading academic and industry institutes and universities.

The main directions of scientific research and applied developments in this direction are reflected in the proceedings of the I and II Forums on industrial safety [12].

The solution of problems of formation and development of industry norms and rules for substantiating the strength, durability, resource, and reliability of pipelines is concentrated in the research institutes of Transneft and Gazprom.

In normative documents [13] that are governing the industry, the following assumptions were made:

- Temporary technological heredity is not explicitly taken into account from the processes of obtaining the parent metal and the production of sheets and pipes in factories and enterprises.

- Mechanical properties (including limits σ_y and σ_u) of structural pipe steels in the process of pipeline transportation, construction, and operation of pipelines are assumed to be unchanged.
- Strength margins n_σ in Eq. (1) and margins n_y and n_u in Eqs. (2), (4), and (5) are accepted unchanged for all stages of the life cycle τ.
- Degradation of pipes and pipelines is associated mainly with a decrease in wall thickness due to corrosion (general and local) and erosion.
- The crucial part in material consumption reduction is in the increase in nominal operating stresses $\sigma^s_{n\max}$, yield strength σ_y, and strength σ_u and reduction of margins n_y and n_u according to the Eq. (1).

The normative approach has an important development element in comparison with [2, 13]—in it, the strength and durability evaluation is carried out not only by nominal stresses $\sigma^s_{n\max}$ but also by local deformations $e^s_{\max\kappa}$ in the concentration zones created by structural, technological, and operational factors (welds, defects, corrosion). This makes the normative calculation of the strength of pipelines comply with both the modern deformation criteria [6, 7] and the norms in nuclear power engineering and rocket and space technology [9–11].

5. Main directions of development of pipeline strength standards

Taking into account parts 1–3, the perspective directions of calculation and experimental analysis of the strength of pipelines in the deterministic interpretation should include:

- Direct quantitative accounting of the degradation and aging in time τ of tube steels at various temperatures t and the number of cycles N, leading to a change in the basic design characteristics—the yield strength $\sigma_т$ and strength $\sigma_в$:

$$\{\sigma_y(\tau,t,\sigma,e,N),\sigma_u(\tau,t,\sigma,e,N)\}=\{\sigma_y,\sigma_u\}\cdot F_c\{\tau,t,\sigma,e,N\}, \tag{6}$$

where $\sigma_y(\tau,t,\sigma,e,N)$ and $\sigma_u(\tau,t,\sigma,e,N)$—kinetically varying yield and strength limits for a given time τ, temperature t, stress σ, and deformation e; $F_c\{\tau,t,\sigma,e,N\}$—generalized functionals describing the change in the basic mechanical properties under the influence of temperature t, time τ, stress σ, cyclic N, and deformation e factors at all stages of the life cycle of the pipeline.

The functional $F_c\{\tau,t,\sigma,e,N\}$ with its parameters τ, t, σ, e, and N essentially reflects the processes of degradation and aging of pipeline steels in the process of sheet and pipe manufacturing, their transportation, construction, testing, and exploitation of pipelines.

Despite of a huge number of studies in factory laboratories; scientific institutes; design, construction, and operation organizations; and powerful industry research centers, in Russia and abroad, it has not yet been possible to obtain and justify this functional F_c with the appropriate statistical and probabilistic equations and parameters. The prerequisites for the formation of a system of initial equations for the functional F_c are presented in [4, 8, 11, 13, 14].

Currently, knowledge on the processes of aging and degradation in time τ of carbonaceous and low-alloy steels is reduced to the following basic provisions (**Figure 5**):

- Natural aging (curve 1) of steels in the initial state ($e = \sigma = 0$) at room temperature t_o is characterized by a slow increase in the yield strength σ_y, reaching values of 1.1–1.25 in about 30–40 years τ; furthermore, the ratio of the yield strengths $\sigma_y(\tau)$ to the tensile strengths decreases.
- Thermal aging (curves 2^{I} and 2^{II}) of steels in the initial state ($e = \sigma = 0$) at elevated temperatures t_1 and t_2 ($t_1 > t_0$; $t_2 > t_1$) leads to an accelerated growth of the yield point $\sigma_T(\tau, t)$ at the initial stages of exposure (up to 10^3–10^4 h) with its subsequent reduction (steel over ageing).
- Deformation aging (curve 3) of steels in the riveted state for $e > 0$ even at room temperature t_0 gives a smaller change of $\sigma_y(\tau, e)$ than the natural one.
- Dynamic aging (curve 4) at elevated temperatures in the plastically deformed state ($e > 0$) under stress conditions ($\sigma > 0$) can be accompanied at first by an insignificant increase, while later there is a fall in yield strength $\sigma_y(\tau, t, e, \sigma)$ and strength $\sigma_u(\tau, t, e, \sigma)$ with a decrease in the degree of hardening of tube steels in the plastic area.

In all cases of aging (curves 1–4), the ratio of the yield strengths σ_y to the tensile strengths σ_u increases (due to a smaller change in the tensile strength σ_u as compared to the yield point σ_y).

In the normative strength calculations [10], it is suggested not to take into account the areas of increase in the yield strength $\sigma_y(\tau, t, e, \sigma, N)$ due to aging, which goes to the margin of safety. In the refined basic and normative calculations of the strength of pipelines, one should take into account [4–9, 14–16]:

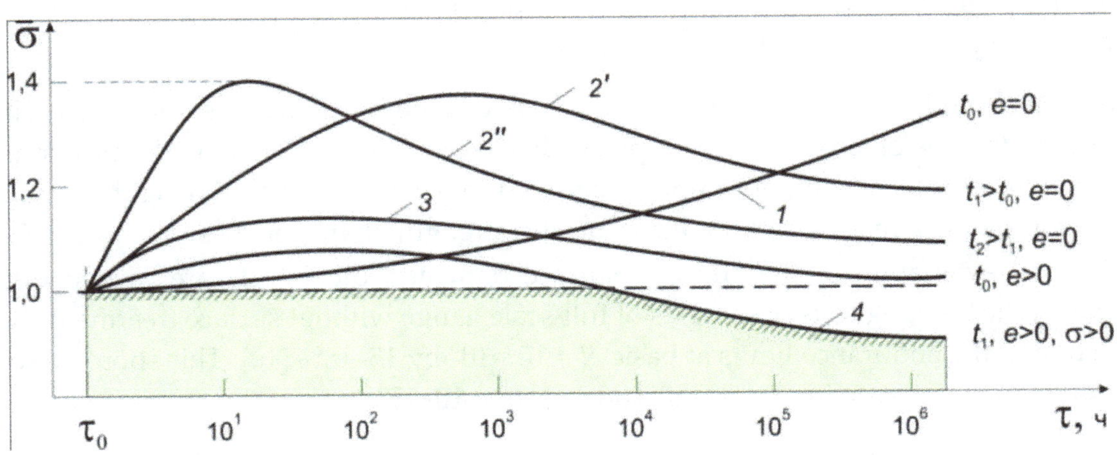

Figure 5. Scheme of aging processes of pipe steels.

- Continuous σ_u under all types of aging and degradation and the change in values σ_y and σ_u in Eq. (6)
- Effects of degradation of mechanical properties—decrease in relative yield point $\overline{\sigma}_y = \overline{\sigma}_y/\sigma_u$; $\overline{\sigma}_y(\tau, t, e, \sigma, N) \leq 1$;
- The decrease in plasticity (δ_K from **Figure 2**), which accompanies aging and degradation, as well as the fracture toughness

In accordance with the above, based on Eqs. (2)–(5), and taking into account **Figures 2–5**

$$n_y = \frac{\sigma_y(\tau, t, \sigma, e, N)}{\sigma^s_{n\max}}, \tag{7}$$

$$n_y/n_u = \sigma_y(\tau, t, \sigma, e, N)/\sigma_u(\tau, t, \sigma, e, N). \tag{8}$$

Equations (7) and (8) mean that the safety margins n_y and n_u are dependent on the aging and degradation processes of the tubular steels, time-dependent τ, temperature t, the cyclicity N, and the stress-strain state $\sigma - e$. This circumstance, which was not explicitly reflected in domestic [1, 2] and foreign [11, 12] regulatory materials, is to be taken into account in promising developments of pipe strength standards.

In [2–5], an experimental analysis was made of the time-dependent change in the characteristics of the mechanical properties of tube steels, primarily the yield strength σ_y and strength σ_y from the tensile tests of samples cut from the pipes in the initial state and after prolonged use. The time τ was varied from $\tau \cong 5 \times 10^{-2}$ to 3×10^5 h, operating temperature from -45 to +50°C, stress σ from 0.6 σ_y to 1.0 σ_y, and deformation e from 0.8×10^{-3} to 3×10^{-3}.

The averaged data from these tests showed that the reduction of the yield strength $\sigma_y(\tau, t, e, \sigma, N)$ during exploitation from the initial τ_0 to the maximum of τ = 2, $3{\cdot}10^5$ h was 10–15% of the yield strength σ_y. Meanwhile, the ratio of the yield strengths to the tensile strengths increased by 1.15–1.2. This means that the margin n_y of the yield strength σ_y can be reduced by 1.1–1.17 times, and the margin n_u of the ultimate strength σ_u by 1.20–1.25 times. This corresponds to the generalized statistical experimental data from Transneft, obtained during tests of laboratory samples from actually operated pipes.

However, it should be borne in mind that the bulk of pipeline damage is associated with the most severe damage of surface layers of pipes (due to corrosion, erosion, mechanical impacts). In the standard tensile testing of samples (with surface layers removed during their manufacture), this type of damage has little effect on the strength characteristics σ_y and σ_u. For the experimental evaluation of the effect of surface damages, other tests are carried out. For example, cyclic bending tests of samples of full-scale gauge without surface treatment showed a reduction in the endurance limits at basic $N = 10^5$–10^6 by 15–18% [16]. This should affect the abovementioned decrease in margins $n_т$ and $n_в$ (up to 10–15%).

For these margins n_y and n_u, the degradation of pipelines is significant due to a decrease in time τ because of corrosion and erosion of the wall thickness δ that is included in Eq. (1) for determining the nominal maximum operating stresses $\sigma^s_{n\max}$. As shown by laboratory tests and observations of the actual processes of metal loss while in the operation due to these mechanisms, the rate of corrosion and erosion reduction of the wall thickness $d\delta/d\tau$ can be from 0.05–0.1 to 0.3 mm/year. With wall thicknesses from 10 to 30 mm, the decrease of margins can reach 10–30%.

Thus, the aging of tubular steels and the degradation of pipes can, in the course of operation, with unfavorable combinations of all the abovementioned damaging factors lead to a substantial reduction in determined margins n_y and n_u and breach of strength as shown by Eqs. (1), (7), and (8). The number of such cases in real operation [3–5, 14] in the period of 1970–2015 gradually decreased from 1.2–1.0 to 0.12–0.14 damages per 1000 km per year.

6. Analysis of resistance to the development of cracks

A special place in the analysis of the pipeline strength is and will be occupied by the problems of their crack resistance and survivability, when formation and development of cracks of technological and operational origin are observed [3–6, 13–19]. In calculating the strength of pipelines with cracks of depth ℓ in thickness and length a over the surface, equations and criteria for linear and nonlinear fracture mechanics are used [3–7, 12, 13]. Then, the local stress-strain state at the crack tip is determined from the solution of the boundary value problem by numerical methods with defining of stresses $\sigma^s_{\max\kappa}$ and deformations $e^s_{\max\kappa}$:

$$\{\sigma^s_{\max k}, e^s_{\max k}\} = \sigma^s_{n\max} \cdot K_{\sigma\ell}, \tag{9}$$

where $\sigma^s_{n\max}$—maximum rated stress in Eq. (1); and $K_{\sigma\ell}$—effective coefficient of stress concentration in the zone of cracks.

The value $K_{\sigma\ell}$ is determined on samples with cracks:

$$K_{\sigma\ell} = F_\ell\{D, \sigma, \ell, a, S_*\}, \tag{10}$$

where $F_\ell\{D, \sigma, \ell, a, S_*\}$—function of pipe geometry (D, δ) and cracks (ℓ, a); and S_*—the structural parameter of the material, determined experimentally when testing samples with cracks.

Since $\sigma^s_{\max\kappa} > \sigma^s_{n\max}$ and $F_k\{D, \delta, \ell, a\} \geq 1$, then safety margins from Eq. (7) for pipes with cracks taking into account Eq. (9) will be further reduced (**Figure 6**):

$$\left\{(n_y)_\ell, (n_u)_\ell\right\} = \{n_y, n_u\}/F_k\{D, \delta, \ell, S_*, a\}. \tag{11}$$

In general, all the parameters of Eqs. (9)–(11) are deterministic, statistical, and probabilistic.

In calculating the strength of pipelines with defects, two basic estimated defect sizes are introduced:

- ℓ_o – Initial size (depth) of the defect, determined by the accepted methods of flaw detection (with their resolving power, sensitivity)

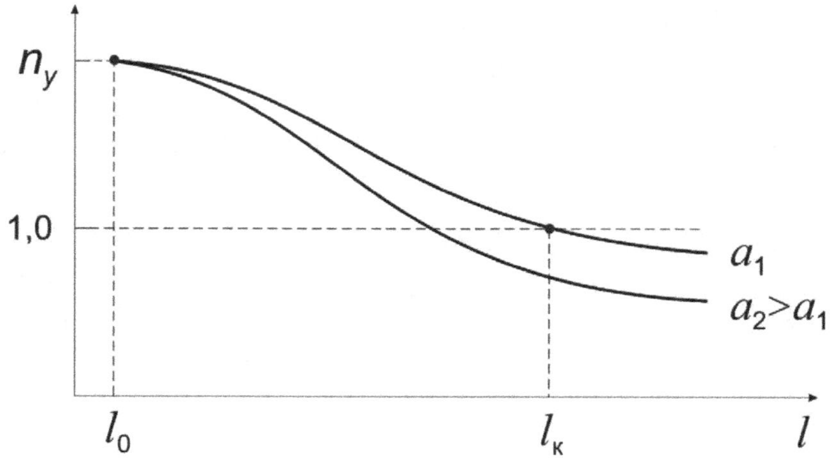

Figure 6. Influence of defects (such as cracks) on safety margins.

- ℓ_K – The critical size (depth) of the defect at which the margin of safety n_y (or n_u) in Eq. (10) becomes less than 1

The calculations ℓ_K take an elliptical ($\ell/a \approx 1/3$) or extended ($\ell/a \to \infty$) fracture shape. Typically, the most dangerous ones are surface cracks, taking into account more intensive accumulation of corrosion, erosion, and mechanical damage in the surface layers.

The second and most common way of assessing the strength of pipelines is to estimate margins $(n_y)_e$ and $(n_u)_e$ according to the equations and criteria of linear and nonlinear fracture mechanics [3, 7, 10, 16]. In this approach, the stress intensity factors are determined by the calculation for the given $\sigma^s_{n\max}$ in Eq. (1) and $F_k\{D, \delta, \ell, a\}$ in Eq. (9):

$$K_I^s = \sigma^s_{n\max}\sqrt{\pi\ell} \cdot F_k\{D, \delta, \ell, a\} \tag{12}$$

When a sample or a pipe with a crack breaks up, a critical value of the stress intensity factor is reached at the crack tip in accordance with the linear fracture mechanics. Then, in calculating the crack, resistance (survivability) of pipes with cracks by analogy with Eq. (2) introduced a margin by the stress intensity factor:

$$n_k = \frac{K_{Ic}}{K_I^s}. \tag{13}$$

By the values of K_I^s and K_{Ic} and Eqs. (9) and (13), the equation below can be obtained:

$$\left\{(n_y)_\ell, (n_u)_\ell\right\} = \left\{n_y, n_u\right\} \cdot \frac{K_{Ic}}{\left\{\sigma_y, \sigma_u\right\}\sqrt{\pi\ell} \cdot F_k}. \tag{14}$$

The difference in margins according to Eqs. (11) and (14) should not be significant.

In the event of plastic deformations, instead of the stress intensity factors K_I and K_{Ic}, the strain intensity factors should be used [4, 6, 8].

A generalized analysis of the strength, resource, reliability, survivability, and safety of complex technical systems of pipeline transport is made in one of the volumes [17] of the multivolume series "Safety of Russia."

7. Statistical characteristics and probabilistic modeling of pipeline systems

Multiparameter pipelines with a wide range of service lives are functioning nowadays in Russia and in various countries across the world, according to parts 1 and 2 (**Figure 7**).

In further analysis of their initial and residual strength, durability, and crack resistance, both statistical data on service life τ and statistical data on changes in the mechanical properties of tubular steels σ_y, σ_u, K_{Ic}, as well as on developing defects ℓ, should be taken into account. This consideration can be performed on the basis of Eqs. (1)–(15) in both deterministic and statistical forms.

According to statistical data [20] on oil pipelines of Russia with a total length of more than 70,000 km (see **Table 1**), about 70% of them have a service life of more than 30 years. Their age structure is shown in **Figure 7**.

Statistical studies of mechanical properties (tensile strength σ_u) of 29 tube steels were carried out in 217 pipe sections manufactured at 14 plants. Upward bias from data on technical conditions was revealed in 8.9% cases and downward bias 2.6%.

Primary and repeated in-tube condition diagnostics on the length of more than 80,000 km of oil and gas pipelines revealed the presence of unacceptable corrosion and mechanical and erosive damage in 0.2–0.3% of pipes. This required repair and restoration works, as well as replacement of pipes or its sections. These works over the past 20 years have made it possible to reduce the frequency of accidents on pipelines from 0.14–0.16 to 0.09–0.10 per 1000 km per year.

The generally recognized statistical characteristic of the technical condition and safety of pipelines with due regard of their period of operation is [1, 3–7, 17–20] the number of system failures (failures $N^o(\tau)$) generated per time unit. The failure of a specific section of the pipeline is a very

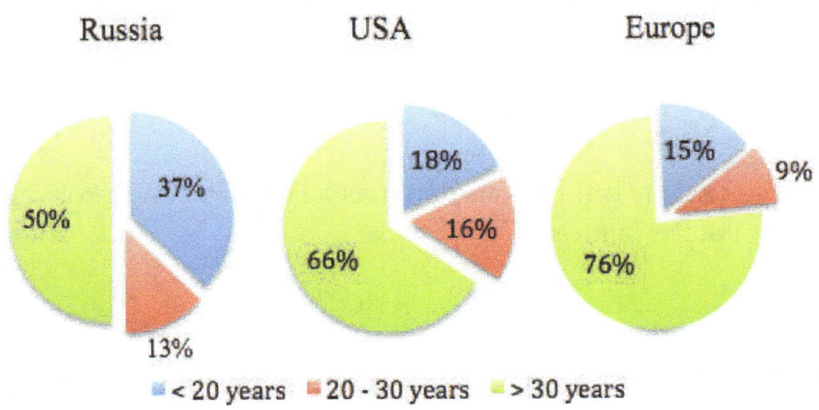

Figure 7. Statistics on the service life of pipelines.

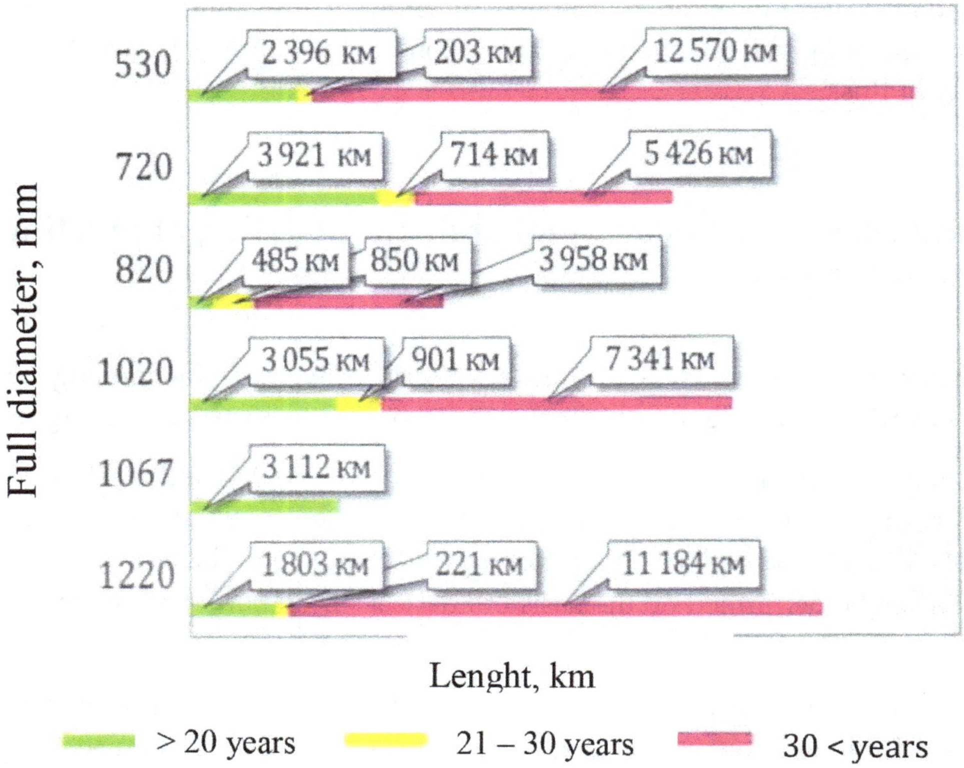

Figure 8. Age structure of long-term running main pipelines of large diameter.

rare event, even for a fairly long period of time τ. But taking into account the considerable length of the whole system (more than 70,000 km), the reduced frequency or failure flow $P^o(\tau)$ at the length L (L = 1000 km) will have a finite value depending on the time of operation τ^s:

$$P^o(\tau) = \frac{dN^o(\tau)/d\tau^s}{L}. \tag{15}$$

The failure flow $P^o(\tau)$, in our country and abroad, of oil and gas pipelines decreases over time —from 0.3 to 0.4 in the 1960s and 1970s to 0.012–0.015 at the present.

According to Eq. (15), the reliability $P_o(\tau)$ of section L at a given time τ can be estimated [1, 4, 6, 7, 18] by the failure flow $P^o(\tau)$:

$$P_o(\tau) = 1 - P^o(\tau). \tag{16}$$

In this case, the value of $P_o(\tau)$ can be considered as a statistical and probabilistic indicator of the technical risk $R_o(\tau)$ of the failure:

$$R_o(\tau) \; = \; 1 - P^o(\tau) \tag{17}$$

On the basis of (16) and (17), the safety $S^o(\tau)$ of the MPS functioning at $\tau = \tau^э$ can be considered as.

$$S^o(\tau) = 1 - R_o(\tau) = P_o(\tau) \quad (18)$$

According to operational statistical data on failures N^o and failure flows $dN^\circ / d\tau$, the standard (permissible) operating time $[\tau]$ can be established—a resource of reliable operation excluding the transition of the MPS to the critical (ultimate) state.

Operational experience shows that the service life of the pipeline, as well as of other complex technical systems, can be conveniently divided into three main periods (**Figure 9**):

- Run-in period (τ_I), when there is a high failure rate (N^o), associated with unacceptable defects in construction and installation works and factory defects in pipes
- Stabilization period (τ_{II}), when the number of failures is minimal and their increase is insignificant
- Wear period (τ_{III}), associated with a steady increase in the number of failures and a decrease in throughput due to the occurrence of damage accumulation processes and the formation and development up to critical dimensions (K) of the initial and operational defects of metal pipes, welded joints, protective coatings, etc.

For mastered deterministic technologies of designing and manufacturing, the following correlations are fulfilled:

$$\tau_k = \tau_I + \tau_{II} + \tau_{III}.$$

$$\tau_I << \tau_{II} < \tau_{III}. \quad (19)$$

The allowed period $[\tau]$ of reliable operation of pipelines based on the allowed failure flow may include periods τ_I and τ_{II} and part of the τ_{III} period:

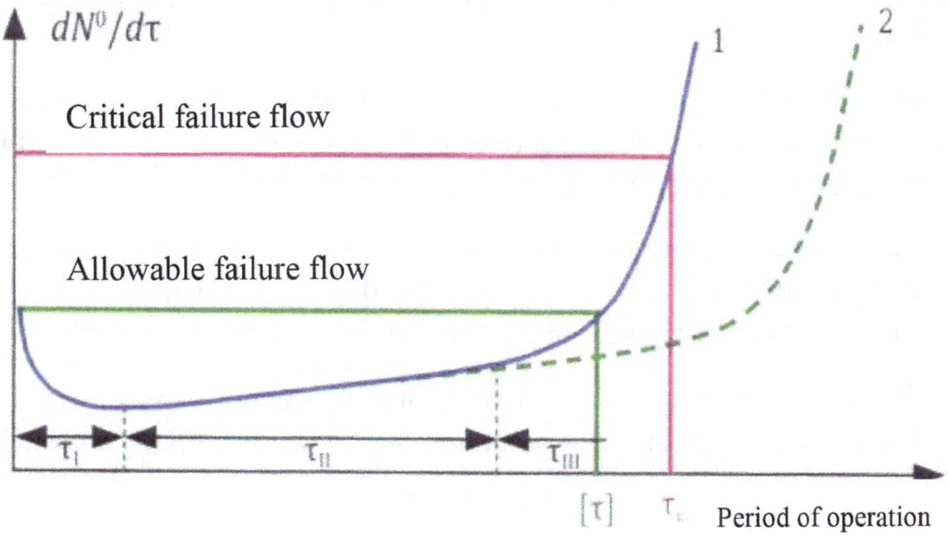

Figure 9. The failure of technical systems in dependence from the period of operation.

$$\tau^s \leq \tau_I + \tau_{II} + k\tau_{III} < \tau_k, [\tau] = \frac{\tau_k}{n_\tau} \tag{20}$$

where k—coefficient of using the pipeline with damages ($k < 1$); and n_τ—service life margin.

Equations (15) and (16) are valid both for MPS and for their individual elements when failures are associated with the development in the length of time of operational defects. At the same time, for the main pipeline transport, the period of operation and margins n_τ with deterministic, statistical, and probabilistic approaches should be taken into account under Eqs. (15)–(20).

The period of stable operation of the pipeline according to Eq. (20) can be increased by carrying out special organizational and technical measures, including the implementation of local or major repairs, diagnostic surveys, efficiency improvement of the corrosion protection system, etc. The most important aim of these measures is the extension of the safe operation period for the entire system (MPS) as well as for individual sections and pipes (the transition from curve 1 to curve 2 in accordance with **Figure 9**), subject to specified safety and reliability parameters.

Considering economic consequences $V_o(\tau)$, failures $N^o(\tau)$, risks $R^o(\tau)$, and costs for improving reliability and safety $Z(\tau)$ allows us to evaluate the economic effectiveness of integrated measures to improve the working capacity of MPS:

$$V_o(\tau) = V^o(\tau)\left[1 - k_p P^o(\tau)\right] = V^o(\tau)P^o(\tau), \tag{21}$$

where $V_o(\tau)$ and $V^o(\tau)$—designed throughput of the system with and without consideration for reliability; and kp—coefficient of influence of failures on the throughput.

Therefore, in accordance with Eqs. (1)–(4), the requirements for MPS operation efficiency are inextricably linked to the high requirements for ensuring reliability $P_o(\tau)$, safety $S_o(\tau)$, and risk management $R_o(\tau)$ in the process of its operation $\tau = \tau^s$, which determines the priority importance of economic, environmental, and industrial safety of transportation of oil, oil products, and gas. These issues are assigned to the scope of strategic planning at the federal, regional, and sectoral levels.

Statistical information on the quantities σ^s and σ_u, σ_y makes it possible to construct the probability density functions $f(\sigma^s)$ and $f(\sigma_u, \sigma_y)$ (**Figure 10**) describing the operational loads (nominal σ^s and strength characteristics) from Eq. (21).

The probability of fracture P_p as an extremely dangerous (critical) failure, accident, and catastrophe will be determined by the overlapping of the distribution density functions $f(\sigma^s)$ and $f(\sigma_u, \sigma_y)$. In general, all the parameters of Eq. (21) are time-dependent $\tau = \tau^s$.

Parameter $P_p(\tau)$ is taken into account when assigning the safety margins $\{n_u, n_y\}$, and Eq. (2) makes it possible to estimate the strength properties in accordance with the following equation:

$$P_{p^o}(\tau) = 1 - P_p(\tau). \tag{22}$$

In the calculations for the permissible stress under codes and rules for building [16], this approach is reflected in the separation from the total factor of margin n factors of homogeneity k_o, overload k_{Π}, and operating conditions m:

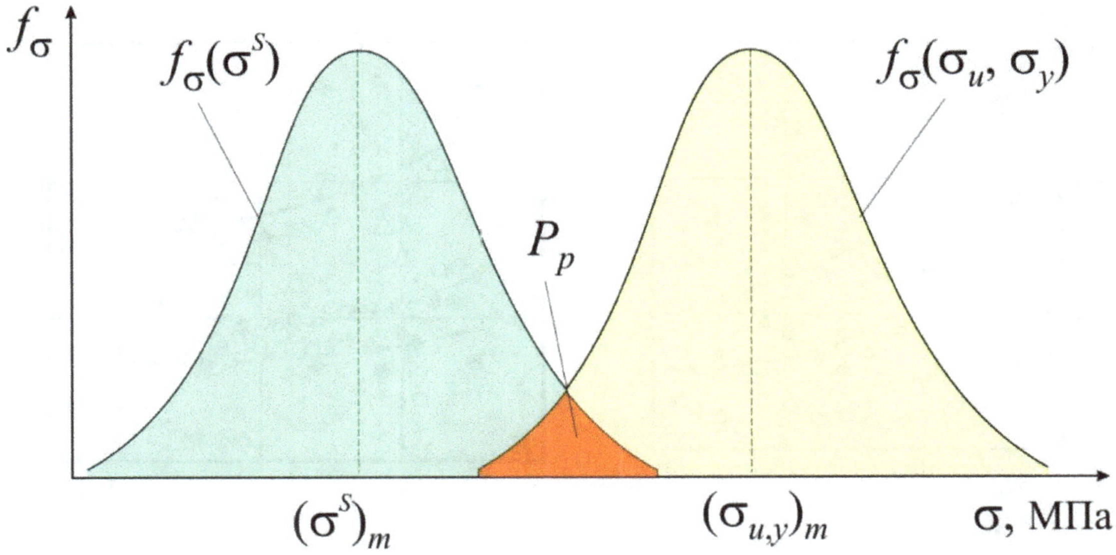

Figure 10. The scheme for determining the probability of fracture P_p by parameters of reliability and durability.

$$n = n(k_o, k_п, m) \tag{23}$$

As a result, the calculated strength (σ_u^p, σ_y^p) and load σ_p^s are calculated by multiplying their mathematical expectation by the corresponding factors:

$$\sigma_{u,y}^p = k_o(\sigma_{u,y})_m; \sigma_p^s = k_п(\sigma^s)_m. \tag{24}$$

The values of the factors in Eqs. (23) and (24) will depend on the assumed probability of fracture P_p, which is determined by the safety characteristic $S_o(\tau)$, the shape of the load distribution curves, and the strength in **Figure 9**:

$$k_o = 1 - z_p v_{u,y}; k_п = 1 + z_p v_s, \tag{25}$$

where $v_{u,y}$, z_p—factors of variability and quantiles of distribution of the strength characteristics of the material; and v_s—variation factor of the operational load.

Statistical analysis [5, 7] of the distribution functions of the mechanical properties of low-alloy steels (type 15XCHD-C 0.12–0.18, Ci 0.4–0.7, Mn 0.4–0.7, Ni 0.3–0 (6%)) on a large number of n = 2500 laboratory samples from a 15-mm-thick sheet showed the acceptability of the use of the normal distribution law.

The generalization (**Figure 11**) of the test results of this steel at n = 22.000 samples with thickness of 5 to 24 mm revealed while increasing thickness δ, decrease of the yield strength for the probabilities P = (1%, 50%, 99%), as well as the variation coefficients v.

In the generally accepted normative calculations for the strength of the MPS, the time parameters τ are not explicitly introduced in Eqs. (1) and (2). They become necessary in the future specified calculations of the strength $\sigma_u(\tau)$ and $\sigma_y(\tau)$, reliability $P_o(\tau)$, safety $S^o(\tau)$, and efficiency $V_o(\tau)$ under Eqs. (15)–(25) in case of assessing the technical condition and extending the life of the functioning facilities and while designing new MPS:

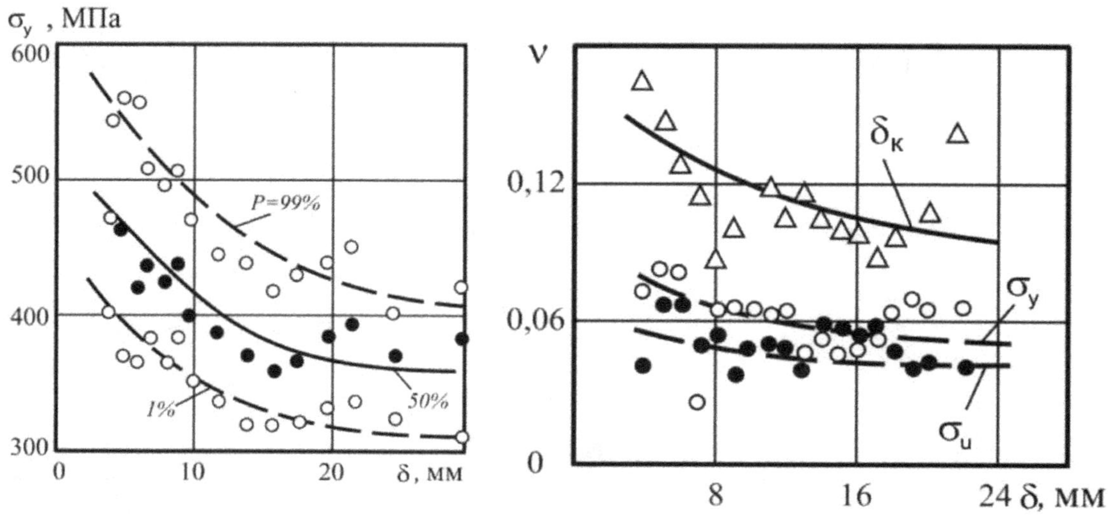

Figure 11. Dependence of yield strength σ_y, factors of variation v of yield strength σ_y, strength δ_u, and elongation δ_k from the rolled thickness δ.

The currently developed combined probability statistical method [4] makes it possible to assess the reliability $P_o(\tau)$ as a function of time τ^s on the basis of analysis of the initial deterministic, statistical, and probabilistic information about the design $K(\tau)$ and technological $T(\tau)$ features of MPS objects, the operating loads $Q(\tau)$ and environmental impacts $\Phi(\tau)$, stress-strain states in the coordinates $\sigma(\tau) - e(\tau)$ and probable mechanisms of accumulation of damage $d(\tau)$, and nucleation and development of defects $l(\tau)$.

The main design parameters will be determined under:

$$\{\sigma_u(\tau), \sigma_y(\tau), S^o(\tau), P_o(\tau)\} = \begin{cases} F_{\text{в}}\{K(\tau), T(\tau), Q(\tau), \Phi(\tau)\}; \\ F_p\{\sigma(\tau), e(\tau)\}; \\ F_n\{d(\tau), l(\tau)\}, \end{cases} \quad (26)$$

The abovementioned basic calculated dependencies in equations (1)–(26) allow [1, 3–7, 18–20] to make the transition from traditional deterministic engineering calculations of strength with the standard characteristics of mechanical properties σ_y, σ_u to calculations of strength, durability, crack resistance, reliability, and safety using new developing statistical and probabilistic methods of mathematical and physical modeling and refined calculations.

8. Conclusion

Ultimately, the problems of functional and strength reliability, resource, and safety of pipeline systems should cover all stages of the life cycle of facilities, representing three interrelated and interdependent processes: design, construction, and operation.

Designing while taking into account the prospects of statistical and probabilistic modeling of reliability and safety criteria should include the development and coordination of the technical assignment with the introduction of basic requirements and criteria for strength, resource, and safety in accordance with applicable standards and development of physical and mathematical models for regular, damaged, and emergency situations. When designing facilities of new generations, strength analysis should be carried out in accordance to the basic standard and additional verification calculations, based on known internal and external influences and object characteristics, parameters of stress-strain state, and damaging factors with justification of initial resources for reliable and safe operation.

In the subsequent stages of design and manufacturing, reliability problems will be addressed, including selection, justification, and development of materials technology and control in accordance with existing norms and rules. Generally, for the manufactured elements of MPS, the actual mechanical properties and their deviations from the technical requirements, the level of real defectiveness, the geometry parameters, and their deviations should be established. On their grounds, the basic design parameters of strength and resource will be refined. At this stage, the issues of stability and safety of the elements require an analysis of possible failures for reasons of technological heredity.

At the operational stage, the system of routine diagnostics of the main characteristics of the MPS facility and the external environment that determine reliability will be specified, and information will be collected on confirming or adjusting design decisions on strength and resource. As the finalized design resource is exhausted, an evaluation of the residual life of safe operation should be carried out. To harmonize all deterministic, statistical, and probabilistic information for all stages of the life cycle of an object, it is necessary to use unified mathematical and physical models, calculation equations, criteria, and computer programs for MPS.

In the future, considering formation of a new legal and regulatory framework, in which the standardized requirements for safety $S^o(\tau)$, risks $R^o(\tau)$, as well as economic efficiency $V_o(\tau)$ will be of decisive importance, reverse solutions will be decided. At the same time, all the scientific and methodological potential accumulated in previous years will be fully utilized in selecting models, methods, design equations, and design parameters to achieve the required values $S^o(\tau)$, $R^o(\tau)$, and $V_o(\tau)$ in engineering design and technological and operational solutions for pipeline systems for oil and gas transportation.

Acknowledgements

This work was financially supported by the Russian Science Foundation (grant #14 19 00776-P).

Author details

Yuriy V. Lisin[1], Nikolay A. Makhutov[1], Vladimir A. Nadein[2]* and Dmitriy A. Neganov[1]

*Address all correspondence to: vladimir_nadein@ogsed.ru

1 The Pipeline Transport Institute (PTI, LLC), Moscow, Russia

2 LLC "Oil and Gas Safety – Energodiagnostika", Moscow, Russia

References

[1] Radionova SG, Lisin YV, Makhutov NA, Revel-Muroz PA, Neganov DA, Zorin NE. Scientific, technical, socio-economic and legal aspects of the reliability of transport of oil and oil products. Science and Technology of Pipeline Transport of Oil and Oil Products. 2016;**6**:20-31

[2] SP 36.13330.2012 Trunk pipelines. Updated Version of SNiP 2.05.06–85*

[3] Mazur II, Ivantsov OM. Safety of Pipeline Systems. Moscow: IC ELIMA; 2004. p. 1104

[4] Makhutov NA, Permyakov VN, et al. Analysis of Risks and Ensuring the Security of Critical Facilities of the Oil and Gas Chemical Complex. Tyumen: TyumGNGU; 2013. p. 560

[5] Lisin YV, Makhutov NA, Neganov DA, Varshitskiy VM. Comprehensive analysis of the pipeline. Pipeline Science and Technology. 2017;**1**:5-16

[6] Makhutov NA. Strength and Safety: Fundamental and Applied Research. Novosibirsk: Science; 2008. p. 528

[7] Makhutov NA, Radionova SG, Zhulina SA, Gadenin MM, Lisin YV, Neganov DA, Nadein VA. Prospects of research in the field of risk analysis for improving state regulation and improving the safety of oil and gas chemical facilities. Safety in Industry. 2017;**9**:5-13

[8] Series "Investigation of Stresses and Strength of Nuclear Reactors". Vol. 1–9. Moscow: Science; 1987–2009

[9] Norms for Calculating the Strength of Equipment and Pipelines of Nuclear Power Plants. Moscow: Energoatomizdat; 1989. p. 525

[10] ASME Boiler and Pressure Vessel Code. 2015. Section III

[11] Makhutov NA, Rachuk VS, Gadenin MM. Strength and Life of a Liquid Rocket Engine. Moscow: IMASh RAS, Nauka; 2011. p. 525

[12] Makhutov N. Industrial Safety is the Responsibility of the State, Business and Society. I and II Forum-Dialogue. Moscow: Rostekhnadzor; 2015–2016

[13] ASME B31.4 Systems for Pipeline Transport of Liquid Hydrocarbons and Other Liquids

[14] Gumerov AG. The Aging of Oil Pipelines. Moscow: Nauka; 1995

[15] Chuvildeeva VN, editor. Problems of Aging and Resource of Steel Trunk Pipelines. Materials of Reports. N. Novgorod: UNN; 2010. p. 560

[16] Makhutov NA, Moskvicheva VV, Gadenina MM, et al., editors. Problems of Destruction, Resource and Safety of Technical Systems. Krasnoyarsk: Association "KODAS" - SPE "SIBERA"; 1997. p. 520

[17] Safety Guide. A Methodology for Assessing the Risk of Accidents at Hazardous Production Facilities of the Oil and Gas Refining and Petroleum-Gas-Chemical Industries. Moscow: RTN; 2004. Series 09, Issue 38

[18] Safety of Russia. Legal, Socio-Economic and Scientific-Technical Aspects. Safety of Pipeline Transport. Moscow: MGOF "Knowledge"; 1998–2015. pp. 1-50

[19] Pluvenage G. Improvement of the failure-assessment diagrams used to check the harmfulness of pipe defects. Pipeline Science and Technology. 2017;**1**:17-23

[20] Lisin YV, Sergaev AA, Neganov DA. Determination of permissible operating pressures for long-term main pipelines based on the results of in-pipe diagnostics. Science and Technology of Pipeline Transport of Oil and Oil Products. 2016;**6**:30-37

9

Probabilistic Modeling Processes for Oil and Gas

Vsevolod Kershenbaum, Leonid Grigoriev,
Petr Kanygin and Andrey Nistratov

Abstract

Different uncertainties are researched for providing safe and effective development of hydrocarbon deposits and rational operation of oil and gas systems (OGS). The original models and methods, applicable in education and practice for solving problems of system engineering, are proposed. These models allow us to analyze natural and technogenic threats for oil and gas systems on a probabilistic level for a given prognostic time. Transformation and adaptation of models are demonstrated by examples connected with non-destructive testing. The measures of counteraction to threats for the typical manufacturing processes of gas preparation equipment on enterprise are analyzed. The risks for pipelines, pumping liquefied natural gas across the South American territory, are predicted. Results of probabilistic modeling of the sea gas and oil-producing systems from their vulnerability point of view (including various scenarios of possible terrorist influences) are analyzed and interpreted.

Keywords: analysis, modeling, operation, probability, process, risk, system, threat

1. Introduction

A history of development of the oil and gas industry all over the world, and in the Russian Federation, is impressive. Since 1930s large oil and gas fields have been opened; a huge number of oil refining and petrochemical factories are constructed. In recent years, the role of the gas branch has essentially increased; pipeline transport, thanks to which the basic part of Russia's territory is provided with gas, oil and mineral oil, has actively developed; export of these products is carried out; there has been development of sea deposits. Hydrocarbon reservoirs, pipeline transport, oil refining and petrochemical factories, various storehouses of oil and gas, sea platforms and terminals and so on are examples of objects of modeling in oil and gas systems (OGS)—see **Figure 1**.

Technological processes of oil and gas branches are various. As a matter of fact, it is all a spectrum of processes from hydrocarbon extraction to end-product production. There are geological and geophysical researches; drilling; developing of hydrocarbon reservoirs (both on land, shelves and on the sea); pipeline transport and oil and gas storage; refining and chemistry. The end production of oil and gas manufacturing is used in majority branches of the modern economy. Unfortunately, up-to-date claims for deposits of hydrocarbons are the reasons for international conflicts.

Features which are necessary for consideration for the creation of control systems by technological processes and at construction are peculiar to the oil and gas branch. So technological processes are continuous, and objects are difficult and demand at the management level of performing synergistic researches. Objects of oil and gas manufacturing are technologically dangerous; therefore, the role of systems' safety and ecological monitoring is significant. The initial information possesses are characterized by the high level of uncertainty generated by natural factors.

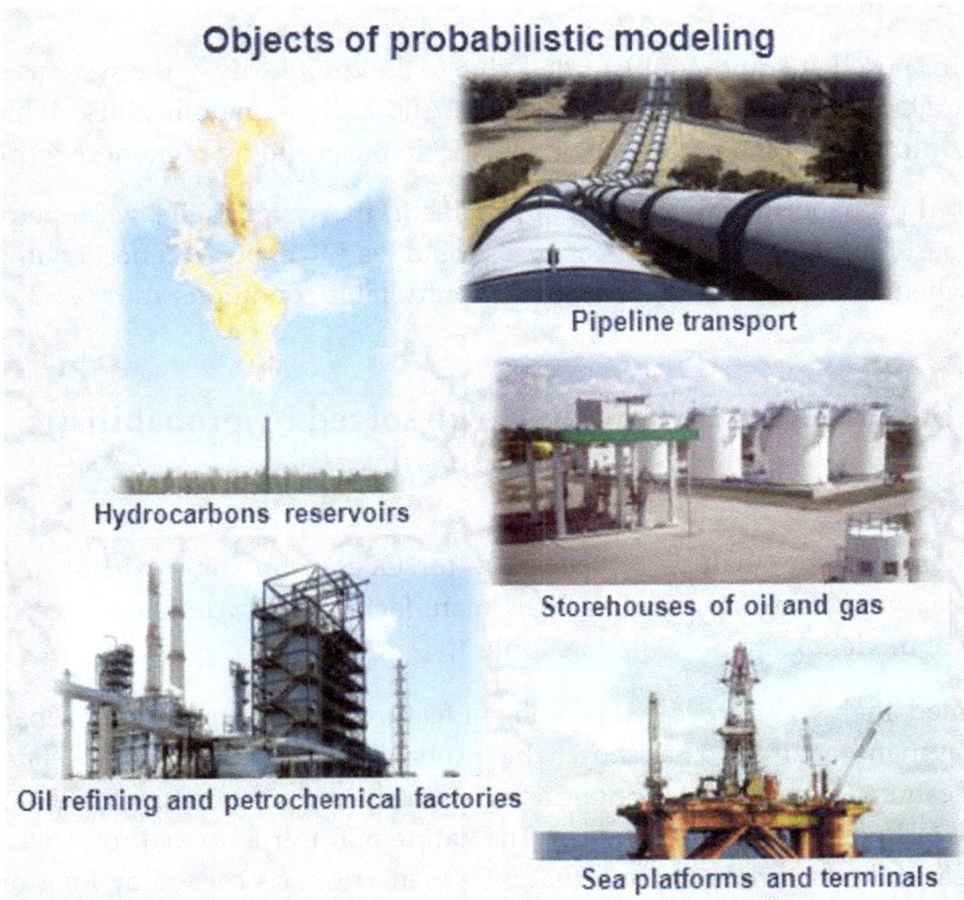

Figure 1. Objects of modeling.

The automated dispatching control (ADC) meets the requirements of continuous technological processes control (ADC is the heterogeneous man–machine control system of the technological process integrating the dispatcher with an information-operating system, providing automatic information gathering, transfer, processing and display [1, 2]).

Theoretical bases for creating heterogeneous control systems are at the formation stage. Effective ACD operation in general depends on the quality of modeling objects and managerial processes. Problems of modeling for oil and gas systems should be considered for two levels:

- problems of technological process control taking into account features of oil and gas manufacture and.
- problems of monitoring and prediction of the integrated metrics, providing safe and competitive development of the OGS enterprises.

In such a manner, the systemic uncertainty inherent in oil and gas technologies due to the specificity of the objects under study leads to the need for modeling oil and gas systems, the goal of which is ultimately to manage risks at all levels of the hierarchy and all stages of the life cycle [1–10].

The problems posed are quite sophisticated, due to the complexity of the systems being studied, the operation of which is clearly non-linear. And at the same time, it is highly an actual one, taking into account the noted role of the OGS in the economy of modern world.

The proposed probabilistic approaches, applicable in the system's life cycle, help to answer the main question: "What rational measures should lead to expected effects without wasted expenses, when, by which controllable and uncontrollable conditions and costs?"

2. About the problems that are due to be solved by probabilistic modeling

Modeling demands the analysis of specificity for OGS, estimating existing uncertainties. Prominent features of objects of oil and gas manufacture, characterizing uncertainties and complexity of modeling are presented in **Table 1**.

The performed analysis has revealed prominent features of uncertainties for separate objects of oil and gas manufacture and has shown that probabilistic modeling, models for estimations and identifications, a method of Monte-Carlo, is widely and successfully applied for solving problems of technological process control. The nature of uncertainty of processes and objects of oil and gas manufacture is various; that is in many respects caused by long processes of hydrocarbon formation. Therefore, the occurrence of technology of evolutionary modeling as often named synergistic analysis (with the theory of non-linear systems and the self-organizing

Object or process	Distinctive factors (uncertainty) and applicable modeling tool
Geological structures (it is the isolated area of the earth shell, differing from the cross-border regions for the tectonic behavior, i.e. specific combination of the geological formations, its bedding and structuring conditions)	The text and cartographical information symbiosis is used. The most popular tasks: correlation analysis, cluster analysis, association theory, interpolation theory. The spatial data modeling with the variograms evaluation (the spatial correlation measure) and other estimations. Natural phenomena modeling – one of the most progressive lines of the modern science. It is based on the digital models of the geological data combined with the spatial databases. The computer modeling tasks in the practical geology are solved with help of the modeling software packages. Actual tendencies of the geostatistics are connected with the development of: spatial analogues of the Monte-Carlo methods; approaches based on the multipoint statistics; hybrid models with artificial intelligence algorithms application; with additional information of the varied type and applications in the images processing and transfer area and others [11].
The Oil (it is typified as the oil disperse system)	In the oil disperse system the behavior principles and physical–chemical properties in the molecular or disperse state can be quite differ from each other, and this is the reason of the non-linear response while changing of the external input character and scale. Then the phase transitions occur, and system properties are changed in a qualitative manner. This field researches show that oil systems are structured at the nanoscale level, what creates the basics for the new technologies development in the oil and gas industry. So the phase transitions are possible with the aggregate state changing. This is the object of the synergistic analysis; also the physical–chemical analysis models are applied; fractal analysis is used.
Hydrocarbons reservoirs	They were formed for millions years. Recently the count of hypothesizes about the hydrocarbon reservoirs origin increased, and generally speaking the self-organizing processes are typical for the oil and gas reservoirs.
Hydrocarbons reservoir rock (it is mine rock containing the voids, i.e. the pores, cavities and others and having the potential to store and filtrate the fluids)	Porosity, permeability are the key indexes for the estimation task solution. The deformations are typical for the reservoir rocks. In certain cases it is needed to take into account the non-Newtonian fluid and to use the rheological models. The most dependencies have the non-linear character. The initial information has the statistical character, then the mathematical statistics and probabilistic modeling apparatus is actively applied. The percolation task is also has its special features.
Processes of the petrochemical industry and oil refining	Non-linear processes with catalysts' application, and its activity, are varied with time. Most wide-spread modeling and engineering evaluations systems are actively used for the project design as well as the calculations providing in management.
Management of the oil production process (Intelligent field, i.e. I-field)	While solving the problem of the hydrocarbons production on the reservoir the task of the adjusted management with Kalman filter application. In classic case the adjusted management task supposes the object model correction with control input generation. Toward to the hydrocarbons reservoir the uncertainty is increased with the changing of the object characteristics while the reservoir development process. The tasks of the estimation and identification are widely used and applied.

Object or process	Distinctive factors (uncertainty) and applicable modeling tool
Management in the emergency situations and accidents	In the oil disperse system occur the negative phenomena, which are connected with the phase transitions in technological processes of the oil refining, hydrocarbons reservoirs development, wells drilling and other processes of the oil and gas industry. These phenomena may include: the asphaltenes, paraffins and salts sedimentations, gas hydrates formation and others. Physical–chemical analysis of the oil disperse systems is supposed to be the key point in development of the decisions support systems in the emergency situations. Combined to the experimental researches practice the computer modeling at the molecular level is carried. Based on the results the type of the catastrophe is evaluated and the order parameter is identified. Finally the recommendations to the management in the emergency situations are developed.

Table 1. Distinctive uncertainties of the objects and processes in the oil and gas industry.

processes, the determined chaos, fractal analysis, etc.) has considerably expanded possibilities of the researches of the natural uncertainty of oil and gas manufacture.

Evolutionary processes as a development basis, actively acted not only the system analysis in a control context, but also for the decision of problems at level of organizational-economic management. At this level, the nature of uncertainty is connected with many criteria. In **Figure 2**, the evolution of risk-oriented criteria is shown: from economic criteria to competitiveness.

In different areas, the heterogeneous threats for complex systems are inevitable. The uncertainties in the system's life cycle are usual. Different problems, connected with evaluations, comparisons, selections, controls, system analysis and optimization, are solved by the probabilistic modeling of processes according to system engineering standards (general–ISO/IEC/IEEE 15288, ISO 9001, IEC 60300, 61,508, CMMI и т.д. and specific for the oil and gas industry ISO 10418, 13,702, 14,224, 15,544, ISO 15663, ISO 17776 etc.). The saved-up experience confirms the high importance of scientific system researches based on probabilistic modeling. For example, in general cases, prediction and optimization are founded on modeling different processes. Any process is a repeated sequence of consuming time and resources for outcomes' receiving in all application areas. From the probability point of view, the moments for any activity beginning and ending are random events on the timeline. In practice, a majority of timed activities is repeated during the system's life cycle (estimations, comparisons, analysis, rationale, etc.) [1–10]. See some problems that are due to be and can be solved by the mathematical modeling of processes according to ISO/IEC/IEEE 15288 "Systems and software engineering. System life cycle processes" in **Figure 3**. The applications of models allows one to manage rationally risks, raise the quality and safety of oil and gas systems and at the expense of them be successful in the market—see in **Figure 4** the example of formalized problems which are solved on the basis of probabilistic modeling in the life cycle [6].

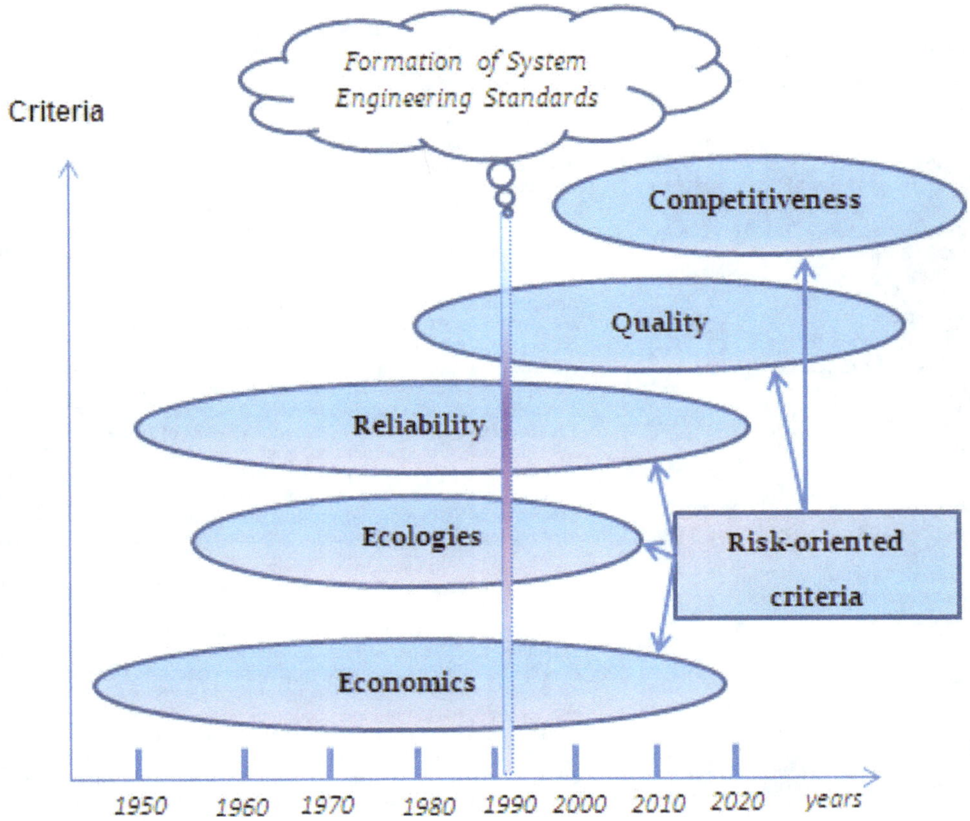

Figure 2. Evolution of risk-oriented criteria.

The summary of the analysis of existing approaches is presented in the next section.

An existing risk control concept tries to consider different uncertainties. But in the application for various areas, the results of information gathering and processing are not used purposefully for modeling, because as the used models of risk prediction that are used in a majority of complex systems, are specific, results and interpretations are not comparable. A universal objective scale of measurement is not established yet. Moreover, the terms "acceptable quality" and "admissible risk" should be defined on a probability scale level only in dependence with corresponding methods and precedents (considering system analogue). For heterogeneous threats, an analytical rationale of the balanced preventive measures of system integrity support at limitations on admissible risks and resources cannot be solved in many cases. The probabilistic modeling, aimed at pragmatic effects, helps to prove probability levels of" acceptable quality" and "admissible risk" for different systems in uniform interpretation, creates techniques to solve different problems for quality and helps in risk optimization. It supports making-decisions in quality and safety and/or helps to avoid wasted expenses in the system's life cycle—see the proposed purposeful way in **Figure 5**, based on dozens of probabilistic models and software tools [6]. There are proposed universal metrics for system processes: probabilities of success or failure during a given period for an element, subsystem and system. A calculation of these metrics within the limits of the offered probability space built on the basis of the theory for random processes allows one to predict quality and risks on a uniform probability scale, quantitatively proving comprehensive levels of acceptable quality and admissible risks from "precedents cases." The prediction of risks can use widely safety monitoring data and statistics. In practice, an application of the proposed model and method

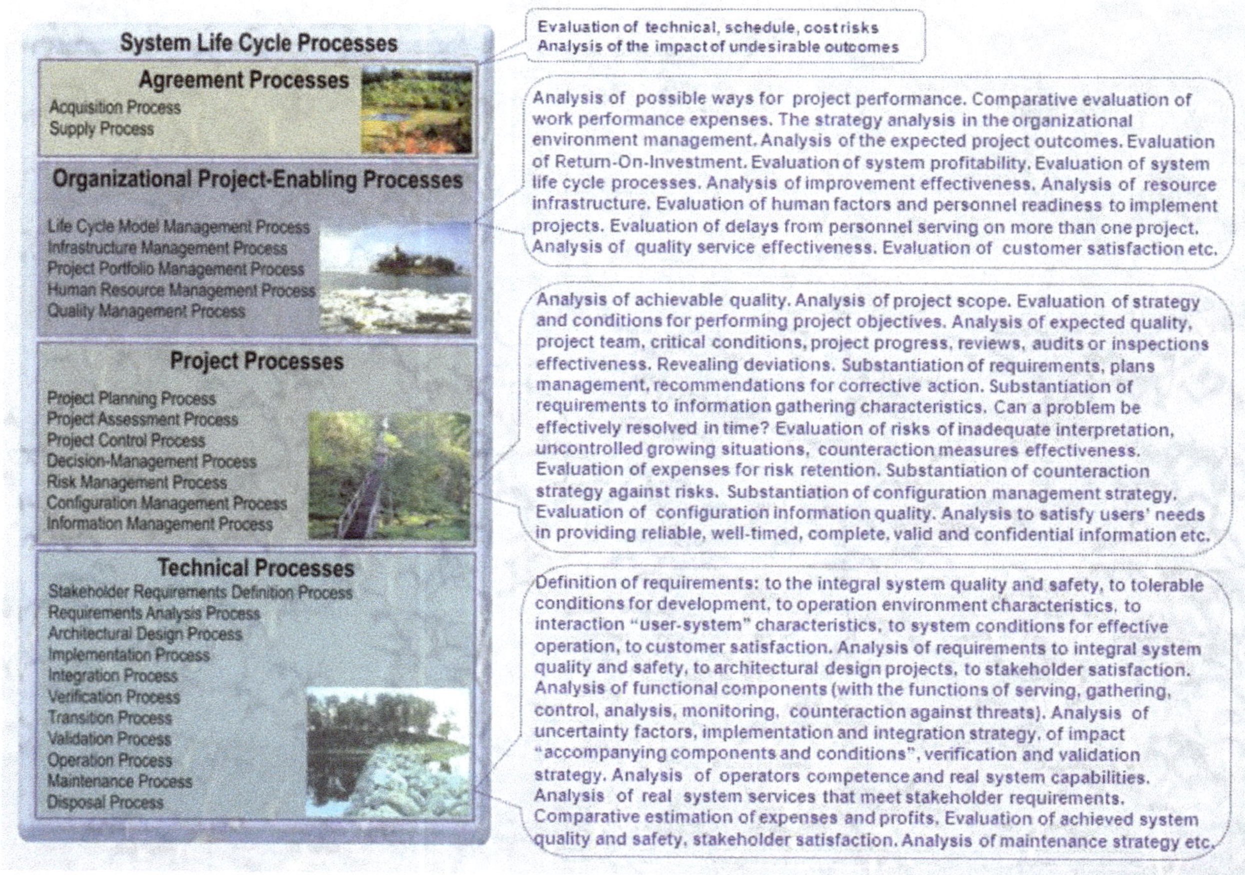

Figure 3. The problems that are due to be and can be solved by probabilistic modeling processes.

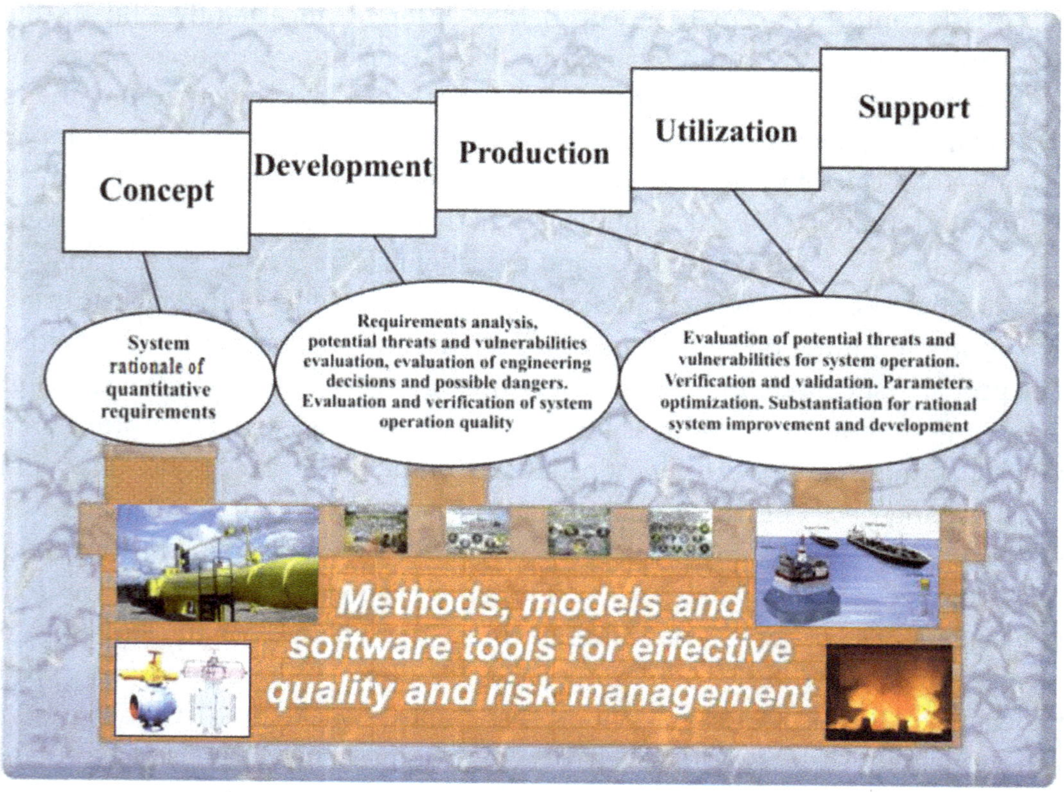

Figure 4. Examples of formalized problems which are solved on the base of probabilistic modeling.

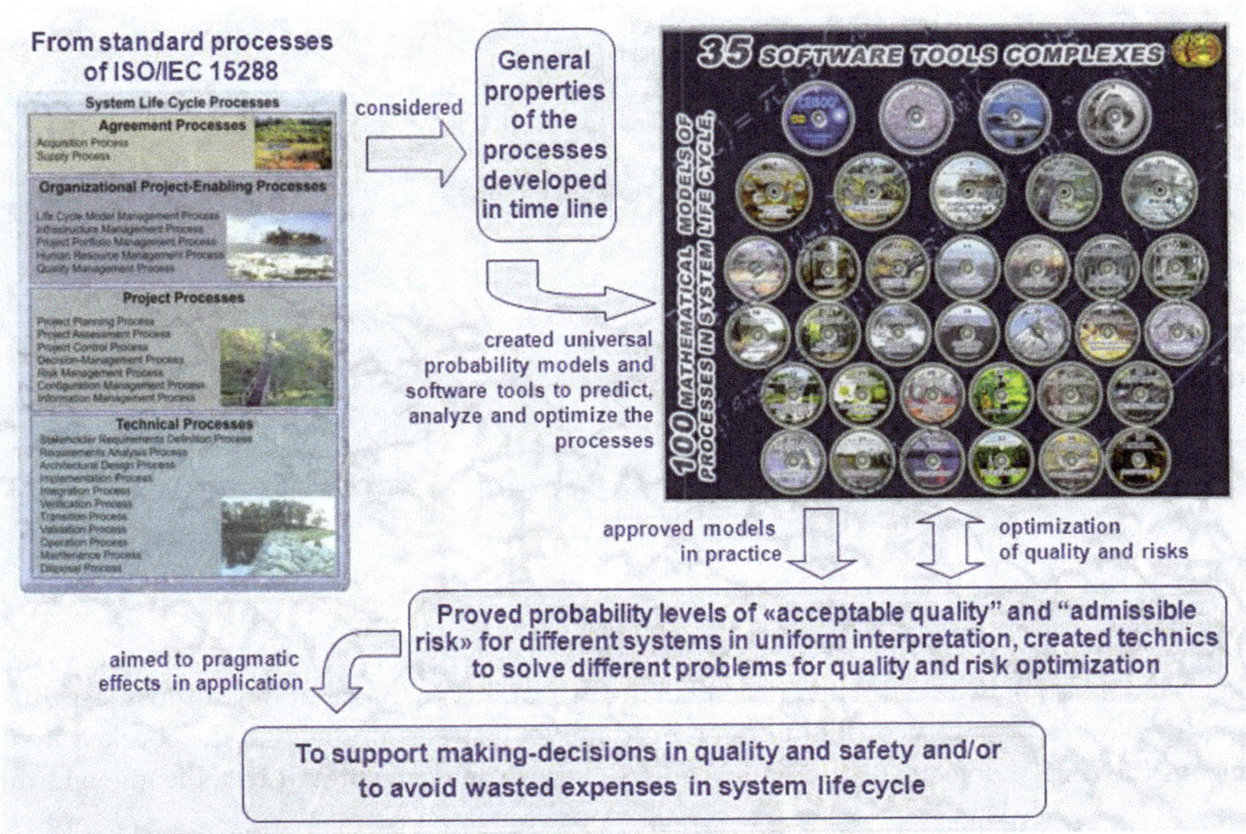

Figure 5. The proposed way to support making-decisions in quality and safety.

allows a customer to formulate better justified requirements and specifications, a developer to implement them rationally without wasted expenses and a user to use the system's potential in the most effective way [1–12].

In a general case, a probabilistic space (Ω, B, P) for the evaluation of system operation processes should be proposed, where Ω is a limited space of elementary events; B is a class of all subspaces of the Ω space, satisfied to the properties of σ algebra; P is a probability measure on the space of elementary events Ω. Because $\Omega = \{\omega_k\}$ is limited, there is enough to establish a reflection $\omega_k \rightarrow p_k = P(\omega_k)$ like $p_k{}^{3}0$ and $\sum_k p_k = 1$.

The descriptions for some from the proposed probabilistic models and methods for their transformations, adaptations, applications and result interpretations are the following.

3. Model for estimating non-destructive testing

Problems of item content analysis are everywhere for any oil and gas systems in their life cycle. Pipes and pipelines, the equipment (e.g., fountain armature, columned heads and welded tanks), monolithic walls of buildings and the constructions, to be checked in the presence of emptiness, can be considered as such items—see **Figure 6**.

For solving some problems of item content analysis, the existing probabilistic model "information faultlessness after checking" may be used by renaming input and output [6]. For example, for estimating non-destructive testing, the probability of soundness of the checked

Figure 6. Examples of item content analysis.

item (renamed) may be estimated instead of the probability of information faultlessness during the required term (according to referenced model [6]). A soundness of the checked item means the zero of defects (or anomalies) after non-destructive testing during the given term.

What about the effectiveness of non-destructive testing methods for some technical items?

Example 1: Let an application of some instruments of non-destructive testing be planned in the applications to check 10,000 conditional items (the items can be meters of pipes, square meters of walls in storehouses and so on). The operator using instruments forms a system for non-destructive testing. Speed of testing equals 5000 items a day. Taking into account the human factor, a frequency of first-type errors (when the absence of defect [anomaly] is accepted as defect [anomaly]) equals one error a week. The mean time between second-type errors for the system (when real defect [anomaly] does not come to light) is equal once a month. The non-destructive testing is performed permanently for 10 days. It needs to estimate the maximum density of defects (anomalies) for which the probability of soundness of the checked 10,000 conditional items is more than 0.90.

Results of probabilistic modeling have shown that the required density is about 0.02%, that is, 2 defects (anomalies) on 10,000 items. In addition it is expedient to notice that since density of defects about 1%, the probability of soundness is stabilized at level 0.88. It does not fall as less, because first-type and second-type errors seldom occur in example 1.

Example 2: Continuing example 1, it needs to prove minimum speed of non-destructive testing, the checked volume for which the probability of soundness of 10,000 conditional items will exceed 0.95 at continuous work within 8 h of working hours.

The results of probabilistic modeling are reflected in **Figure 7**.

The analysis shows that the found rational speed is about 1100 items per hour. And the part of defects after the control in the checked-up volume of 10,000 items will be 0.0008% against the primary 0.02%. It can be interpreted: at the checked volume of 1,00,000 items (i.e., in 10 times more primary 10,000, when quantity of defects is 20), the average residual quantity of defects

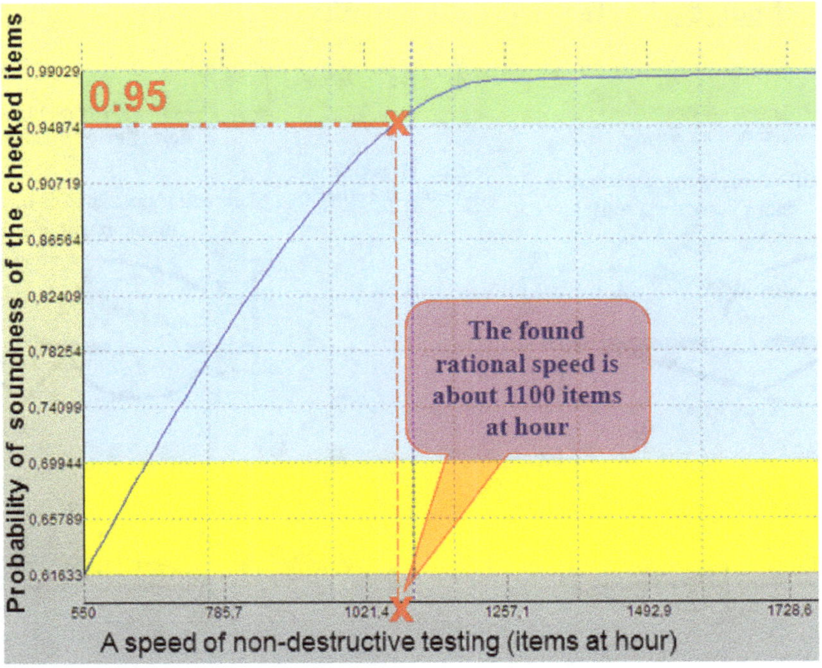

Figure 7. The way for rationale speed of non-destructive testing.

will not exceed 1. It means that under the second example conditions, 19 from 20 defects will be revealed in time with probability 0.95 and more.

4. Models for "black box" and for complex structures

The probabilistic approaches for modeling "black box" and complex structures operating in conditions of heterogeneous threats are proposed.

4.1. "Black box"

There are two general technologies proposed of providing protection from critical influences on the system: technology 1 is the periodical diagnostics of system integrity (without the continuous monitoring between diagnostics) and technology 2 is the continuous monitoring between periodical diagnostics added to technology 1—see **Figure 8**.

Technology 1 is based on periodical diagnostics of system integrity, which is carried out to detect danger source penetration from threats (destabilizing factors) into a system or the consequences of negative influences. The lost system integrity can be detected only as a result of diagnostics, after which system recovery starts. Dangerous influence on a system is acted upon step by step: at first, a danger source penetrates into a system and then after its activation begins to influence. System integrity cannot be lost before a penetrated danger source is activated. Danger from threats (destabilizing factors) is considered to be realized only after a danger source has influenced a system.

Technology 2, unlike the previous one, implies that operators alternating each other trace system integrity between diagnostics (the operator may be a man or a special device or their

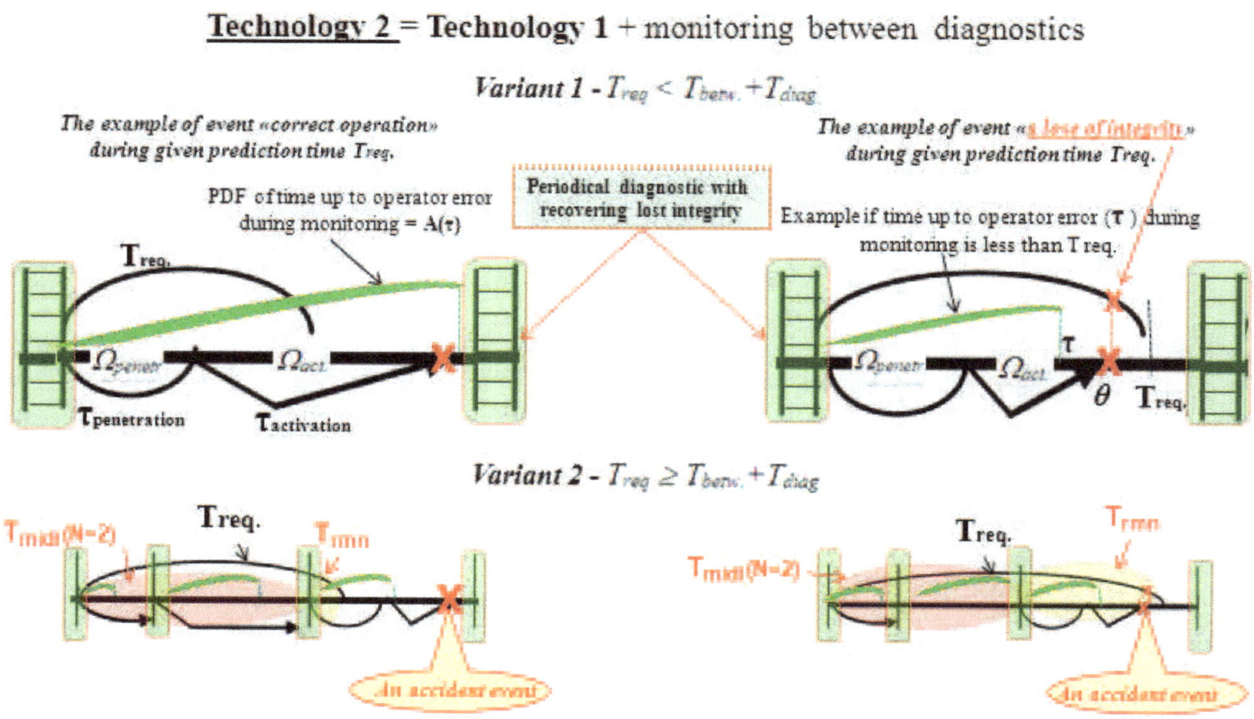

Figure 8. Some accident events for technology 2 (left – "Correct operation", right – "a loss of integrity" during $T_{req.}$).

combination). In case of detecting a danger source an operator recovers system integrity. The ways of integrity recovering are analogous to the ways of technology 1. Faultless operator's actions provide the neutralization of a danger source trying to penetrate into a system. When operators alternate a complex diagnostic is held. A penetration of a danger source is possible only if an operator makes an error but a dangerous influence occurs if danger is activated before the next diagnostic. Otherwise the source will be detected and neutralized during the next diagnostic.

It is supposed for technologies 1 and 2 that the used diagnostic allows to provide necessary system integrity recovery after revealing danger source penetration into a system or consequences of influences. Assumption: for all time input characteristics, the probability distribution function (PDF) exists. Thus, the probability of the correct system operation within the given prognostic period (i.e., the probability of success) may be computed as a result of the use of models. For identical damage risk, to lose integrity is an addition to 1 for the probability of correct system operation, R = 1–P [3–4].

There are possible next variants for technologies 1 and 2: variant 1 in the given prognostic period T_{req} is less than the established period between neighboring diagnostics ($T_{req} < T_{betw.} + T_{diag}$); variant 2 in the assigned period T_{req} is more than or equals to the established period between neighboring diagnostics ($T_{req}\ {}^{3}T_{betw.} + T_{diag}$). Here, $T_{betw.}$ is the time between the end of the diagnostic and the beginning of the next diagnostic, T_{diag} is the diagnostic time.

4.2. Integration of probabilistic models for complex structures

The main output of integration modeling is the probability of the correct system operation or risk to losing system integrity during the given period of time. If probabilities for all points $T_{req.}$ from 0 to ∞ are computed, it means a trajectory of the PDF, depending on the characteristics of threats, periodic control, monitoring and recovery. And the building of PDF is the real base to prediction metrics P and R for given time $T_{req.}$. In analogy with reliability, it is important to know a mean time between neighboring losses of integrity like mean time between neighboring failures in reliability (MTBF), but in application to quality, safety, etc.

For complex systems, parallel or serial structure existing models with known PDF can be developed by usual methods of probability theory. Let's consider the elementary structure from two independent parallel or series elements. Let PDF of time between losses of the ith element of integrity be $B_i(t)$, that is, $B_i(t) = P(\tau_i \leq t)$; then:

1. Time between losses of integrity for the system combined from series-connected independent elements is equal to a minimum from two times τ_i: failure of first or second elements (i.e., the system goes into a state of lost integrity when either the first or second element integrity is lost). For this case the PDF of time between losses of system integrity is defined by expression:

$$\begin{aligned} B(t) &= P(\min(\tau_1, \tau_2) \leq t) = 1 - P(\min(\tau_1, \tau_2) > t) = 1 - P(\tau_1 > t)P(\tau_2 > t) \\ &= 1 - [1 - B_1(t)]\,[1 - B_2(t)] \end{aligned} \tag{1}$$

2. Time between losses of integrity for system combined from parallel-connected independent elements (hot reservation) is equal to a maximum from two times τ_i: failure of first and second elements (i.e., the system goes into a state of lost integrity when both first and second elements have lost integrity). For this case the PDF of time between losses of system integrity is defined by the expression:

$$B(t) = P(\max(\tau_1, \tau_2) \leq t) = P(\tau_1 \leq t)P(\tau_2 \leq t) = B_1(t)\,B_2(t) \tag{2}$$

Applying recurrently expressions (1), (2), it is possible to build PDF of time between losses of integrity for any complex system with parallel and/or series structures.

All these ideas for analytical modeling operation processes are supported by the software tools "Mathematical modeling of system life cycle processes"—"know how" (registered by Rospatent №2,004,610,858), "Complex for evaluating quality of production processes" (registered by Rospatent №2,010,614,145) and others [1–4].

5. Optimization

By using the models and software tools above the problems of optimization for an element, subsystem and system can be solved through calculating probabilities of success or failure during a given period on the timeline. This approach considers the threats, conditions of counteractions and the given admissible risk established by the precedent principle. Thus, the final choice of integrated measures is allocated on a payoff to the customer in view of specificity of the created or maintained system.

For example, the next general formal statements of problems for optimization can be used [6]:

1. For the stages of system concept, development, production and support: System parameters, technical and management measures, presented in terms of time characteristics of threats, control and/or monitoring of conditions and comprehensible recovery of lost integrity are the most rational for the given period if the minimum amount of expenses for the creation of the system is reached at limitations on an admissible level of risk to lose integrity and/or probability of an admissible level of quality and expenses for operation under other developments, operations or maintenance conditions.

2. On an operation stage: System parameters, technical and management measures, presented in terms of time characteristics of threats, control and/or monitoring of conditions and comprehensible recovery of lost integrity, are the most rational for the given period of operation if the minimum of risks to system integrity loss is reached at limitations on the admissible level of risk and/or probability of an admissible level of quality and expenses for operations under other operations or maintenance conditions.

The combination of these formal statements also can be used in the system's life cycle.

The approach for using the developed models, methods and software tools to analyze and optimize system processes is illustrated in **Figure 9**.

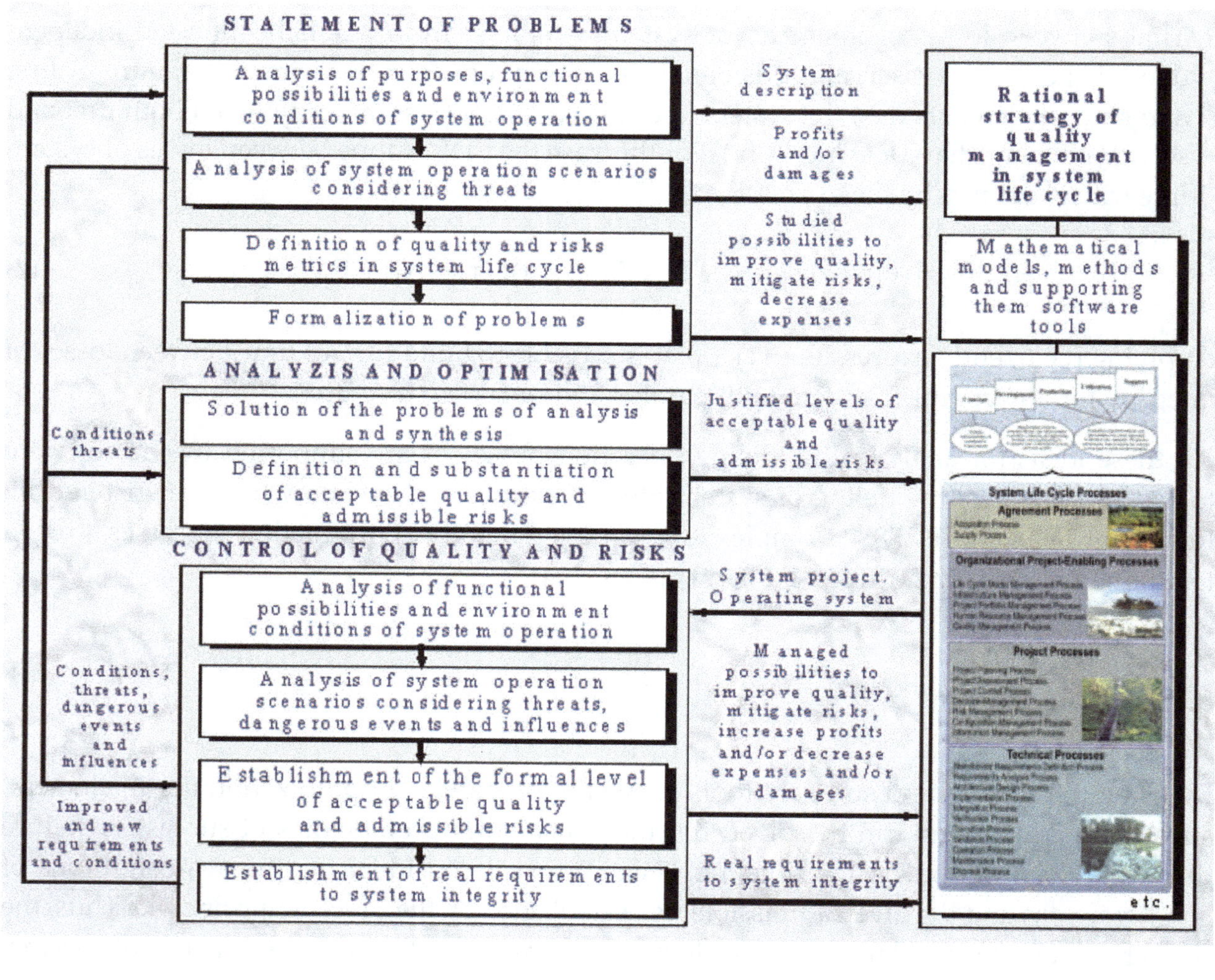

Figure 9. The approach to analyze and optimize system processes.

6. Examples for complex structures: quality prediction for manufacturing processes

A typical set of manufacturing processes of gas preparation equipment (GPE) on the enterprise includes:

- processes connected with operation of entrance threads;
- processes of low temperature gas separations;
- process of economical measure of gas;
- processes of gas heating and reduction;
- processes of candle and torch separation;
- processes connected with storage and use methanol;
- processes connected with storage, supply and drainage dumps of the weathered condensation and diesel fuel;
- managing processes in the engineering division;
- managing processes in the manufacturing division;
- managing processes in booster compressor station division;
- managing processes in the administrative department.

Not to tire the attentive reader, we will not state results of modeling for all processes—in examples 3 and 4, there are only results for processes connected with the operation of entrance threads and managing processes.

Example 3: It is required to predict the quality of the production processes and reliability of equipment connected with the operation of entrance threads.

Input data for modeling are formed as an analysis result of the average statistical data and requirements for production processes of the enterprise. A separate quality of each group of processes is estimated; then, quality of productions for GPE as a whole is predicted. Let an average time of recovery of each group of the processes earlier is equal to the duration of work of one shift, that is, 8 h. The predicted period is 1 month, 1 year and 5 years at observance of set modes for processes.

Note: For a pre-emergency condition, input data can essentially differ; that will cause also change of modeling results.

For the decision the models above are used. The results of modeling of the productions connected with the operation of entrance threads are analyzed in **Figure 10**.

Results of modeling: Owing to the recovery in time technological and production processes as a result of periodic control, the mean time between failures (MTBF), affecting quality,

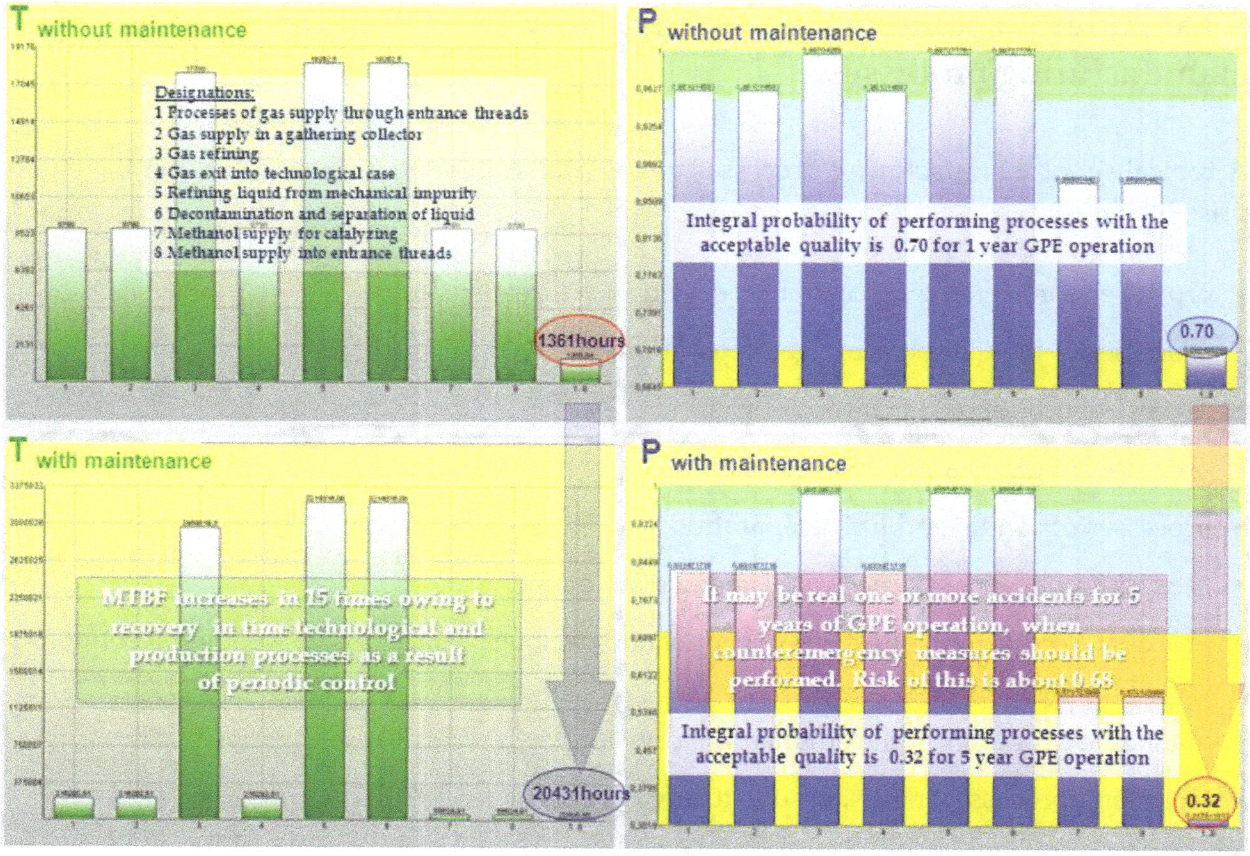

Figure 10. Prediction of quality of production processes connected with operation of entrance threads.

increases from 1361 to 20,431 h, that is, by 15 times. It is reached at the expense of timely reaction in process control. The integral probability of the performing processes, connected with the operation of entrance threads with an acceptable quality, is 0.97 for a month of GPE operation, 0.70 for GPE operation in a year and 0.32 for GPE operation in 5 years. The last probability (0.32) means that it may be a real one or more accidents or failures for 5 years of GPE operation, when counter-emergency measures should be performed. Risk of this is about 0.68, that is, twice more than the probability of success.

And what about reliability? The maintenance and diagnostic measures are performed every half a year according to recommendations of equipment suppliers. How much it is effectively for real operation conditions on the level of predicted reliability?

Results of predicting reliability of equipment connected with the operation of entrance threads are demonstrated in **Figure 11**. Expected integral MTBF is equal to 5770 h. It is 3.5 times less in comparison with 20,431 h owing to daily periodic control (see the earlier section).

Summary: The account of daily results of control and measurements is necessary. Otherwise, if it is to be guided by only guarantee recommendations of equipment suppliers' occurrence, at least one accident or failure demanding counter-emergency measures of protection annually is possible and for 5 years it is inevitable.

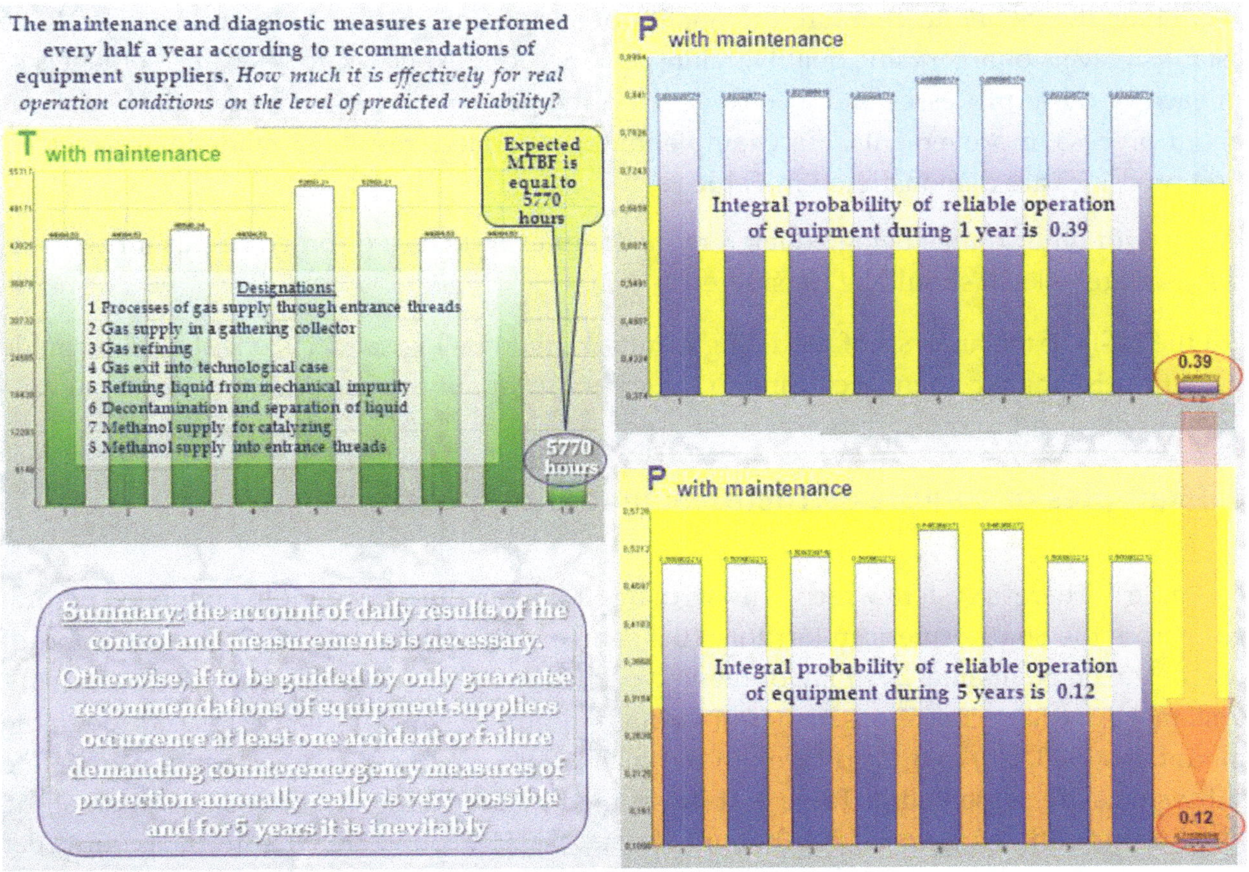

Figure 11. Predicted reliability of equipment connected with operation of entrance threads.

7. General estimation of predicted quality

Example 4: The next system question is very important: What about the benefit for enterprise "the prediction of complex quality" based on the probabilistic modeling of processes?

№	Processes	Probability of providing acceptable quality during a year
1	Processes connected with operation of entrance threads	0.70–0.90
2	Processes of low temperature gas separations	0.62–0.87
3	Processes of economical measure of gas	0.9999
4	Processes of gas heating and reduction	0.94
5	Processes of candle and torch separation	0.82
6	Processes connected with storage and use methanol	0.63
7	Processes connected with storage, supply and drainage dumps of the weathered condensation and diesel fuel	0.60
8	Managing processes in engineering Division, manufacturing Division, booster compressor station Division, administrative Department	0.67

Table 2. Comparative results of production processes modeling.

Modeling allows one to compare the quality of various productions on a uniform scale, to establish levels of acceptable quality, taking into account expenses, to allocate "bottlenecks" in each of these processes and also to develop the general and separate recommendations about process improvements. For example, the comparative results of modeling of production processes are demonstrated in **Table 2**.

Thus, with other things being equal, a more complex structure of processes, as a rule, possesses more risks. It should be considered.

On the basis of the analysis of modeling results, numerous logical decisions should be made by enterprise management according to the criterion "quality-risks-cost."

8. Examples of complex structures: modeling pipelines

Example 5: There is system which consists of a 560-km pipeline for pumping liquefied natural gas across the South American territory (the source of modeling data is a technical report of one of the oil companies). All lay of the line conventionally is divided into three parts (subsystems) by service conditions: first part through the jungle (200 km), second part through the mountains (300 km) and third through the plains (60 km). These characteristics of pipeline subsystems are presented in **Table 3**. It is assumed that the annual profit of operation of the pipeline in the first 5 years is 1500.000 and after is 2500.000 conventional units of accounts per year. It is required to predict the risks taking into account profits and the estimated costs (in conventional units of account) for the construction and maintenance of various sections of the pipeline between 10 and 50 years of its operation.

Characteristics	Part through the jungle (200 km)	Part through the mountains (300 km)	Part through the plains (60 km)
The frequency of potentially hazards impacts on 100 km lay of the line (technical, natural, human or criminal, etc.)	15 times a year	10 times a year	50 times a year
The period between system controls the integrity of area	1 month	1 month	1 week
The mean time to failure of monitoring tools at the area (without using or using existing/ prospective monitoring tools)	1 day/1 year	1 day/1 year	1 day/1 year
The resistance of areas (the average time of preserving the integrity) for the dangerous influences statistically and in comparison with analogues	228.1 days	331.8 days	1217 days
The average cost of construction and maintenance of the area, over 1 km,	1000 c.u. per year	2000 c.u. per year	200 c.u. per year
Average recovery time pipeline integrity after occurrence of the fault		10 days	

Table 3. Characteristics of hazards, measures of control, monitoring and maintaining pipeline integrity.

The solution of a problem: The traditional approach to risk analysis is limited by the obtainment values of the frequency of potential hazard impacts on a 100-km lay of the line— see the first row of characteristics in the table. The proposed solution allows not only for obtaining risks from frequency but also implies how security will change as a result of management. Traditional approaches are not possible to feel the effectiveness of the measures taken for measures of control, monitoring and maintaining integrity. Measures should not seem effective but should be really effective! It is necessary to understand their influence on securing ultimate security. The correct understanding of the possibilities of the impact on safety from measures of control, monitoring and maintaining integrity will allow rationally managing their parameters. The proposed approach provides for the use of these and other data of **Table 3** as input data for subsequent mathematical modeling using the models of Section 3. The results of predictive modeling for 10 and 50 years showed the following (see **Figure 12**).

As a result of applying technologies, which had been developed in 2008, the average time achievable of the safe operation is approximately 3000–5000 h. At the same mean time, failure in the jungle is 5767–8745 h; in the mountains it is 8255–12,676 h; and on the plain it is 29,500–1,22,145 h. Note that the upper estimate was inherent for the systematic maintaining of pipeline integrity (when all failures and critical areas with potential danger are identified)

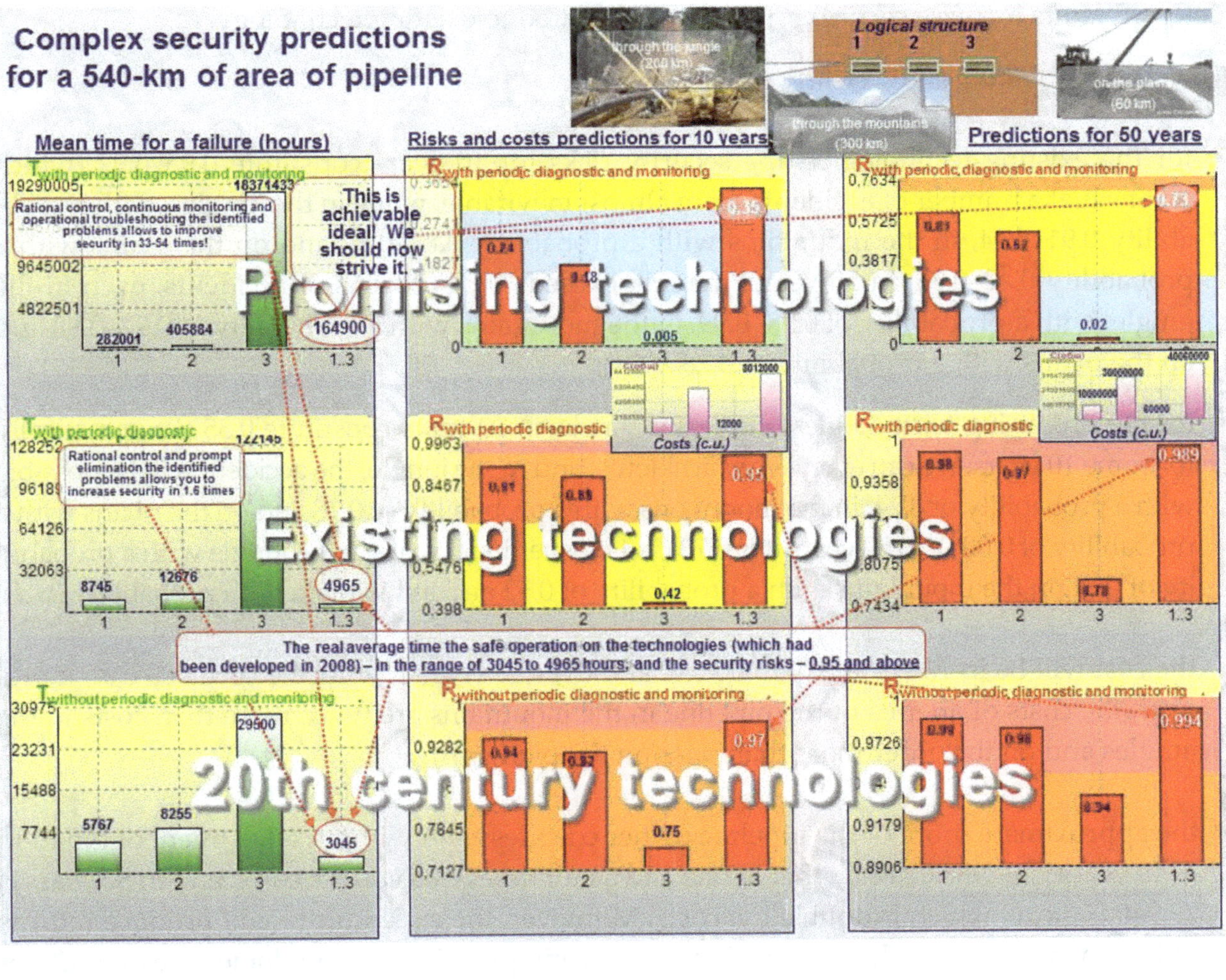

Figure 12. Predicted risks taking into account monitoring possibilities.

in the jungle and in the mountains every month and at the plains weekly. Subsystems' state monitoring is tracked mainly in the days of control. The analysis of the results of the calculations shows that systematic monitoring allows one to increase the safety of operations of the pipeline in the jungle and in the mountains by 1.5 times, in the plain by 4 times, but throughout the 560-km stretch of the pipeline it is by 1.6 times! This is a real job for pre-emption as compared with the case of the absence of any control; when troubleshooting, it is only after the accident that cannot be overlooked. It is assumed that operative repair with restore the integrity follows after the failure detection immediately.

Promising technologies will implement a continuous monitoring of the pipeline at any point. For example, it may be a scan of the air situation using electronic locator fighters of the fourth and fifth generations with the smart cover. Similarly, intellectual filling of the pipeline will signal the dangers of the results with relevant coordinates and diagnosis. If we know the location and cause of the potential failure of the restoration of the integrity it becomes a routine "matter of technique." Under these conditions there's real will be the mean time of safe operation of the pipeline of about 165,000 h, which is achieved when the mean time of failure of monitoring tools is about a year. The mean time of failure in the jungle will be more than 28,0000 h, in the mountains more than 40,000 h and on the plains more than 18 million h (as in the engines of space vehicles!). The analysis of calculation results shows that rational frequency of periodically controlled, continuous monitoring and prompt removal of the detected faults increase security in 33–54% compared to existing technology.

The results obtained show clearly the following:

- for the existing technologies, the security risks for 10 years constitute 0.95–0.97 (which means that a number of accidents seem almost inevitable, while in the jungles, with a probability 0.91–0.94, in the mountains with a probability 0.88–0.92 and on the plains with a probability 0.42–0.75; in 50 years, the risk exceeds 0.99 (dozens of accidents, even in the jungle, with a probability of 0.98–0.99, in the mountains with a probability of 0.97–0.98 and on the plains with a probability of 0.78–0.94);
- for the promising technologies, the security risks in 10 years constitute 0.35 (i.e., practically for some 10 years, we can even avoid accidents, and in the jungle, the accident will be possible with a probability of 0.24, in the mountains with a probability of 0.18 and on the plains with a probability of 0.005), in 50 years it is 0.73 (2–4 crashes in 50 years in the jungle with a probability of 0.61, in the mountains with a probability of 0.52 and in the plain with probability 0.02);
- the cost will be in 10 years 8.012.000 c.u. and over 50 years it will be 40.060.000 c.u.; moreover, the costs of an area of the pipeline in the mountains are twice more than costs in the jungles and on the order more than ones on the plains;
- the approximate profit of the pipeline owner costs less and without adjustment of inflation in 10 years is 11.988.000 c.u.; that in one and a half times exceeds the costs, and in 50 years is 79.940.000 c.u., which is double the costs. Moreover, the expenditure will produce returns in less than in a year. That means that when using promising technologies the quantity of accidents may be reduced on the matter; even these accidents happen either in the jungle or

in the mountains. It's quite a profitable and secure project. It must be admitted that the level of security obtained—the risks are 0.35 in 10 years and 0.73 in 50 years—can be considered as normative "acceptable."

Thus, the examples of forecasting the security operation of the pipelines have illustrated the ability to proactively manage risk. The effectiveness is not just using the universal models but also in the justification of the necessary system requirements for new materials (pipes should be intelligent with the ability of continuous monitoring and mean time of failure for at least a year) and in technologies of restoring functional integrity, in minimizing risks on the basis of the control parameters of the processes of control, monitoring and restoring even before promising technologies have appeared! It is therefore proposed to manage the risks for pipelines of the future even before their creation and based on this, to justify the technical requirements to the system and their components.

Summary for Example 5:

1. Rational control, continuous monitoring and prompt elimination of the revealed accidents and failures allow one to increase the safety tens of times compared to the lack of a systematic control and monitoring!

2. With using advanced technology accidents and failures on plains it is possible to virtually excludes, and in the jungles and mountains to reduce of their number many times.

9. Examples of modeling sea gas-and-oil producing system (GOPS) processes

There are many standards used in the oil and gas industry (ISO 10418 "Basic surface safety systems", ISO 13702 "Control & mitigation of fire & explosion", ISO 14224 "Reliability/maintenance data", ISO 15544 "Emergency response", ISO 15663 "Life cycle costing", ISO 17776 "Assessment of hazardous situations" etc.), but they focus on technical aspects and do not consider terrorist threats.

The principal difference of GOPS consists of the fact that safety problems should be resolved in the sea because long distances from the shore and probable ice conditions in northern regions exclude any help from the outside—see **Figure 13**.

Oil and gas are usually produced on stationary stills and concrete platforms located up to 200 km from the shore at the depth from several dozens to several hundred meters. There are nearly 5000 sea platforms dispersed all over the world. Dozens of thousand oil wells are drilled from these platforms. Produced oil is delivered to the buyers by tankers or directly through pipelines.

Produced gas before transportation goes into the liquefied natural gas terminals. After the liquefaction its volume reduces by 600 times, that makes its transportation profitable. Statistics shows that during the time of sea field development, emergencies are distributed in the following ways [10, 11]: drilling—32% (including 23% at survey and 9% at production

Figure 13. Some explanation of conditions for examples 6 and 7.

drilling); gas-and-oil production—19%; ship collision and towing of floating drilling rigs and blocks for platform construction —14%; storms—11%; floating drilling rigs' delivery to the point of drilling—6%; and other kinds of works—18%.

A safety policy concerning sea GOPS safety includes accident prevention and drawing, plans concerning failure consequences of liquidation and actions taken in case of emergencies. Special brigades are formed and trained to prevent and liquidate failure consequences. High-quality materials and pipes and application of computer diagnostics for pipe integrity monitoring provide safety of the GOPS operation.

All safety measures undertaken nowadays provide system protection from inexperienced personnel (because according to statistics about 80% of all failures are connected with the human factor) or from the natural causes and "cataclysms" which are of an unpremeditated character. However, the attitude to safety cardinally varies in case of terrorist threats because terrorist actions are malicious and aimed at damaging the system through its vulnerable "bottlenecks." As a result the existing risks of system safety violation essentially grow.

The examples 6 and 7 are devoted to modeling processes of possible terrorist influence and GOPS safety provision (including platforms, coastal technological complexes including terminals for floating storage and offloading, liquefied natural gas terminals, pipelines, tubing stations) and to withdraw quantitative evaluations of their vulnerability in various scenarios.

Example 6: connected with an estimation of effectiveness of a safety monitoring system for sea GOPS. Before we start the analysis of possible terrorist threats, let us consider the basic dangers that can arise on sea platforms in case of failures. They are explosions of fuel-air mixed clouds; generation and burning of fire balls; oil spill and burning; separation and spread of technological equipment parts; and others. Each of these dangers can aggravate consequences of failures, that is, lead to the "dominoes" effect. To control risks the following measures are taken: application of safe technologies; measures preventing dangerous situations; applications of systems providing early detection of emergencies; control over operating parameters of the technological process, the signal system and the notification about emergency technologies; measures directed on mitigation of emergency consequences; and preparation of a platform staff to react immediately.

The analysis shows that the basic preventive mechanism of risk reduction is safety monitoring in various variations of its application. Let us estimate commonly used safety technologies, technology 1 (periodical diagnostics of system integrity without the continuous monitoring between diagnostics) and technology 2 (continuous monitoring between periodical diagnostics is added to technology 1)—see Section 4.

Let's estimate efficiency of sea GOPS safety technologies used in the case of emergencies for dozens of years. Thus we take into account that the basis of safety systems consists of automatic facilities' mean time where failures of which are estimated for several years.

To form inputs for probabilistic modeling the arising of basic dangers for sea platforms in the case of failures is considered. There is generally one of the abovementioned technologies to provide safety of GOPS components (platforms, coastal technological complexes including floating storage and offloading terminals, liquefied natural gas terminals, pipelines, tubing stations). The script of emergency development provides frequency of danger source appearance equal to 1 time per 24 h with a mean time of activization within an hour. Time between the termination of the previous and the beginning of the next diagnostics taking into account broken integrity recovery is 2 h. Let us suppose that monitoring is performed by automatic means of tracking the integrity of system components. To such means systems of fire and gas detection, systems of water fire-fighting and foaming, circled fire mains, systems of platform irrigation, pressure relief systems, emergency switching of systems, various locking device and so on may be related. Let the mean time between failures of these means be not less than 2 years.

It needs to estimate a safety of the sea platform operation in such scenarios within several hours, a day, several weeks and a month.

The integrated results of calculations prove that at the realization of technology 1, the required safety is provided only for several working days—what is inadmissible in practice. If the most effective technology, technology 2 is realized, the probability that a dangerous influence does not occur within 24 h is above 0.99997, that is, the probability of emergency is about 0.00003. At the same time provision of the required safety within a month in conditions of daily failure danger this risk increases up to 0.001 that also appears to be a practically admissible result.

Summary for example 6: An effectiveness of the existing safety systems of sea GOPSs appears to be rather enough or high if the frequency of danger source appearances is about once a day. The high level of GOPS protection in emergencies is mainly provided by application of approved automatic safety technologies.

Against the background of proved measures of counteraction to sources of emergency the situation concerning the struggle against terrorist threats appears to be cardinally worse because this problem is still at the initial stage. Other things being equal, let us estimate the expected sea GOPS protection from terrorist threats with differences in abilities of a security service operator to reveal suspicious actions and objects which can be a means of terrorist purpose realization.

Example 7: Let the deliberately formal conditions of a terrorist influence scenario be similar to the emergency dangers in example 6. Let us suppose that for providing protection of sea GOPS platforms from terrorist threats, any of the protecting technologies are used. Let the scenario of the potentially dangerous influence of terrorists provide the frequency of a danger source appearance from air, the water table or from under water equal to 1 time per 24 h with the mean time of activation after penetration onto a platform equal to 1 h. The time between the termination of the previous and the beginning of the following diagnostics taking into consideration the broken integrity recovery is 2 h. Let us assume the mean continuous time of potentially faultless operators' works in each shift to be 6 hours. It is required to evaluate the safety of the sea platform operation in such scenarios within several hours, days, weeks and a month.

The integrated calculation results prove that without any additional protection the system remains in safety with the probability of 0.9 only within 2–3 h. It is explained by a comparative rarity of a danger source appearance. If the 1st technology of counteraction to terrorists is applied (this technology exists on the most of platforms and implies visual tracking of air conditions, the working hours what is an equivalent to the case if there are no measures of counteraction to terrorists on a platform at all.

If the most effective second technology is used, the probability of GOPS integrity within a day is more than 0.92, that is, the risks of latent introductions of a terrorist danger source into a system and the overcoming of all technological protection barriers preventing realization of terrorist threats in the conceived volume approximate to 0.08. At the same time to provide the required safety within a month in conditions of daily danger that a sudden terrorist attack happens, this risk runs up to 0.93. The main cause of this is insufficient preparedness of operators to recognize terrorist threats at the background of other technical threats. That's why it is necessary to increase the mean time between failures of a safety service operator to tens and hundreds of hours what requires creation of special "smart subsystems" in order to support operator functions (radar-tracking, optical, acoustic, electromagnetic means etc.). Compare the results of examples 6 and 7.

Pragmatic interpretation of example 7 results: If the characteristics of terrorist dangers growing are similar to the characteristics of emergency danger, the risks of terrorist threats' realization in the conceived volume are incommensurably higher. Owing to insufficient preparedness and technical equipment of operators for timely and valid recognition of terrorist threats at the background of other technical threats, a variety of GOPSes are completely helpless in case of terrorist dangers that are growing.

10. Instead of conclusion

The presented probabilistic approaches allow us to research different problems for providing safe and effective development of hydrocarbon deposits and rational operation of oil and gas systems. Their application in the system's life cycle helps to answer the main question. "What rational measures should lead to expected effects without wasted expenses, when, by which controllable and uncontrollable conditions and costs?" The efficiency from implementation is commensurable with expenses for system creation.

The probabilistic modeling, comprehensive and systematic studies on the competitiveness of OGS have been carried out in Gubkin Russian State University of Oil and Gas (National Research University) over a period of several years. These researches are concerned with the most important economic branch, which largely determines the country's energy security and efficiency. Certainly, the integral view of OGS competitiveness seamlessly includes the most important components of quality, safety, energy efficiency, environmental compatibility, economic aspects and so on. In turn, these components are also complex, integral and affect a wide range of activities. Moreover, competitiveness as a complex integral metric characterizes the studied considered systems and objects (as living organisms) that have the property of changes in the life time. The proposed probabilistic models and methods are widely used in the practice of education and research.

Author details

Vsevolod Kershenbaum[1,2], Leonid Grigoriev[2], Petr Kanygin[3] and Andrey Nistratov[3,4*]

*Address all correspondence to: andrey.nistratov@gmail.com

1 National Institute of Oil and Gas, Moscow, Russia

2 Gubkin Russian State University of Oil and Gas, Moscow, Russia

3 Chamber of Commerce and Industry of Russian Federation, Moscow, Russia

4 The Federal State Organization, Russian Energy Agency of the Ministry of Energy of the Russian Federation, Moscow, Russia

References

[1] Kostogryzov AI, Stepanov PV. Innovative Management of Quality and Risks in Systems Life Cycle. Moscow: APC; 2008. 404 p

[2] Grigoriev LI, Kershenbaum VY, Kostogryzov AI. System Foundations of the Management of Competitiveness in Oil and Gas Complex. Moscow: National Institute of Oil and Gas; 2010. 374p

[3] Kolowrocki K, Soszynska-Budny J. Reliability and Safety of Complex Technical Systems and Processes. Springer-Verlag London Limited; 2011. 405p. DOI: 10.1007/978-0-85729-694-8

[4] Kostogryzov A et al. Mathematical models and applicable technologies to forecast, analyze and optimize quality and risks for complex systems. Proceedings of the 1st Intern. Conf. on Transportation Information and Safety (ICTIS), June 30–July 2, 2011, Wuhan, China. 2011. p. 845-854

[5] Kostogryzov A, Nistratov A, Nistratov G. Some Applicable Methods to Snalyze and Optimize System Processes in Quality Management. InTech; 2012. ISBN 979-953-307-778-8 http://www.intechopen.com/books/total-quality-management-and-six-sigma/some-applicable-methods-to-analyze-and-optimize-system-processes-in-quality-management

[6] Kostogryzov A, Grigoriev L, Nistratov G, Nistratov A, Krylov V. Prediction and optimization of system quality and risks on the base of modelling processes. American Journal of Operations Research. 2013, 3, p. 217-244. DOI: 10.4236/ajor.2013.31A021. http://www.scirp.org/journal/ajor

[7] Grigoriev L, Kostogryzov A, Tupysev A. Automated dispatch control; problems and details of modeling. Proceedings of IFAC Conference on Manufacturing modeling, management and control. Saint Petersburg, Russia; June 19-21, 2013. pp. 1157-1161

[8] Kostogryzov A, Nistratov A. George Nistratov the innovative probability models and software technologies of risks prediction for systems operating in various fields. International Journal of Engineering and Innovative Technology (IJEIT). September 2013;**3**(3):146-155. http://www.ijeit.com/archive.php

[9] Demyanov VV, Savelyeva EA. Geostatistics: Theory and Practice. Under the editorship of R.V. Arutyunyan; Nuclear Safety Institute of the Russian Academy of Sciences. The Science. 2010. 327 p

[10] Leonid G, Chingiz G, Vsevolod K, Andrey K. The methodological approach, based on the risks analysis and optimization, to research variants for developing hydrocarbon deposits of Arctic regions. Journal of Polish Safety and Reliability Association. Summer Safety and Reliability Seminars. 2014;**5**(1-2):71-78. http://jpsra.am.gdynia.pl/archives/jpsra-2014-contents/

[11] Guseinov C, Tagiev R. The Safety Fundamentals to Design the Objects for Development to Commercial Level of the Production Capacity of a Hydro-Carbon Deposit on a Shelf of Arctic Seas. Manual book. Moscow University of Oil and Gas; 2001

10

Probabilistic Methods and Technologies of Risk Prediction and Rationale of Preventive Measures by Using "Smart Systems": Applications to Coal Branch for Increasing Industrial Safety of Enterprises

Vladimir Artemyev, Jury Rudenko and George Nistratov

Abstract

Abilities of "smart systems" for processing information, adaptation to conditions of uncertainty, and performance of scientifically proven preventive actions in real time are analyzed. Basic probabilistic models and technologies for the analysis of complex systems, using "smart systems," ways of generation of probabilistic models for prognostic researches of the new systems projected, modernized, or transformed, are proposed. The proposed methods are described to predict risks to lose integrity for complex structures on the given prognostic time and rationale of preventive measures considering admissible risk, estimate "smart system" operation quality, and predict in real time the mean residual time before the next parameter abnormalities. The methods and technologies are implemented on the level of the remote monitoring systems. The application is illustrated on the examples of the joint-stock company "Siberian Coal Energy Company."

Keywords: analysis, method, model, prediction, probability, quality, risk, safety, smart system, technology

1. Introduction

All next years and decades form an epoch of using smart systems. What about the usefulness of smart systems for prediction and rationale of preventive measures against possible threats? To answer this question, we address to some definitions.

According to ISO Guide 73, in general, case risk is defined as the effect of uncertainty on objectives. An effect is a deviation from the expected—positive and/or negative. Objectives can have different aspects (such as financial, health and safety, and environmental goals) and can be applied at different levels (such as strategic, organization-wide, project, product, and process). Risk is often characterized by reference to potential events and consequences or a combination of these. Risk may be estimated by a probability of potential events, leading to effects considering consequences. The chapter, including examples, is focused on events leading to losses of system integrity (often with negative consequences). But it does not limit a generality of proposed approaches.

According to ISO/IEC/IEEE 15288 "Systems and software engineering—System life cycle processes," a system is a combination of interacting elements organized to achieve one or more stated purposes. An enabling system is a system that supports a system of interest during its life cycle stages but does not necessarily contribute directly to its function during operation. A system of systems (SoS) is a system of interest whose elements are themselves systems. A SoS brings together a set of systems for a task that none of the systems can accomplish on its own. Each constituent system keeps its own management, goals, and resources while coordinating within the SoS and adapting to meet SoS goals. The research covers systems defined in itself as "smart" system or using "smart" systems (see **Figure 1**).

For modern or perspective system or for a system of systems from the point of view of prediction and rationale of preventive measures against possible threats, the "smart" systems are and will be used as the systems in itself or as system elements or enabling systems. In a

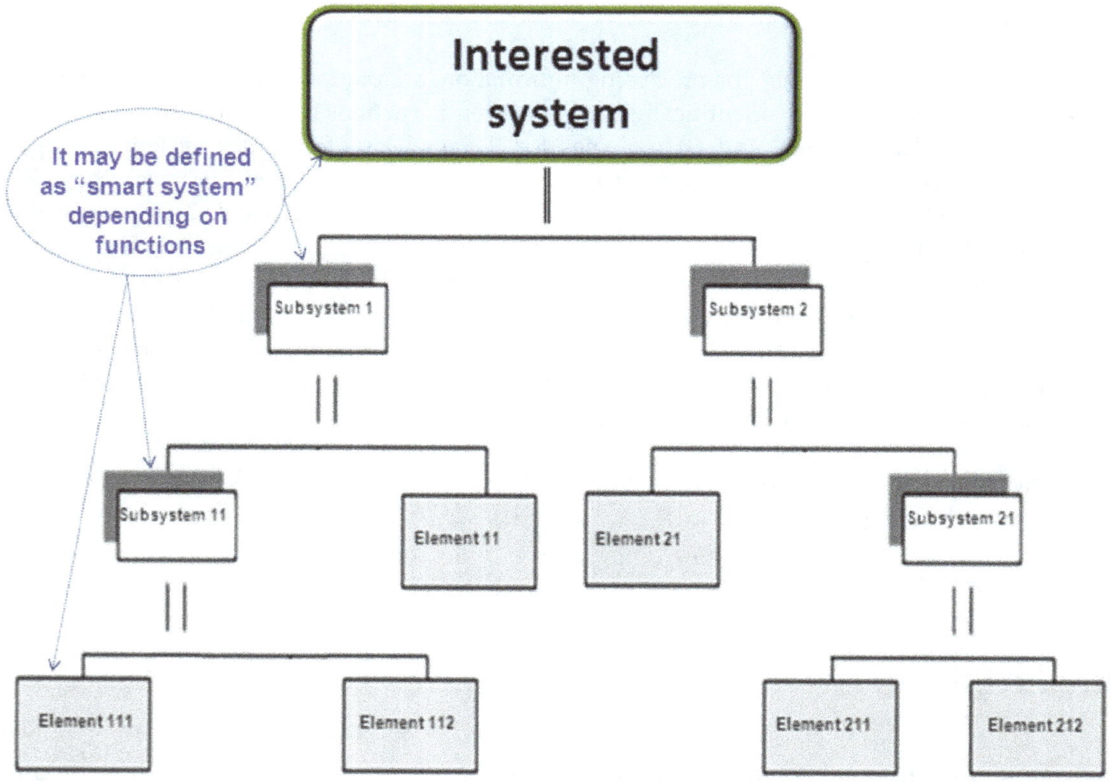

Figure 1. To the definition of "smart system" in a system.

general case, "smart" is a mnemonic acronym, giving criteria to guide in the setting of objectives, and "smart systems" are defined as miniaturized devices that incorporate functions of sensing, actuation, and control (www.wikipedia.org, www.thefullwiki.org).

Developing existing researches [1–17], this manuscript includes correct probabilistic interpretation of risk prediction effectively using "smart" systems, some original basic probabilistic models for risk prediction, the improvement of existing risk control concept, and approaches for solving some problems of industrial safety for coal branch.

2. Probabilistic interpretation of risk prediction for effective using "smart" systems

Because "smart" possibilities allow to forecast a future, we should view probabilistic vision of event prediction, its scientific interpretation, and, unfortunately, some existing illusory vision. Here, from the scientific point of view for anticipating dangerous development of events, it is difficult to construct an adequate probability distribution function (PDF) [1–4] of time between losses of system integrity. Damage may be to some extent estimated on practice (we will consider that the deviations in estimations can reach 100%). Therefore, leaving an estimation of a possible damage out of the work, we will stop on researches of a probabilistic component of risk. What deviations in risk predictions are possible here? To answer this question, it is necessary to understand typical metrics and engineering methods of risk predictions, in definition and concept to use "admissible risk," and then to compare various variants.

In practice probabilistic estimations of system integrity losses are quite often carried out by the frequency of emergencies or any adverse events. For example, with reference to safety, it can be frequencies of different danger threats influences, leading to a damage. That is, frequency replaces estimations of probability (risk to lose integrity of system during prognostic period). It is correct? From probability theory it is known that for defined PDF one of its characteristics is the mathematical expectation ($T_{exp.}$). In turn, for PDF of time between losses of system integrity, the mathematical expectation is the mean time between neighboring losses of system integrity $T_{exp.}$, and moreover the frequency λ of system integrity losses is equal to $1/T_{exp.}$ If to be guided only by frequency λ (with ignoring PDF) in practice, a large deviation may take place. Indeed, a probability that event has occurred till moment $T_{exp.}$ can be equal to 0.00 for approximation by deterministic (discrete) PDF and 0.36 for exponential approximation (see **Figure 2**). That is, as a result of erroneous choice of PDF, characterized by identical λ, the enormous difference may take place! On the one hand, it means ambiguity of a probabilistic estimation of events, being guided only on frequency λ, and on the other hand, a necessity of search (or creations) of more adequate PDF of time between losses of system integrity is very high.

Often today, engineers prefer exponential PDF: $R(t, \lambda) = 1 - \exp. (-\lambda \cdot t)$. If, for example, for 1 year of prognostic period to put λ about 10^{-3} times in a year or less, then under Taylor's expansion $R(t, \lambda) \approx \lambda \cdot t$ with accuracy $o(\lambda^2 \cdot t^2)$. And, if t = 1 year, the absolute value of frequency practically coincides with the value of probability. But if value $\lambda \cdot t$ increases, it is capable to exceed 1 and by definition generally cannot be perceived as probability. Resume: focusing on

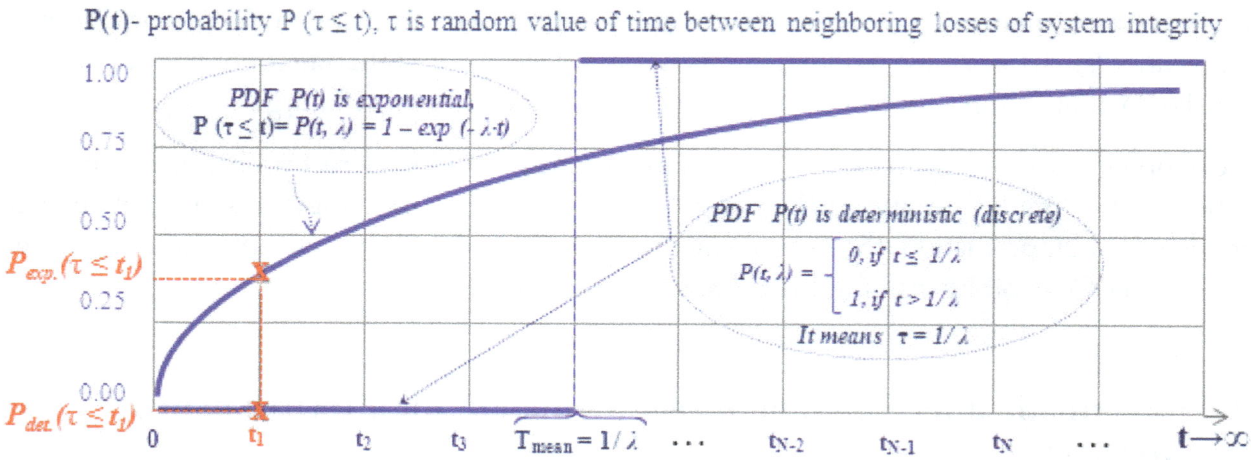

Figure 2. For the same λ, a probability that event has occurred can be equal to 0.00 for approximation by deterministic PDF and 0.36 for exponential PDF approximation.

probability is correct from the point of view of universal risk metric. And, focusing on frequency may be incorrect if $\lambda \cdot t$ is approximately more than 10^{-3}.

The special importance has the concept of "admissible risk." The matter is that there should be a result of the consent of all parties involved in unsafe business on condition that it does not ruin business; by all it is unequivocally estimated and interpreted (not excluding emergencies) and is scientifically proven. In practice frequently the "admissible risk" is interpreted as "border strip," i.e., it is supposed that if it does not cross this "border strip," the system integrity cannot be lost. But in reality it is not so! The residual risk always remains. In operation research the similar restrictions are considered as a starting point for the decision of synthesis problems, connected with searching effective preventive measures of system integrity in life cycle. The complex use of these measures promotes in retaining the risk on the admissible level. It is the typical approach which should work correctly. And how does it work in practice?

Here, it is to quite pertinently address the developed form of the quantitative requirements, connected with the level of admissible risks. The elementary forms of requirements are:

- "A frequency λ of system integrity losses should not exceed admissible level $\lambda_{adm.}$"
- "Probability to lose integrity of system during time T_{req} should not exceed admissible level $R_{adm.}$ (T_{req})."
- Their combination.

What engineering explanations occur in practice? They are as follows:

- If the limitation on the admissible level of probability $R_{adm.}$ (T_{req}) is set, it means that crossing "border strip" should not occur on an interval of time from 0 to T_{req}. For exponential approximations there is an unequivocal functional dependence: $\lambda_{adm.} = -\ln(1 - R_{adm.}(T_{req}))$. That is, this dependence means that a given value of admissible probability $R_{adm.}$ (T_{req}) corresponds unequivocally with a value of the maximum frequency of system integrity losses.

- If the limitation on the admissible level of maximum frequency of system integrity losses $\lambda_{adm.}$ is set, it means that for exponential approximations the function of probability from time t is considered: $R(t, \lambda_{adm.}) = 1 - \exp.(-\lambda_{adm.} \cdot t)$. That is, this is the same "border strip" but in the form of the function from t and without an obvious binding to value T_{req}. This level of limitation by function $R_{adm.}(T_{req})$ is logically to interpret also as "admissible" for the period of time from 0 to t. Admissible risk in the point of probability $P_{adm.}(t_{req.})$ for time $t_{req.}$ May be prolonged on the level of PDF by exponential distribution and the admissible frequency of system integrity losses $\lambda = -\ln(1 - P_{adm.}(t_{req.}))$. It is convenient, but is it adequate? In reality a vision about exponential PDF for behavior of "smart" system may be roughly erroneous (see **Figure** 3).

Despite obvious incompleteness of the elementary forms of requirements to "admissible risks" (in reality, only the limitations in one or several points) and the absence of interrelations with a kind of real PDF of time between losses of system integrity (depending from many parameters: structure of system, heterogeneity of threats, different measures of counteraction to threats, etc.), these forms have got accepted by engineering community. In the further statement of the work, we will be guided by these elementary forms of requirements to "admissible risks." They also allow to extract latent knowledge from the results of adequate probabilistic modeling.

Today, specifications of safety in different fields characterize a frequency **λ** of system integrity losses at the level 10^{-3}–10^{-7} times a year. As a matter of fact, it is one danger event for 1000 years, i.e., cannot be tested in system life! In practice it can be estimated by means of mathematical and/or physical modeling. And, from statistics we know only that at the Russian systems of oil and gas industry, thousand emergencies are annually. But, the number of incidents with a comprehensible result (with prevented emergencies) is usually a hundred times more!

Accordingly, there is an important question: what frequencies of system integrity losses should be used for risk predictions and where does it take? If these are only the frequencies of emergencies, the predicted risks will be essentially underestimated! These final frequencies are output

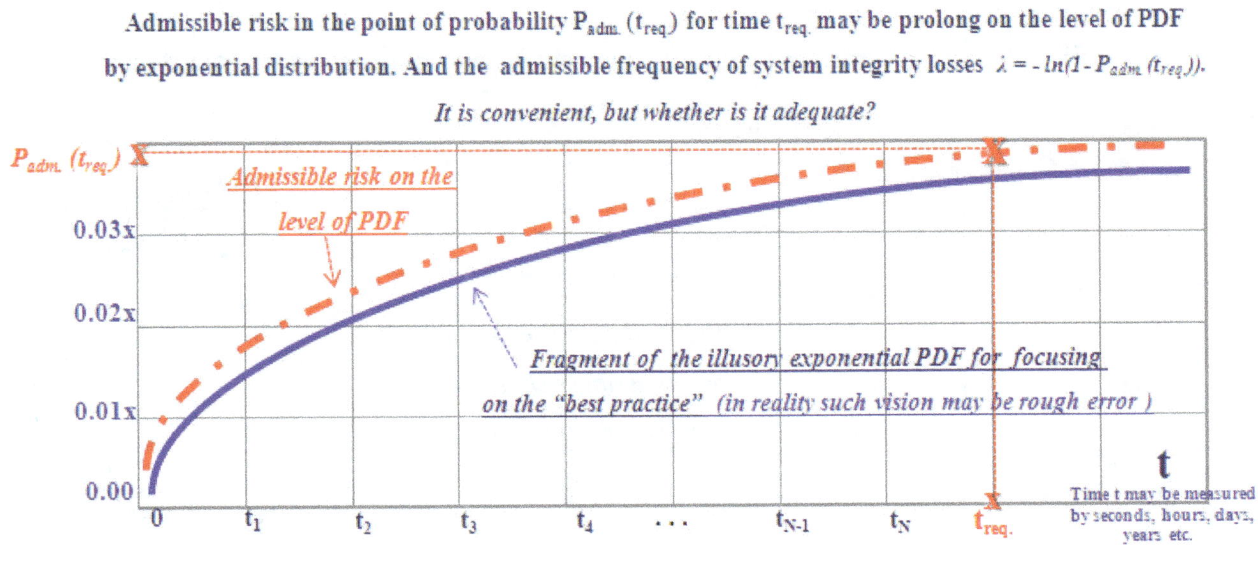

Figure 3. About erroneous vision of exponential PDF approximation instead of more adequate approximation.

instead of input data for modeling. Estimate, please: if to be guided by these frequencies and to consider that 50–70% of failures are the result of "human factor," it should mean that the frequency of critical errors from "human factor" on systems is about one time in thousand years! However, that is not so in real life! Errors are committed much more often. But they are under control, and the majority of them is in due time corrected. As consequence of these counteraction measures, required system integrity (including safety) is reached. The answer arises obviously: the frequency λ of system integrity losses used at risk predictions itself should pay off by the results of probabilistic modeling. Indeed, for adequate risk prediction, there is an important frequency of all the primary incidents (including neutralized incidents at the expense of control measures, maintenance, and timely reaction on initial signs of threat development).

Consideration of "smart" system possibilities for proactive diagnostics of system integrity, monitoring of conditions, and recovering the lost integrity allows to create more adequate

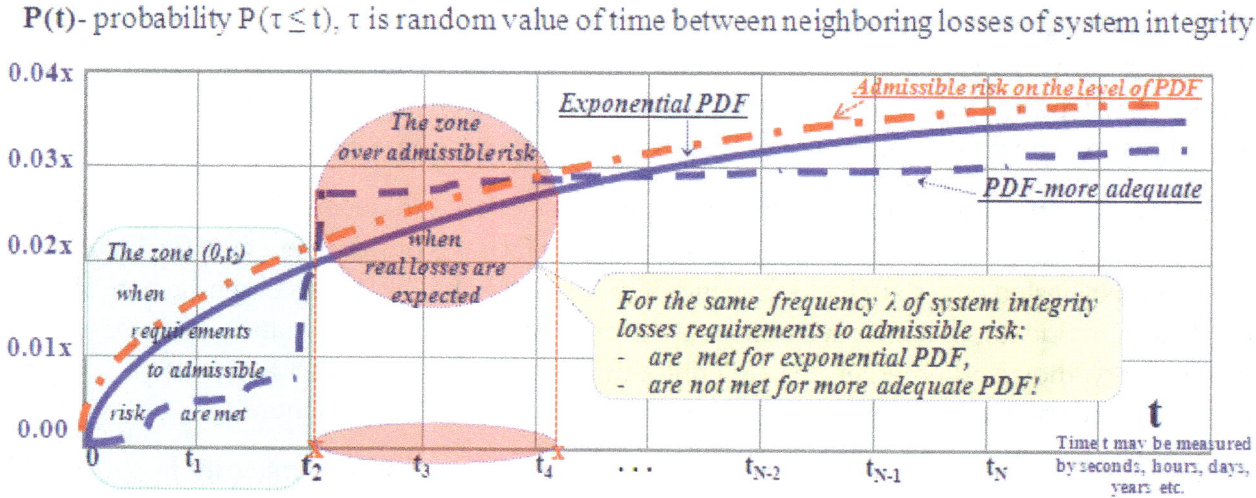

Figure 4. The possible variants of correlations of the limitations to admissible risks, exponential, and an adequate PDF of time between losses of system integrity with identical frequency of system integrity losses λ.

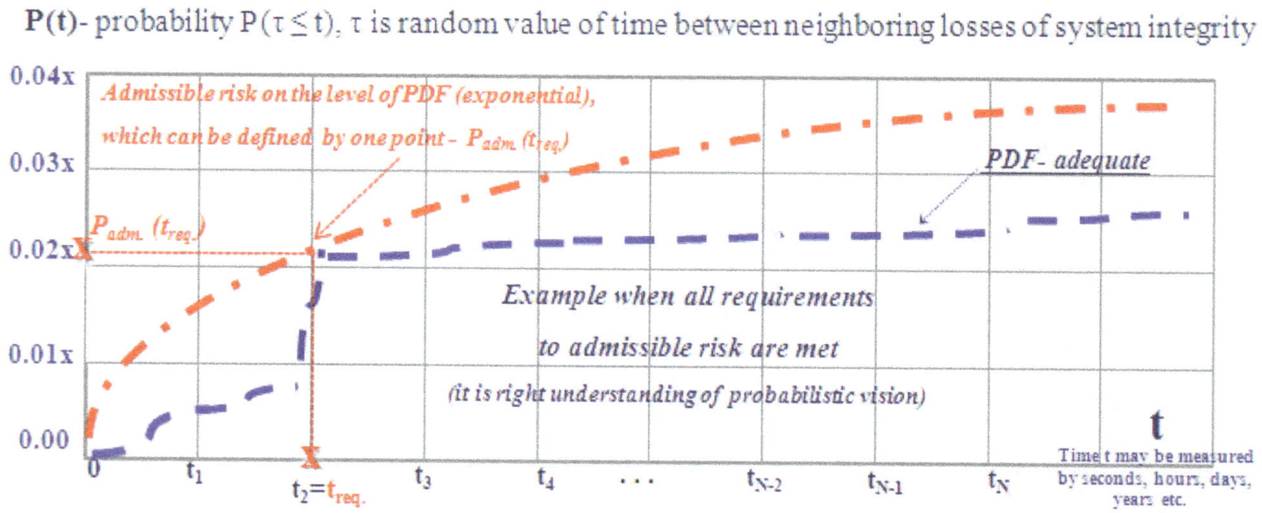

Figure 5. All requirements to admissible risk are met for an adequate PDF of time between losses of system integrity.

PDF for risk predictions. In **Figure 4** the limitations to admissible risks, fragment of exponential, and an adequate PDF of time between losses of system integrity with identical frequency of system integrity losses are demonstrated. The errors in comparison with vision in **Figure 3** are noted.

An example when all requirements to admissible risk are met is presented on **Figure 5**. It is the right understanding of probabilistic vision of event prediction with scientific interpretation considering situations in **Figure 4**.

3. Some basic probabilistic models for risk prediction

Considering possibilities of "smart" systems, two general technologies of providing protection in different spheres are described: proactive periodical diagnostics of system integrity (technology 1) and additionally monitoring between diagnostics (technology 2) including recovery of integrity [2–3, 6–10]. These models allow to create more adequate PDF of time before the next event of the lost integrity.

3.1. The models for the systems that are presented as one element ("black box")

Technology 1 is based on proactive diagnostics of system integrity that are carried out periodically to detect danger occurrences into a system or consequences of negative influences. The lost system integrity can be detected only as a result of diagnostics, after which the recovery of integrity is started. Dangerous influence on system is acted step by step: at first a danger occurrence into a system and then after its activation begins to influence. System integrity cannot be lost before an occurred danger is activated. A danger is considered to be realized only after a danger has activated and influenced on a system. Otherwise, the danger will be detected and neutralized during the next diagnostic.

Note: it is supposed that used diagnostic tools allow to provide system integrity recovery after revealing of danger occurrences into a system or consequences of influences.

Technology 2, unlike the previous one, implies that operators alternating each other trace system integrity between diagnostics. In case of detecting a danger, an operator recovers system integrity (ways of dangers removing and system recovery are the same as for technology 1). Faultless operator's actions provide a neutralization of a danger. When a complex diagnostic is periodically performed, this time operators are alternated. An occurrence of a danger is possible only if an operator makes an error, but a dangerous influence occurs if the danger is activated before the next diagnostic.

The probability of system operation with required safety and quality within the given prognostic period (i.e., probability of success) may be estimated as a result of using the next models for technologies 1 and 2. Assumption: for all time input characteristic, the probability distribution functions exist. Risk $R(T_{req})$ to lose integrity (safety, quality, or separate property, e.g., reliability), i.e., to be though one time in "red" range during period T_{req}, is addition to 1 for probability $P(T_{req})$ of providing system integrity ("probability of success," i.e., to be in "green" or "yellow" ranges all period T_{req}). $R(T_{req}) = 1 - P(T_{req})$ considering consequences.

The next variants for technologies 1 and 2 are possible: variant 1—the given prognostic period T_{req} is less than established period between neighboring diagnostics ($T_{req} < T_{betw.} + T_{diag}$); variant 2—the prognostic period T_{req} is more than or equals to established period between neighboring diagnostics ($T_{req} \geq T_{betw.} + T_{diag}$). Here, $T_{betw.}$ is the time between the end of diagnostic and the beginning of the next diagnostic, and T_{diag} is the diagnostic time.

The next formulas for PDF of time between the losses of system integrity are proposed [2–3].

PDF for the model of technology 1 (variant 1): Under the condition of independence for characteristics, the probability of providing system integrity for variant 1 is equal to

$$P_{(1)}(T_{req}) = 1 - \Omega_{occur}{}^{*}\, \Omega_{activ}(T_{req}), \tag{1}$$

where $\Omega_{occur}(t)$ is the PDF of time between neighboring occurrences of danger (from the "green" to the "yellow" range), mathematical expectation $T_{occur} = \sigma^{-1}$; $\Omega_{activ}(t)$ is the PDF of activation time of occurred danger (threat: from the first input at the "yellow" range to the first input in the "red" range), and mathematical expectation is β. The PDF $\Omega_{occur}(t)$ and $\Omega_{activ}(t)$ may be exponential (see rationale in [6]). For different threats a frequency of dangers for these PDF is the sum of frequencies of every kind of threats.

PDF for the model of technology 1 (variant 2): Under the condition of independence for characteristics, the probability of providing system integrity for variant 2 is equal to

$$P_{(2)}(T_{req}) = N((T_{betw} + T_{diag})/T_{req})\, P_{(1)}{}^{N}(T_{betw} + T_{diag}) + (T_{rmn}/T_{req})\, P_{(1)}(T_{rmn}), \tag{2}$$

where $N = [T_{req.}/(T_{betw.} + T_{diag.})]$ may be real (for PDF) or the integer part (for estimation of deviations).

$$T_{rmn} = T_{req} - N(T_{betw} + T_{diag}).$$

The probability of providing system integrity within the given time $P_{(1)}(T_{given})$ is defined by Eq. (1).

PDF for the model of technology 2 (variant 1): Under the condition of independence for characteristics, the probability of providing system integrity for variant 1 is equal to

$$P_{(1)}(T_{req}) = 1 - \int_{0}^{T_{req.}} dA(\tau) \int_{\tau}^{T_{req.}} d\Omega_{penetr}{}^{*}\, \Omega_{act.}(\theta) \tag{3}$$

Here, $A(\tau)$ is the PDF of time between operator's errors. $A(\tau)$ may be exponential PDF (see rationale in [6]).

PDF for the model of technology 2 (variant 2): Under the condition of independence of characteristics, the probability of providing system integrity for variant 2 is equal to

$$P_{(2)}\ (T_{req}) = N((T_{betw} + T_{diag})/T_{req})\ P_{(1)}{}^{N}(T_{betw} + T_{diag}) + (T_{rmn}/T_{req})\ P_{(1)}(T_{rmn}), \quad (4)$$

where the probability of providing system integrity within the given time $P_{(1)}(T_{req.})$ is defined by Eq. (3).

The final clear analytical formulas for modeling are received by Lebesgue integration of expression (3).

The models are applicable to the system presented as one element. The main result of such system modeling is a probability of providing system integrity or of losses of system integrity during the given period of time. If a probability for all points $T_{req.}$ from 0 to ∞ will be calculated, a trajectory of the PDF for each combined element depending on threats, periodic control, monitoring, and recovery time is automatically synthesized.

3.2. Probabilistic approach to estimate "smart" system operation quality

In general case "smart" system operation always aims to provide reliable and timely producing the complete, valid and, if needed, confidential information for its proper further pragmatic use, including incorporate functions of sensing, actuation, and control. And, potential threats to "smart" system operation are influencing the used information (**Figure 6**).

In general case a probabilistic space (*Ω*, *B*, *P*) for an evaluation of system operation processes is proposed, where *Ω* is the limited space of elementary events; *B* is the class of all subspace of *Ω* space, satisfied to the properties of *σ*-algebra; and *P* is the probability measure on a space of elementary events *Ω*. Because $\Omega = \{\omega_k\}$ is limited, there is enough to establish a reflection $\omega_k \rightarrow p_k = P(\omega_k)$ like that $p_k \geq 0$ and $\sum_k p_k = 1$. Such space (*Ω*, *B*, *P*) is built [6] and proposed for use because "smart" system may be considered as specially focused information system, incorporating functions of sensing, actuation, and control. The proposed analytical models and calculated measures are as follows [6]:

"The model of function performance by a complex system in conditions of unreliability of its components" (the measures: T_{MTBF} is the mean time between failures; $P_{rel.}(T_{given})$ is the probability of reliable operation of IS, composed by subsystems and system elements, during the

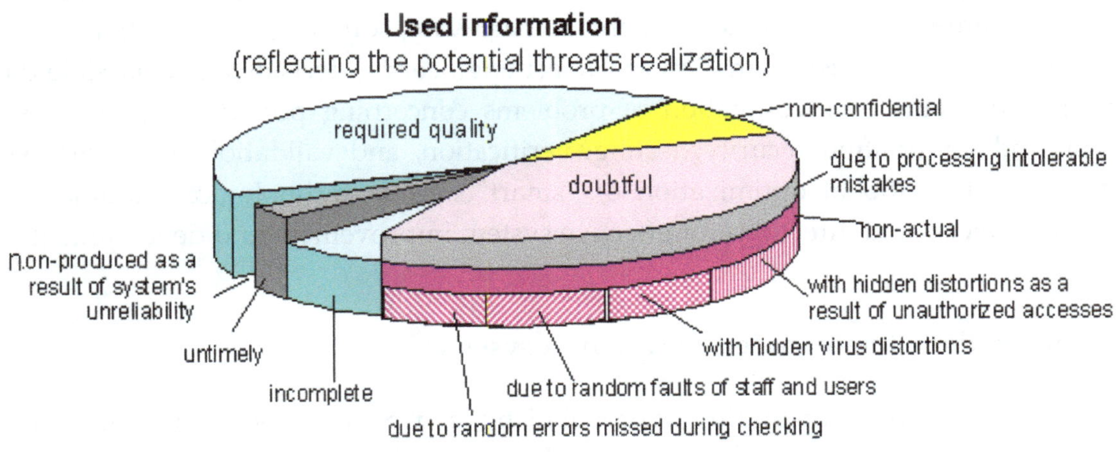

Figure 6. Potential threat realization to "smart" system operation on the level of used information.

given period T_{given}; and $P_{man}(T_{given})$ is the probability of providing faultless man's actions during the given period T_{given}).

"The model complex of call processing" (the measures for the different dispatcher technologies (for unpriority call processing in a consecutive order for single-tasking processing mode, in a time-sharing order for multitasking processing mode; for priority technologies of consecutive call processing with relative and absolute priorities; for batch call processing; for combination of technologies above): the mean wait time in a queue; the mean full processing time, including the wait time; P_{tim} is the probability of well-timed processing during the given time; the relative portion of all well-timed processed calls; the relative portion of well-timed processed calls of those types for which the customer requirements are met C_{tim}).

"The model of entering into IS current data concerning new objects of application domain" (the measure: P_{compl} is the probability that IS contains complete current information about the states of all objects and events).

"The model of information gathering" (the measure: $P_{actual.}$ is the probability of IS information actuality on the moment of its use).

"The model of information analysis" (the measures: P_{check} is the probability of error absence after checking; the fraction of errors in information after checking; $P_{process}$ is the probability of correct analysis results obtained; the fraction of unaccounted essential information).

"The model complex of dangerous influences on a protected system" (the measures: $P_{inf.l.}(T_{given})$ is the probability of required counteraction to dangerous influences from threats during the given period T_{given}).

"The model complex of an authorized access to system resources" (the measures: P_{prot} is the probability of providing system protection from an unauthorized access by means of barriers; $P_{conf.}$ (T_{given}) is the probability of providing information confidentiality by means of all barriers during the given period T_{given}).

These models, supported by different versions of software Complex for Evaluation of Information Systems Operation Quality, patented by Rospatent №2,000,610,272 (CEISOQ+), may be applied and improved for solving such system problems in "smart" system life cycle as rationale of quantitative system requirements to hardware, software, users, staff, and technologies; requirement analysis; estimation of project engineering decisions and possible danger; detection of bottlenecks; investigation of problems concerning potential threats to system operation and information security; testing, verification, and validation of "smart" system operation quality; rational optimization of "smart" system technological parameters; and rationale of projects and directions for effective system improvement and development.

3.3. The generation of new models for complex systems

The basic ideas of correct integration of probabilistic metrics are based on a combination and development of the offered models [2–3, 6–10]. For a complex system estimation with parallel or serial structure, new models can be generated by methods of probability theory. For this purpose in analogy with reliability, it is necessary to know a mean time between losses of integrity for each element. Let's consider the elementary structure from two independent parallel elements that means logic connection "OR" or series elements that means logic connection "AND" (see **Figure 7**).

The PDF of time between neighboring losses of ith element integrity is $B_i(t) = P(\tau_i \le t)$; then:

(1) Time between losses of integrity for system combined from series connected independent elements is equal to a minimum from two times τ_i: failure of the first or second elements (i.e., the system goes into a state of lost integrity when either the first or second element integrity will be lost). For this case the PDF of time between losses of system integrity is defined by expression

$$B(t) = P(\min(\tau_1,\tau_2)\le t)=1- P(\min(\tau_1,\tau_2)>t)=1-P(\tau_1>t)P(\tau_2> t)= 1 - [1-B_1(t)]\,[1- B_2(t)], \quad (5)$$

(2) Time between losses of integrity for system combined from parallel connected independent elements (hot reservation) is equal to a maximum from two times τ_i: failure of the first or second elements (i.e., the system goes into a state of lost integrity when both the first and second element integrity will be lost). For this case the PDF of time between losses of system integrity is defined by expression

$$B(t)=P(\max(\tau_1,\tau_2)\le t)=P(\tau_1 \le t)P(\tau_2 \le t)=B_1(t)B_2(t). \quad (6)$$

Note: The same approach is studied also by Prof. E. Ventcel (Russia) in 80th who has formulated the trying tasks for students.

Thus, an adequacy of probabilistic models is reached by the consideration of real processes of control, monitoring, and element recovery for complex structure. Applying recurrently expressions (5)–(6), it is possible to create PDF of time between losses of integrity for any complex system with parallel and/or series structure.

The known kind of the more adequate PDF allows to define accordingly mean time between neighboring losses of system integrity $T_{exp.}$ (may be calculated from this PDF as mathematical expectation) and a frequency λ of system integrity losses $\lambda = 1/\,T_{exp.}$.

Risk to lose integrity (safety, quality, or separate property, e.g., reliability) is an addition to 1 for probability of providing system integrity (correct system operation or "probability of success") $R = 1-P$. The formulas for probabilistic modeling technologies 1 and 2 and the proofs of them are proposed in [2–3, 6].

All these ideas are implemented by the software technologies of risk prediction for complex systems, for example, the "mathematical modeling of system life cycle processes," "know-how"

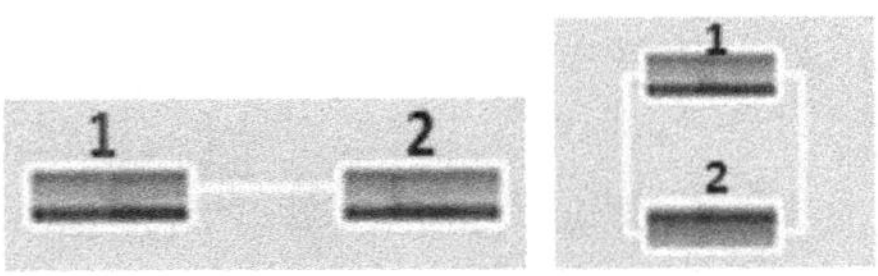

Figure 7. Illustration of system, combined from series (left) or parallel (right) elements.

(registered by Rospatent №2,004,610,858), and "complex for evaluating quality of production processes" (patented by Rospatent №2,010,614,145) [8–9].

4. The improvement of existing risk control concept

The purposed approach to improve existing risk control concept includes (see **Figure 8**) [11–17]:

- Creation and perfection of probabilistic models for problem decision
- Automatic combination and generation of new probabilistic models
- Forming the storehouse of risk prediction knowledge
- For storehouse, dozens of variants of the decision of typical industrial problems for risk control

For example, *system*, combined from complex interested system and "smart" system (also it may be SoS), can be analyzed by the formula (5) and probabilistic models described above (see **Figure 9**). The correct operation of this *system of system* during the given period means during the given period of prediction both the first and the second complex systems should operate correctly according their destinations. That is, integrated system is in the state "Integrity (correct operation)" if "AND" the interested system left and "AND" the "smart" system right are in the state "Integrity (correct operation)."

Thus, the proposed improved that risk control concept can be useful to perform effectively functions: risk prediction; rationale of quantitative system requirements to hardware, software, users, staff, and technologies; requirement analysis; estimation of project engineering decisions and possible danger; detection of bottlenecks; investigation of problems concerning potential threats to operation of complex systems; validation of system operation quality; rational optimization of system parameters; and rationale of plans, projects, and directions for effective system utilization, improvement, and development. The expected pragmatic effect is as follows: it is possible to provide essential system quality and safety rise and/or avoid wasted expenses in system life cycle bases on the rational application of improved concept.

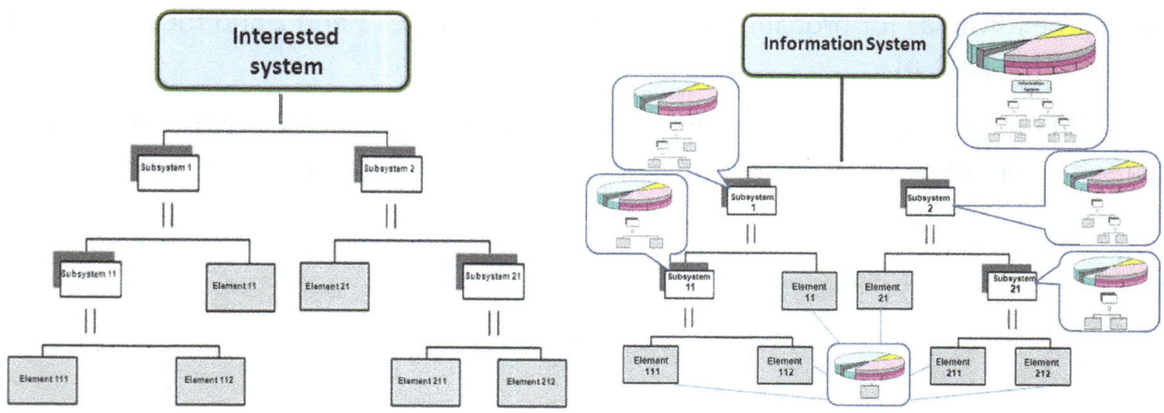

Figure 8. The purposed approach to improve existing risk control concept.

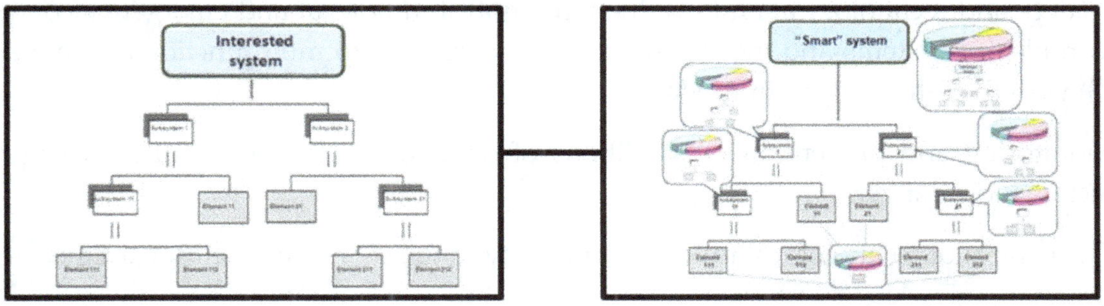

Figure 9. *System* of two different complex systems (serial combination) for modeling integrated system.

5. About some problems of industrial safety for coal branch

As an example of effectively solving the problems of industrial safety, we consider an experience of the joint-stock company "Siberian Coal Energy Company (SUEK)" (see www.suek.com). SUEK is one of the world's largest coal companies with production assets in Russia and a global trading network. SUEK delivers long-term value to shareholders at every stage of the value chain—mining, processing, transportation, and shipment—through port facilities, sales, and distribution (**Figure 10**). This value chain includes different SoS. In practice many SoS of SUEK use "smart" systems [11, 14].

Below are the aspects researched:

- Probabilistic analysis of the remote monitoring system (RMS) possibilities for increasing industrial safety of critical infrastructure safety (CIS).
- Estimating in real time the mean residual time before the next parameter abnormalities considering the results of the control of equipment and technological process conformity to the set normative in real time.

5.1. Probabilistic analysis of the remote monitoring system possibilities

For coal branch the developments of mine, buildings, and constructions should be equipped by a complex of systems and means that provide the organization and implementation of coal

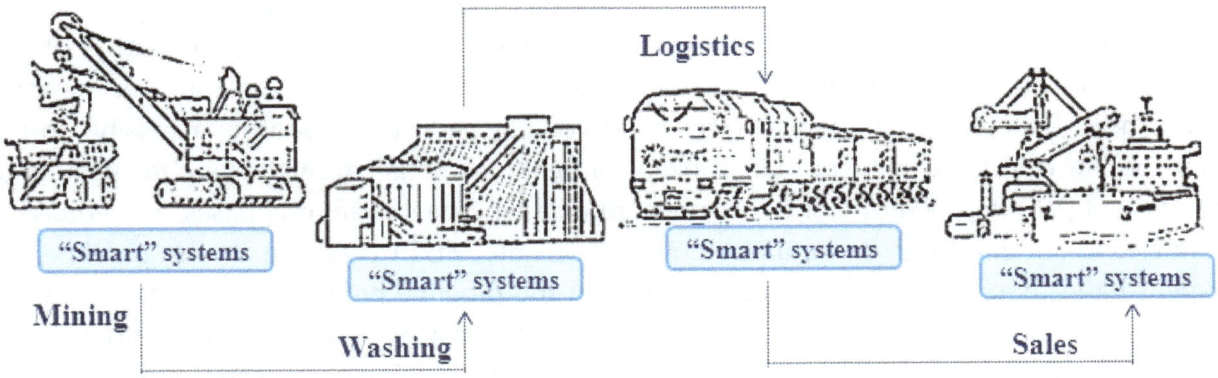

Figure 10. The SUEK value chain includes "smart" systems everywhere.

work safety and technological and productions control in normal and emergency conditions. This complex of systems and means should be integrated into multifunctional safety system (MFSS) with the following main functions:

- Monitoring and prevention of conditions of occurrence of geodynamic, aerologic, and technogenic danger.
- The control of technological process conformity to the set normative in real time.
- Application of counteremergency protection systems.

The usual approaches to critical infrastructure safety (CIS) which have been developed in last dozen years, based on many respects on subjective safety estimations "on places", have reached a high but not sufficient level of efficiency. For the account of interests of all interested parties and the further business development today, rethinking system possibilities of applied information technologies for increasing safety and extracting the innovative effects are not used fully till now.

Search of cardinal directions of improving CIS, favorable to business and the state, has led to comprehension of sharp necessity and expediency of creation and implementation of remote monitoring system (RMS) that is an important part of MFSS. RMS transforms an internal information support of separate CIS in a mode of a needed transparency and wide availability of CIS state in real time for all interested and responsible parties. Along with it on the basis of rational RMS implementation, the transition from the existing subjective expert approach to the risk-based approach for critical infrastructure safety receives necessary information filling.

The proposed probabilistic analysis of RMS operation in their influence on integral risks to lose system integrity is based on researching real remote monitoring systems implemented in Russia for oil and gas CIS. In application to composed and integrated CIS with RMS and without RMS, the earlier models, developed by authors, are used [1–10]. The received results are applicable for an analytical rationale of system requirements to RMS, system definition of the balanced preventive measures of systems, and subsystem and element integrity support at limitations on resources and admissible risks.

Requirements to monitoring and prognosis for critical systems are established at the level of many international standards, for example, ISO 17359, ISO 13381–1, ISO 13379, IEC 61508–1 [18–21], etc. Today, a monitoring of parameter conditions is carried out to increase reliability and industrial safety of critical systems, improve their health management, and provide predictive maintenance and operation efficiency. Here, critical systems are understood as objects of dangerous manufacture and the equipment, energy objects, power and transport systems, and others. Different data about current conditions of parameters become accessible in real time. So, for coal mine some of many dozens of heterogeneous parameters are for ventilation equipment (VE) (temperature of rotor and engine bearings, a current on phases, and voltage of stator) and for modular decontamination equipment (MDE) (vacuum in the pipeline, the expense and temperature of a metano-air mix in the pipeline before equipment, pressure in system of compressed air, etc.). Effects from RMS may be reached on the basis of gathering and analytical processing in real time the information on controllable parameters of objects monitored (see **Figure 11**). RMS is intended for a possibility of prediction, the prevention of possible

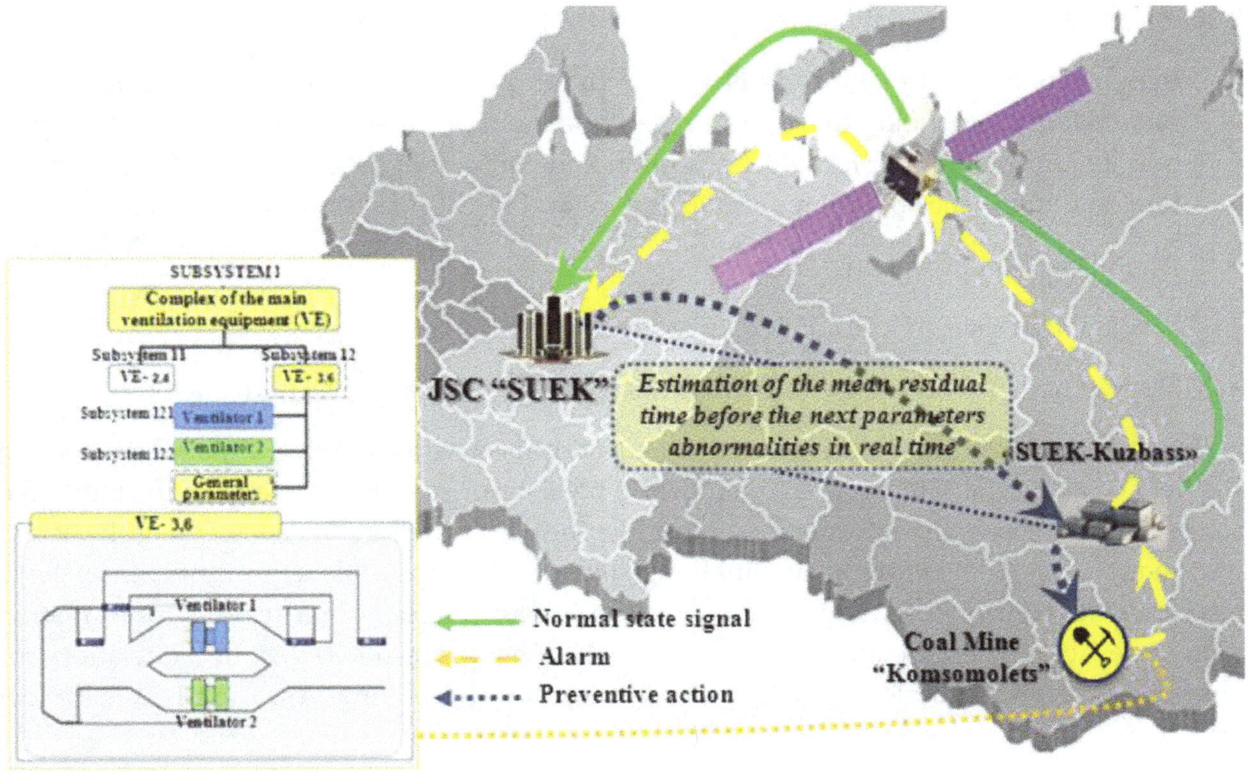

Figure 11. Example of reaction in real time.

emergencies, minimization of a role of human factor regarding control, and supervising functions. The role of RMS is defined by their functions, to the basic of which concern:

- Remote continuous monitoring of CIS condition in real time (gathering data about key parameters of technological processes; gathering and processing data of industrial inspection, the information of technical condition and equipment diagnostics, and the information on the presence of failures and incidents; and results of system recovery, etc.)
- Analytical data processing
- Prediction of risks to lose object integrity
- Display of parameter conditions and predictions with the necessary level of details

In this subsection analytical decomposition and the subsequent integration of complex systems are used according to propositions above in Sections 1–4. Admissible conditions (ranges) of traced parameters for each element, the reservation possibilities, implemented technologies of the control, and recovery of integrity are considered.

RMS is intended for a possibility of prediction, the prevention of possible emergencies, minimization of a role of human factor regarding control, and supervising functions. It may be reached on the basis of gathering and analytical processing in real time the information on controllable parameters of objects monitored. For example, objects monitored for oil and gas CIS are the technological equipment and processes of extraction, transportation, refining, the personnel, systems, and means of safety support.

The role of RMS is defined by their functions, to the basic of which concern:

- Remote continuous monitoring of CIS condition in real time (gathering data about key parameters of technological processes; gathering and processing data of industrial inspection, the information of technical condition and equipment diagnostics, and the information on the presence of failures and incidents; and results of system recovery, etc.)
- Analytical data processing
- Prediction of risks to lose CIS integrity
- Display of CIS conditions and predictions with the necessary level of details

Unlike the usual control which is carried out at enterprises (when the state supervising body in the field of industrial safety and frequently also the enterprise/holding bodies of the industrial safety control receive the information only upon incident or failure, not possessing the actual information about deviations at an initial stage when still it is possible to prevent failure), RMS translates the control, a transparency of CIS conditions, the important real-time information (about the facts and predictions), and also necessity of proper response to critical deviations for absolutely new time scale characterized as the scale of real time, measured by seconds-minutes.

Effects from the remote control can be reached only when quality of RMS operation is provided. It means that it is reliable and timely producing the complete, valid, and, if needed, confidential information by RMS.

Generally, system analysis of RMS operation consists in evaluation of reliability, timeliness, completeness, validity, and confidentiality of the used information. In special cases for compound subsystems and system elements, not all measures may be used. For example, for a subsystem of information security enough to use the measures to evaluate protection from an unauthorized access and information confidentiality during the given time period. Dependence of the purposes of researching RMS can be decomposed to the level of compound subsystems and separate elements (see **Figure 6**).

In this case according to the system engineering principles, the operation quality of every component should be evaluated.

For evaluating integral RMS operation quality, the next measure is proposed: the probability of providing reliable and timely representation of the complete, valid, and confidential information during the given time ($P_{RMS}(T_{given})$).

In general case

$$P_{RMS}(T_{given}) = P_{rel.RMS}(T_{given}) \bullet C_{tim\ RMS} \bullet P_{compl..RMS} \bullet P_{actual..RMS} \bullet P_{check\ RMS} \bullet P_{process.RMS} \bullet P_{inf.I.\ RMS}(T_{given}) \bullet P_{man\ RMS}(T_{given}) \bullet P_{prot.RMS} \bullet P_{conf.RMS}(T_{given}),$$

where all measures are calculated by the models proposed in Section 2.

For complex structures the ideas of combination of the models is proposed in [15]. It allows in an automatic mode to generate new models at the expense that there is possible evaluation of the measures above.

When not all system elements and subsystems are captured by RMS capabilities, two subsystems, operated in different time scales, are cooperated in the CIS. A part of CIS, captured RMS, is served in real time, and the other part is in a usual time scale (with information gathering by a principle "as it is possible to receive"). In many critical situations, this usual time scale cannot be characterized as adequate to a reality. With the use of the offered approach, the system with usual control (UC), used for CIS, i.e., without RMS application, can be estimated. Generally, the analyzed critical infrastructure is presented as a combination "System+RMS" and usual "System without RMS." And, "System+RMS" is a combination of "Structure for RMS" and "RMS" (see **Figure 12**). For these systems some measures of the information delivery may not answer requirements of real time—"System+RMS" because RMS operation quality is inadequate and "System without RMS" without RMS.

All the great number of the factors characterizing threats to analyzed critical infrastructure is considered as 100%, and total frequency of dangerous deviations is designated through λ_{Σ}. Frequency of potentially dangerous deviations traced by "System + RMS" is designated (λ_{RMS}). Frequency of occurrence of other potentially dangerous deviations which are not traced by RMS (i.e., for "System without RMS") is designated ($\lambda_{\Sigma} - \lambda_{RMS}$).

For "System + RMS" the RMS operation quality during the time of prediction T_{given} is evaluated by probability $P_{RMS}(T_{given})$. And, the risk of critical deviation for safety during the time of prediction T_{given}, designated as $R_{RMS}(T_{given})$, can be evaluated by the earlier methods [2–3, 5–17]. For the usual "System without RMS," the same measures $P_{UC}(T_{given})$ and $R_{UC}(T_{given})$ can be used with specified value of input for probabilistic modeling.

Then, in general form, the risk $R(T_{given})$ to lose integrity for analyzed critical infrastructure during the time of prediction T_{given} can be evaluated by the formula:

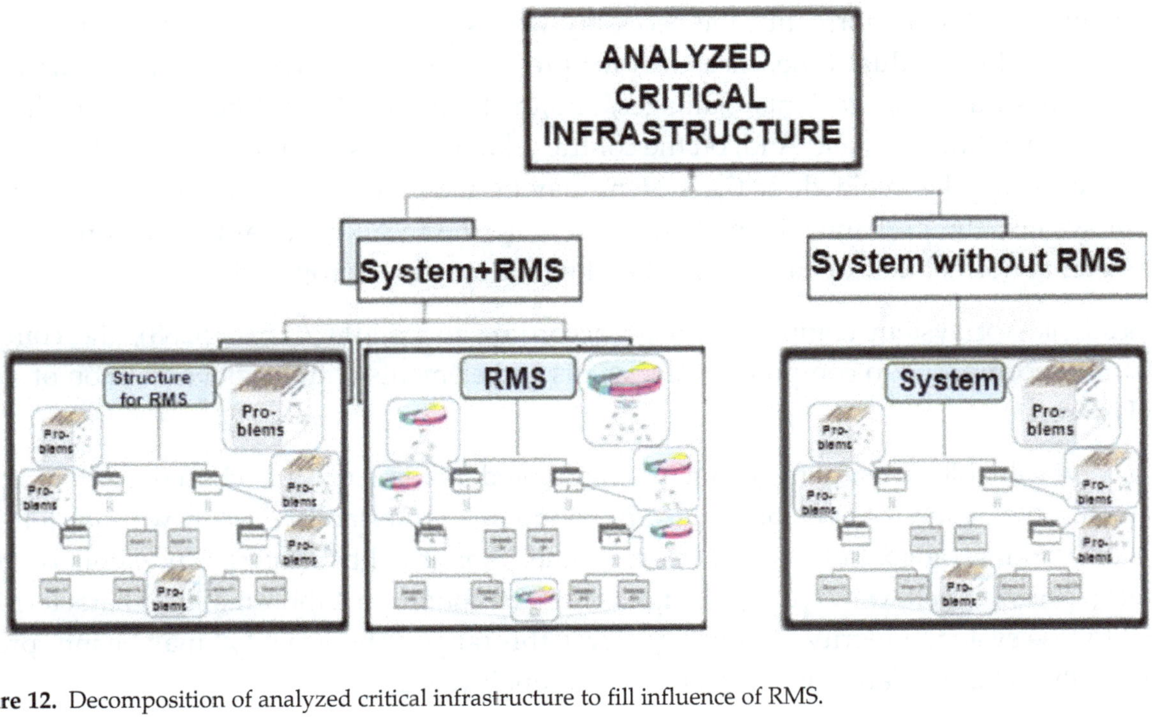

Figure 12. Decomposition of analyzed critical infrastructure to fill influence of RMS.

$$R(T) = 1- [(\lambda_{RMS} / \lambda_{\Sigma}) P_{RMS} (T_{given}) (1- R_{RMS} (T_{given})) + ((\lambda_{\Sigma} - \lambda_{RMS})/ \lambda_{\Sigma}) P_{UC} (T_{given}) (1- R_{UC} (T_{given}))],$$

where expression in square brackets is a probability of successful operation of analyzed critical infrastructure. Depending on the made risk definition in special cases, it can be interpreted as probability of safe or reliable operation or probability of norm observance for critical parameters of the equipment or others in the conditions of associated potential threats. The case $\lambda_{\Sigma} = \lambda_{RMS}$ means full capture of critical infrastructure by RMS capabilities.

5.2. Estimating the mean residual time before the next parameter abnormalities

Unfortunately, in the world the universal approach to adequate prognosis of the future parameter conditions on the basis of current data is not created yet. The uncertainty level is too high. Nevertheless, in practice for each concrete case, subjective expert estimations, regression analysis of collected data, and simulation are often used. And, probabilistic models applied in some cases contain many simplifications, and they frequently do not consider an infrastructure of complex systems, heterogeneity of threats, distinctions in technologies of the control, and recovery of integrity for various elements of these systems [2–3]. The same aspects and also rarity of many random events (with some exceptions) do an ineffective statistical estimation of residual time before the next parameter abnormalities. At the same time, scientifically proven prognosis of a residual time resources is necessary for acceptance of preventive measures on timely elimination of the abnormality reasons. The above-stated characterizes an actuality of this and similar researches for different industrial areas [11–17].

Traced conditions of parameters are data about a condition before and on the current moment of time, but always the future is more important for all. With the use of current data, responsible staff (mechanics, technologists, engineers, etc.) should know about admissible time for work performance to maintain system operation. Otherwise, because of ignorance of a residual time resource before abnormality, the necessary works are not carried out. That is, because of ignorance of this residual time, measures for prevention of negative events after parameter abnormalities (failures, accidents, damages, and/or the missed benefit because of equipment time out) are not undertaken. And, on the contrary, knowing residual time before abnormality, these events may be avoided, or the system may be maintained accordingly. For monitored critical system, the probabilistic method to estimate the mean residual time before the next parameter abnormalities for each element and whole system is proposed.

By principles of system engineering (e.g., according to ISO/IEC/IEEE 15288), the complex system is decomposed to compound subsystems and elements with formal definition of states (see **Figure 13**).

For every valuable subsystem (element), monitored parameters are chosen, and for each parameter, the ranges of possible values of conditions are established: "In working limits," "Out of working range, but inside of norm," and "Abnormality" (interpreted similarly light signals ("green," "yellow," "red") (see **Figure 14**). The condition "Abnormality" characterizes a threat to lose system integrity (on the logic level, this range "Abnormality" may be interpreted analytically as failure, fault, unacceptable risk or quality, etc.).

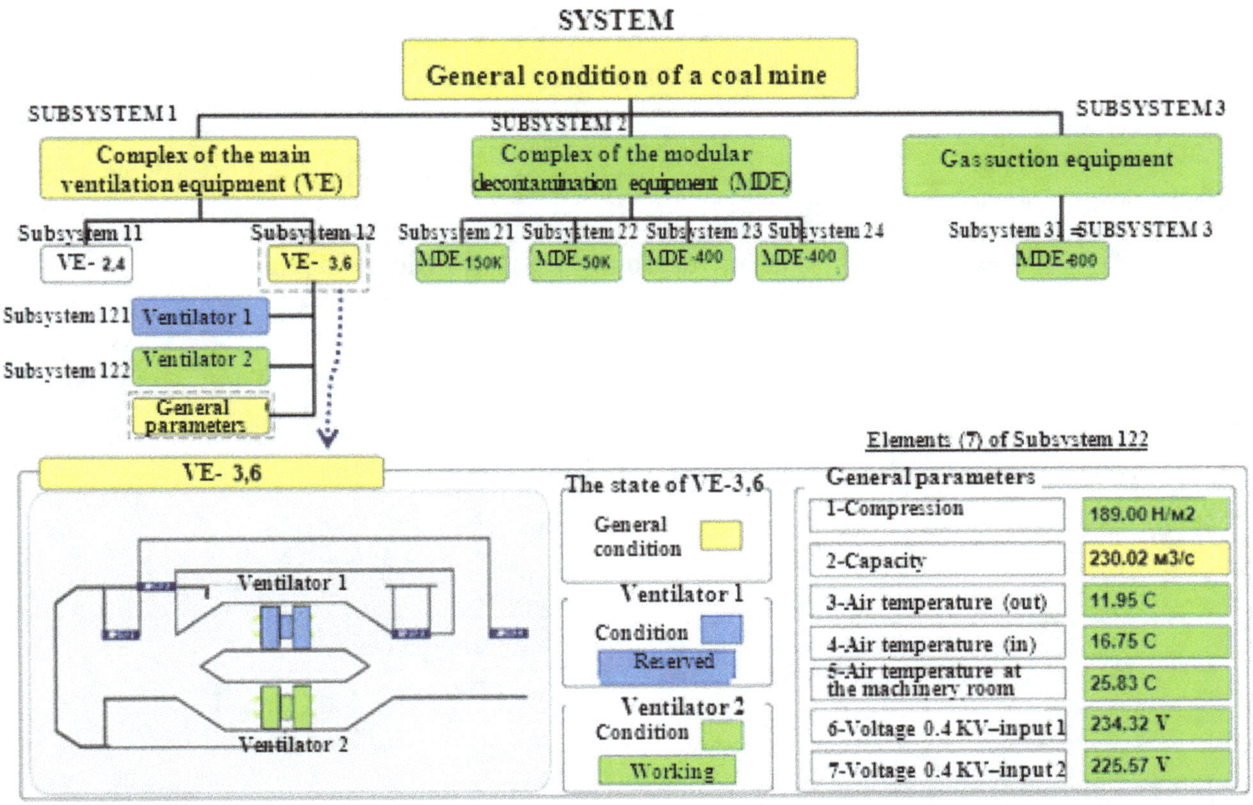

Figure 13. Example of system decomposition.

For avoiding the possible crossing of a border of "Abnormality," a prediction of residual time, which is available for preventive measures, according to gathered data about parameter condition fluctuations considering ranges, should be carried out. For prediction the following are proposed: (1) a choice of probabilistic models for construction (PDF of time before the next abnormality for one element ("black box")), (2) development of the algorithm of generation (PDF of time before the next abnormality for complex system), and 3) formalization of calculative methods of estimating the mean residual time before the next parameter abnormalities for monitored critical system.

The method allows to estimate residual time before the next parameter abnormality (i.e., time before the first next coming into "red" range) [14].

The method allows to estimate residual time before the next parameter abnormality $T_{resid(1)}$ for a given admissible risk $R_{adm.}(T_{req})$ to lose integrity. The estimated $T_{resid(1)}$ is the solution t_0 of equation:

$$R(T_{occur}, t, T_{betw}, T_{diag}, T_{err.}, T_{req.}) = R_{adm.}(T_{req}) \tag{7}$$

concerning of unknown parameter t, i.e., $T_{resid(1)} = t_0$.

Here, $R(T_{occur}, t, T_{betw}, T_{diag}, T_{err.}, T_{req.})$ is the risk to lose integrity; it is addition to 1 for probability $P(T_{req})$ of providing system integrity ("probability of success"), and for calculations formulas (1)–(7) are used (see SubSection 3.1 of this article). So, for exponential PDF, formula (1) transforms into formula.

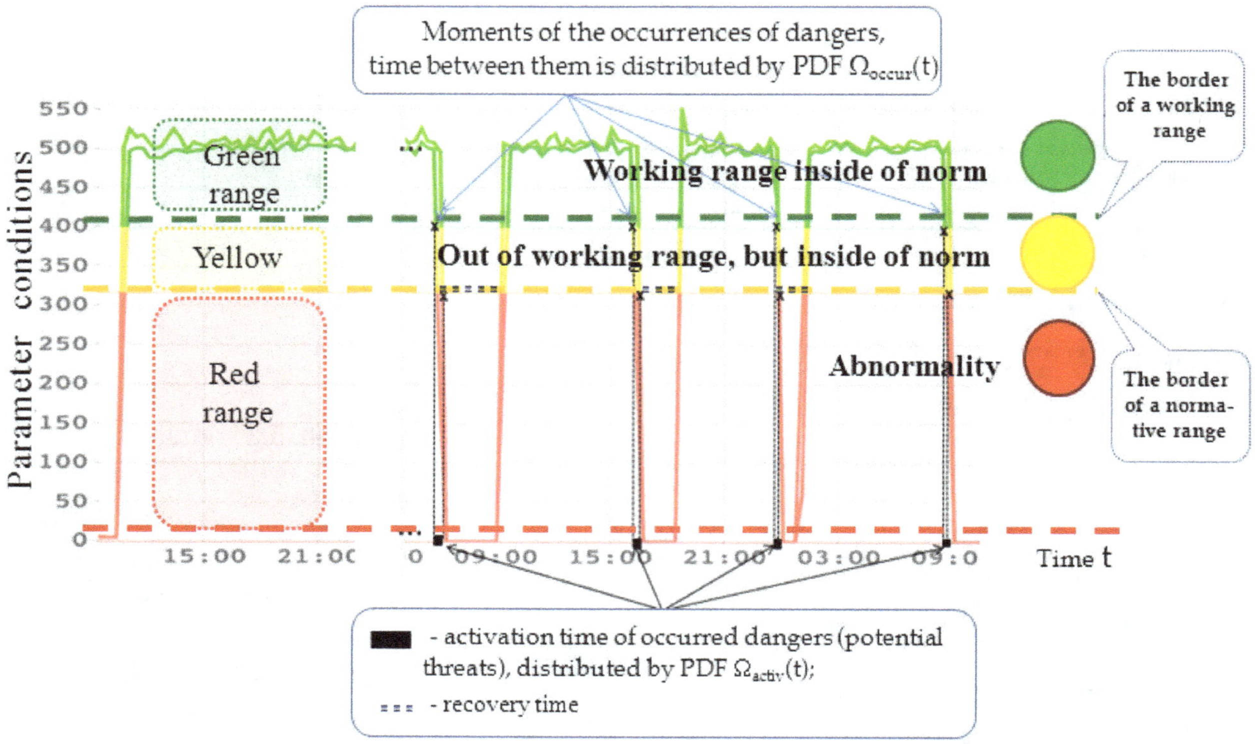

Figure 14. Elementary ranges for parameter conditions.

This formula is used for Eq. (7).

T_{occur} is the mathematical expectation of PDF Ω_{occur} (τ); it is defined by parameter statistics of transition from "green" into "yellow" range (see **Figure 3**). The other parameters T_{betw} and T_{diag} in formula (7) are known. The main practical questions are as follows: what about $T_{req.}$ and what about the given admissible risk $R_{adm.}(T_{req})$? For answering we can use the properties of function $R(T_{occur}, t, T_{betw}, T_{diag}, T_{err.}, T_{req.})$:

- If parameter t increases from 0 to ∞ for the same another parameters, the function R(…, t, …) is monotonously decreasing from 1 to 0, i.e., if the mean activation time of occurred danger (threat: from the first input at the "yellow" range to the first input in the "red" range) is bigger, to lose integrity is less.
- If parameter T_{req} increases from 0 to ∞ for the same other parameters, the function $R(\ldots,T_{req})$ is monotonously increasing from 0 to 1, i.e., for large T_{req} risk approaches to 1.

It means that the such maximal x exists when t = x and $T_{req.}$ = x and $0 < R(T_{occur}, x, T_{betw}, T_{diag}, T_{err.}, x) < 1$. That is, the residual time before the next parameter abnormality (i.e., time before the first next coming into "red" range) is equal to the defined x with the confidence level of admissible risk $R(T_{occur}, x, T_{betw}, T_{diag}, T_{err.}, x)$.

For example, if T_{occur} = 100, T_{betw} = 8 hours, T_{diag} = 1 hour, $T_{err.}$ = 0, and $R_{adm.}$ = 0.05, unknown x is defined from equation, considering (1), (7):

So, if T_{occur} = 100 days, for $R_{adm.}$ = 0.01 residual time x ≈ 2.96 weeks (considering decisions of recovery problems of integrity every 8 hours).

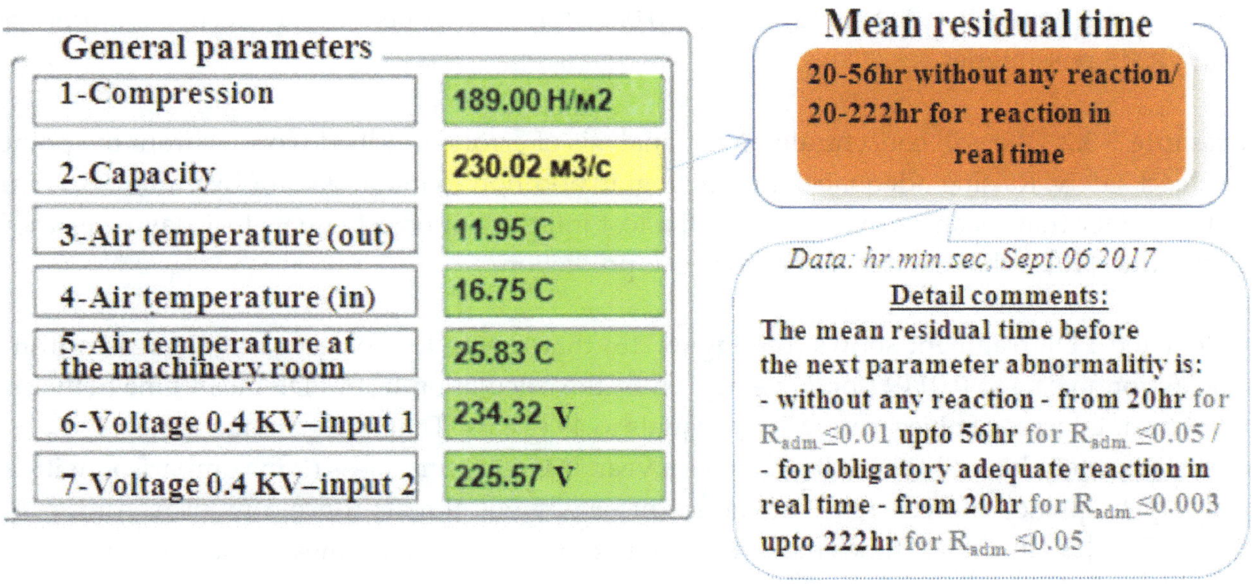

Figure 15. Example of residual time and comments.

The method is implemented by RMS. At once after crossing "yellow" border from "green," the automatic prediction of the mean residual time before the next parameter abnormalities (from the first input at the "yellow" range to the first input in the "red" range) is displayed (see **Figure 15**).

Adequate reaction of responsible staff in real time is transparent for all interested parties.

5.3. About some effects from adequate probabilistic methods and technology applications

Some effects from the proposed adequate probabilistic methods and technologies of RMS are estimated on the level of predicting risks to lose object safety (integrity) by PDF [16].

Example 5.3.1. According to statistics from multifunctional safety system (MFSS), a frequency of occurrence of the latent or obvious threats is equal to once a month, and an average time of development of threats (from occurrence of the first signs of a critical situation up to failure) is about 1 day. A work shift is equal to 8 hours. The system control is used once for work shift, and a mean duration of the system control is about 10 minutes (it is supposed that recovery of object integrity is expected also for 10 minutes). The workers (they may be mechanics, technologists, engineers, etc.) of medium-level and skilled workers are capable to revealing signs of a critical situation after their occurrence, and workers of the initial level of proficiency are incapable. Medium-level workers can commit errors on the average not more often once a month, and skilled workers are not more often once a year. How consideration of the qualification level influences on predicted risks to lose object safety for a year and for 10 years?

The results of modeling. For workers of the initial level of proficiency, risks to lose object safety are near 1 (losses of integrity are inevitable). For workers of medium-level of proficiency, risk to lose object safety for a year is about 0.007 and for 10 years is about 0.067, and for skilled workers, risk equals to 0.0006 for a year and 0.0058 for 10 years because of effective monitoring using RMS possibilities.

Example 5.3.2. We will concentrate on the analysis of errors of skilled workers from the point of object safety. Raising adequacy of modeling, in addition to initial data of Example 5.3.1, we will consider that mean recovery time of the lost integrity of object equals to 1 day instead of 10 minutes [10]. What effect may be from risk prediction?

Calculated PDF fragment shows (see **Figure 16**) that risk to lose object safety increases from 0.0006 (for a year) to 0.0119 (for 20 years). Thus, the calculation from PDF mean time between neighboring losses of object safety T_{mean} equals to 493 years. That is, the frequency $\lambda = 1/T_{mean}$ of system safety losses is about 0.002 times a year. It is 6000 times less (!) in comparison with a primary frequency of occurrence of the latent or obvious threats (once a month). And, estimated T_{mean} is almost 500 times more in comparison with a primary mean time between errors of skilled workers (once a year). And, such effect can be reached at the expense of undertaken control measures, monitoring, and system recovering in case of revealing in time the signs of threat development. To the point, the frequency λ of system safety losses is extracted latent knowledge from PDF, built in a calculated form.

If to compare with exponential approximation of PDF with the same frequency λ, the risk to lose object safety will grow from level 0.002 (for a year) to 0.04 (for 20 years). These are also extracted latent knowledge considering Taylor's expansion $R(t, \lambda) \approx \lambda \cdot t$ (see Section 2). Difference is in 3.3–3.4 times more against adequate PDF. To feel, it is enough to ascertain that for created PDF the border of admissible risk 0.002 will be reached for 3 years, not for 1 year as for exponential PDF. That is, the real duration of effective object operation (i.e., without losses of safety) is three times more!

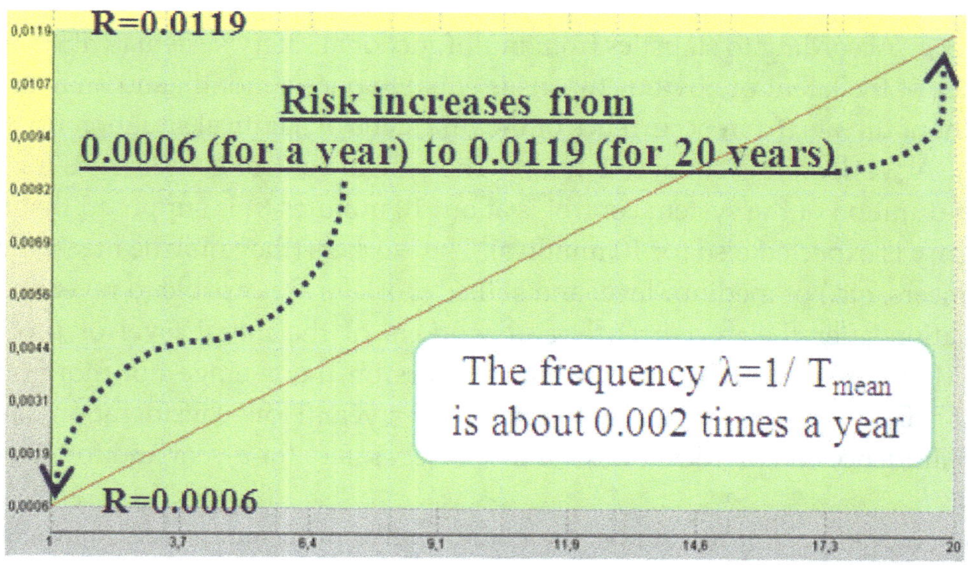

Figure 16. Calculated PDF fragment for Example 5.3.2.

Example 5.3.3. This allowed to estimate operation of object as "black box," described by characteristics of skilled workers. On dangerous manufacture critical operations are carried out by skilled workers in interaction with RMS (including reservation and supports of another). Formally, they operate as parallel elements with hot reservation. Thereby, the consideration of such interaction allows to increase adequacy of modeling. Let's estimate risk to lose object safety for this variant (all input data for each from two parallel elements are the same that in Example 5.3.2).

Calculated PDF fragment shows (see **Figure 17**) that risk to lose object safety increases from 0.0000003 (for a year) to 0.00014 (for 20 years). Thus, the mean time between neighboring losses of object safety T_{mean}, calculated from known PDF, equals to 663 years. That is, the frequency λ of system safety losses is about 0.0015 times a year. It is 8000 times less (!) in comparison with a primary frequency of occurrence of the latent or obvious threats (once a month). And, at the expense of reservation estimated, T_{mean} is 34.5% longer in comparison with T_{mean} from Example 5.3.2.

If to compare with exponential approximation of PDF with the same frequency λ, the risk to lose object safety will grow from level 0.0015 (for a year) to 0.03 (for 20 years). Difference is in 200–5000 times more against adequate PDF. The border of admissible risk 0.0015 will be reached for 195 years, not for 1.3 year as for exponential PDF. That is, the real duration of effective object operation (i.e., without losses of safety) is 150 times more! Such effect can be reached at the expense of mutual aid (reservation and supports) of skilled workers using RMS.

Example 5.3.4. Come back to the SUEK value chain (see **Figure 10**). According to system engineering principles (see ISO/IEC/IEEE 15288 and **Figure 1**), we decompose logically this chain into nine serial components. Components from 1 to 6 are united by MFSS of mine, component 7 is associated with washing factory, component 8 is associated with transport, component 9 is associated with port (see **Figure 18**). For every element of this chain, a specific

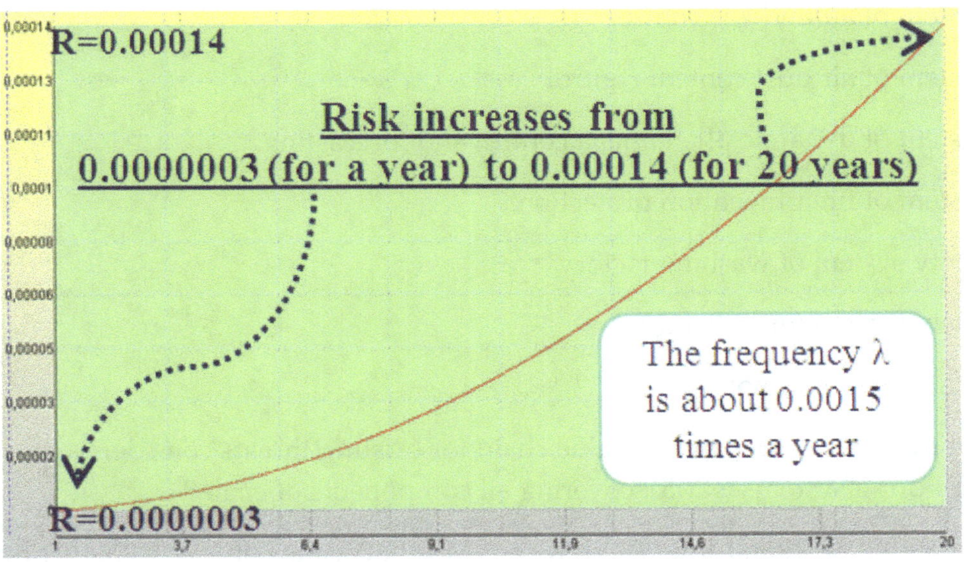

Figure 17. Calculated PDF fragment for Example 5.3.3.

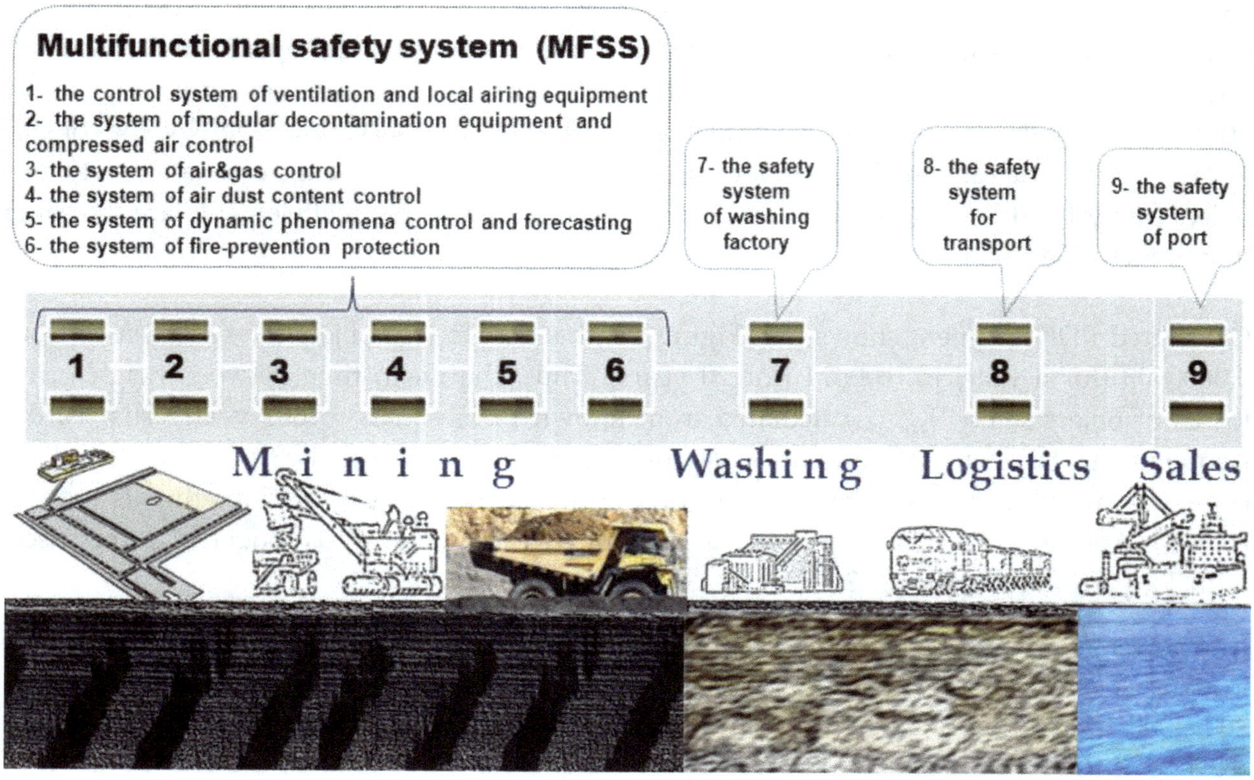

Figure 18. Illustration of system, combined from parallel and series subsystems.

set of threats exists. Let us analyze a system of such value chain. The typical systems of this value chain, including MFSS, are:

1. The control system of ventilation and local airing equipment.
2. The system of modular decontamination equipment and compressed air control.
3. The system of air and gas control.
4. The system of air dust content control.
5. The system of dynamic phenomena control and forecasting.
6. The system of fire prevention protection.
7. The safety system of washing factory.
8. The safety system for transport.
9. The safety system of port.

What about the safety for analyzed value chain for existing threats considering possibilities of remote monitoring systems (RMS), covering all components of chain?

Let's put that the workers, interacted with RMS, participate in each chain process. Their activity is modeled by the models of Section 3, considering examples above. The high adequacy is reached by decomposition of chain system to nine logical subsystems, each of which implements corresponding typical functions of Systems 1–9. Safety of whole value chain system is provided, if "AND" the first subsystem, "AND" the second, ..., and "AND" the ninth subsystem safety are provided (see **Figure 18**). Reservation of elements for every subsystem is explained by RMS possibilities. Those input data for every element are the same as in Example 5.3.3.

Calculated PDF fragment shows (see **Figure 19**) that risk to lose safety increases from 0.000003 (for a year) to 0.0013 (for 20 years). Thus, the mean time between neighboring losses of safety T_{mean} equals to 283 years. That is, the frequency λ of system safety losses is about 0.0035 times a year. It is 2.3 times more often against the results of Example 5.3.3. In comparison with a primary frequency of occurrence of the latent or obvious threats (once a month), the frequency λ is 3430 times lower!

For exponential approximation of PDF with the same frequency λ, the risk to lose safety will grow from level 0.0035 (for a year) to 0.07 (for 20 years). Difference is in 54–1167 times more against adequate PDF.

The border of admissible risk 0.002 will be reached for 24 years, not for 7 months as for exponential PDF (see Section 2). That is, the real duration of effective operation (i.e., without losses of safety) is 41 times more!

Example 5.3.5. How much risks will increase, if in a system of value chain from Example 5.3.4 only medium-level workers are used?

Calculated PDF fragment shows (see **Figure 20**) that risk to lose safety increases from 0.0009 (for a year) to 0.25 (for 20 years). Thus, the mean time between neighboring losses of safety T_{mean} equals to 24 years. That is, the frequency λ of system safety losses is about 0.04 times a year. It is 11.4 times less often against the results of Example 4 for skilled workers. In

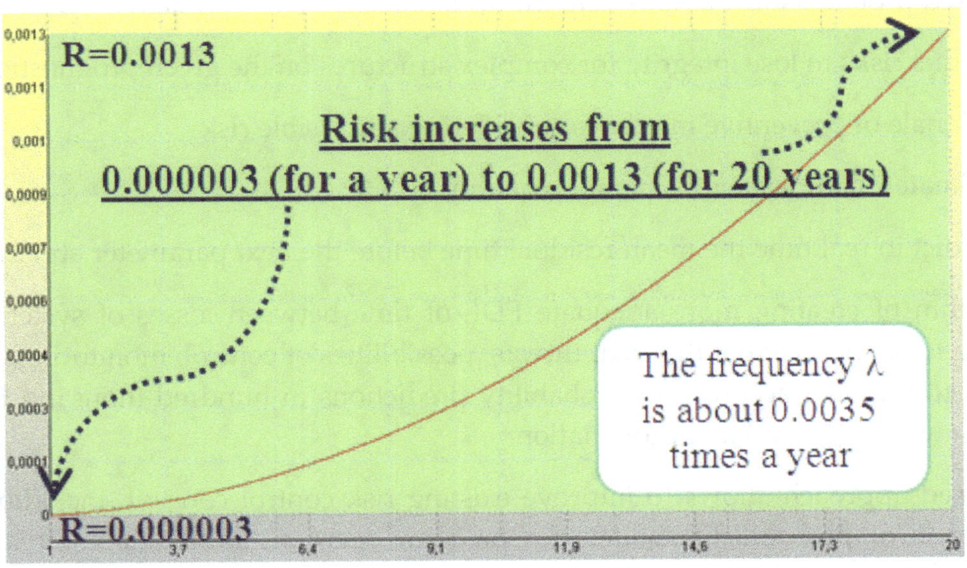

Figure 19. Calculated PDF fragment for Example 5.3.4.

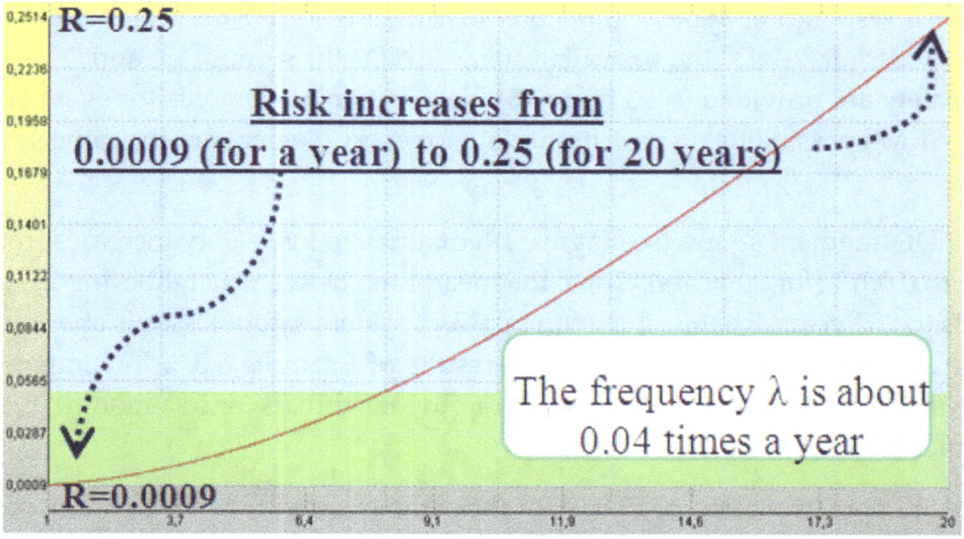

Figure 20. Calculated PDF fragment for Example 5.3.5.

comparison with a primary frequency of occurrence of the latent or obvious threats (once a month), the frequency λ is 21 times lower!

For exponential approximation of PDF with the same frequency λ, the risk to lose safety will grow from level 0.04 (for a year) to 0.55 (for 20 years). Difference is 2.2–44.4 times more against adequate PDF. The border of admissible risk 0.002 will be reached for 2 years, not for one month as for exponential PDF. That is, the real duration of effective operation (i.e., without losses of safety) is 24 times more!

6. Instead of conclusion

The proposed probabilistic methods help the system using "smart systems":

- To predict risks to lose integrity for complex structures on the given prognostic time
- To rationale of preventive measures considering admissible risk
- To estimate "smart system" operation quality
- To predict in real time the mean residual time before the next parameter abnormalities

The algorithm of creating more adequate PDF of time between losses of system integrity, considering for every element different threats, possibilities of control, monitoring, and recovery, allows to improve accuracy of probability predictions in hundred-thousand times (!) in comparison with exponential approximation.

The purposed approach allows to improve existing risk control concept, including creation and perfection of probabilistic models for problem decision, automatic combination, and generation of new probabilistic models, forming the storehouse of risk prediction knowledge; for storehouse, dozens of variants of the decision of typical industrial problems for risk control.

The application of the methods and technologies by the joint-stock company "Siberian Coal Energy Company," implemented on the level of the remote monitoring systems, allowed to rethink system possibilities for increasing reliability and industrial safety, improve multifunctional safety systems, decrease risks, and provide predictive maintenance and operation efficiency in company value chain.

Author details

Vladimir Artemyev[1], Jury Rudenko[1] and George Nistratov[2*]

*Address all correspondence to: george.icie@gmail.com

1 JSC "SUEK", Moscow, Russia

2 Scientific Research Institute of Applied Mathematics and Certification, Moscow, Russia

References

[1] Feller W. An Introduction to Probability Theory and its Applications. Vol. II. Willy; 1971

[2] Kostogryzov A, Nistratov G. Standardization, mathematical modeling, rational management and certification in the field of system and software engineering, Moscow. Armament Policy Conversion. 2004. 395 p

[3] Kostogryzov AI, Stepanov PV. Innovative management of quality and risks in systems life cycle, Moscow. Armament Policy Conversion. 2008. 404p. (in Russian)

[4] Zio E. An introduction to the basics of reliability and risk analysis. World Scientific. 2006. 222 p

[5] Kostogryzov A, Nistratov A, Nistratov G. Applicable technologies to forecast, analyze and optimize reliability and risks for complex systems. Proceedings of the 6st International Summer Safety and Reliability Seminar, Poland. September 2012;**3**(1):1-14

[6] Kostogryzov A, Nistratov G, Nistratov A. Some Applicable Methods to Analyze and Optimize System Processes in Quality Management. Total Quality Management and Six Sigma: InTech. 2012. pp. 127-196. Available from: http://www.intechopen.com/books/total-quality--management-and-six-sigma/some-applicable-methods-to-analyze-and-optimize-system-processes-in-quality-management

[7] Kostogryzov A, Grigoriev L, Nistratov G, Nistratov A, Krylov V. Prediction and optimization of system quality and risks on the base of modeling processes. American Journal of Operations Research, Special Issue;3(1A):217-244

[8] January 2013, Available from: http://www.scirp.org/journal/ajor/

[9] Kostogryzov A, Nistratov G, Nistratov A. The innovative probability models and software Technologies of Risks Prediction for systems operating in various fields. International Journal of Engineering and Innovative Technology (IJEIT). September 2013;**3**(3):146-155. http://www.ijeit.com/archive.php

[10] Kostogryzov AI et al. Security of Russia. Legal, Social&Economic and Scientific & Engineering Aspects. The Scientific Foundations of Technogenic Safety. Machutov N, editor. Moscow, Znanie. 2015. 936 p

[11] Kostogryzov AI, Kosterenko VN, Timchenko AN, Artemyev VB. Osnovy protivoavariynoy ustoychivosty ugolnych predprijatiy (The Foundations of Counteremergency Stability for Coal Enterprises). V. 6 "Industrial safety". Book 11. - Moscow: "Gornoje delo" Kimmerijsky Center Ltd. – 336 p

[12] Kostogryzov AI, Stepanov PV, Nistratov GA, Nistratov AA, Grigoriev LI, Atakishchev OI. Innovative management based on risks prediction. In: Zheng, editor. Information Engineering and Education Science. London: Taylor & Francis Group; 2015. pp. 159-166. ISBN 978-1-138-02655-1

[13] Kostogryzov A, Stepanov P, Nistratov A, Nistratov G, Zubarev I, Grigoriev L. Analytical modeling operation processes of composed and integrated information systems on the principles of system engineering. Journal of Polish Safety and Reliability Association. Summer Safety and Reliability Seminars. 2016;**7**(1):157-166. Available from: http://jpsra.am.gdynia.pl/archives/jpsra-2016-contents/

[14] Artemyev V, Kostogryzov A, Rudenko J, Kurpatov O, Nistratov G, Nistratov A. Probabilistic methods of estimating the mean residual time before the next parameters abnormalities for monitored critical systems. Proceedings of the 2nd International Conference on System Reliability and Safety (ICSRS- 2017), Milan, Italy, pp. 368-373

[15] Svetlana J, Tatiana K, Andrey K, Oleg K, Andrey N, George N. The probabilistic analysis of the remote monitoring systems of critical infrastructure safety. Journal of Polish Safety and Reliability Association. Summer Safety and Reliability Seminars. 2017;**8**(1):183-188. http://jpsra.am.gdynia.pl/archives/jpsra-2017-contents/

[16] Andrey K, Oleg A, Pavel S, George N, Andrey N, Leonid G. Probabilistic modelling processes of mutual monitoring operators actions for transport systems. Proceedings of the 4th International Conference on Transportation Information and Safety, ICTIS 2017. pp. 865-871

[17] Kostogryzov A, Stepanov P, Grigoriev L, Atakishchev O, Nistratov A, Nistratov G. Improvement of Existing Risks Control Concept for Complex Systems by the Automatic Combination and Generation of Probabilistic Models and Forming the Storehouse of Risks Predictions Knowledge. Proceedings of the 2nd International Conference on Applied Mathematics, Simulation and Modelling (AMSM 2017), August 6–7, Phuket, Thailand. DEStech Publications, Inc. pp. 279-283

[18] ISO 17359. Condition monitoring and diagnostics of machines - General guidelines

[19] ISO 13381–1. Condition monitoring and diagnostics of machines - Prognostics - Part 1: General guidelines

[20] ISO 13379. Condition monitoring and diagnostics of machines - General guidelines on data interpretation and diagnostics techniques

[21] IEC 61508–1. Functional safety of electrical/electronic/programmable electronic safety-related systems — Part 1: General requirements

11

Decision-Making Model for Offshore Offloading Operations Based on Probabilistic Risk Assessment

C. E. Patiño Rodriguez

Abstract

To explore offshore oil fields in deepwater, the use of a floating production storage and offloading (FPSO) unit coupled to a shuttle tanker is economically and technically feasible. Shuttle tankers like system for oil transportation are increasingly being accepted as a preferred transportation method for remote and deepwater offshore developments. The offloading operation is considered one of the riskiest operations in offshore environment. The chapter presents a risk-based analysis method aiming at defining the risk profile associated with an offloading operation. For offloading operations, the risk profile is usually evaluated considering that the offloading operation has an approximate duration of 24 hours. The method follows three basic steps: identification of hazard, definition of failure scenarios and their probability of occurrence, and evaluation of failure consequences. The decision-making theory is used to evaluate the possibility of emergency disconnection during the operation. The method is applied to evaluate the risk profile of an offloading operation in Campos Basin, Brazil, considering a FPSO moored with Differentiated Complacent Anchoring System (DICAS). The method is used to model the risk scenario associated with shuttle tanker main engine failure as initiating event. The changes in environmental conditions have great influence in risk profile and increase the probability of disconnection.

Keywords: probability risk assessment (PRA), risk profile, offloading operation, Markovian process, Bayesian techniques

1. Introduction

The occurrence of accidents in complex systems, such as offshore and onshore oil and gas processing plants, power plants, and chemical process industries, is financially expensive because the accidents can cease plant operations and even can cause harm to people, property, and environment. For this reason, to identify vulnerable factors that become unacceptable operating scenarios is a challenge in the risk assessment of complex systems. The risk assessment seeks to minimize undesirable event probability and their impact both for the environment and for the people involved in the operations. The impact in the operation can be measured as economic consequences based on the extension of equipment damage and on reduction of plant performance.

The search for oil fields no longer occurs exclusively onshore, but includes the oceans of the world. This fact has contributed to the development of rigs for drilling and production offshore in deepwater.

The current method for crude oil export in deepwater is using floating production storage and offloading (FPSO). The FPSO is a floating vessel, in that it is equipped with internal or external turret, and equipment to refine crude oil, and storage capacity. Therefore, FPSO have an offloading system to transfer the crude oil to shuttle tankers. As you can see in [1, 2], the shuttle tankers are increasingly being accepted as a preferred transportation method for remote and deepwater offshore developments, for example, according to ONIP (Programa Nacional de Mobilização da Indústria Nacional do Petróleo e Gás Natural) in 2002, Brazil had 46.0% of the total oil production of Petrobras located in deepwater (400–1000 m) and 29.9% in ultra-deepwater, with water depth greater than 1000 m [3]. More recently, shuttle tankers have become the main way to distribute the crude oil produced offshore on Brazilian fields [4]. The options for methods of offloading from a FPSO and shuttle tanker include remote single point mooring, tandem offloading, and alongside configuration.

The tandem offloading operation is frequently a complex and difficult marine operation. FPSO may rotate due to waves and wind actions, and this rotates according to the weather that generates linear motions of a ship (surge, sway, and yaw). To stay connected for loading and at the same time maintain a safe separation distance, shuttle tanker must position itself aligned with the FPSO position. As we show in [5], the situation is dramatically changed in the tandem offloading operation in terms of positioning complexity and damage potential [5], due to the significant amount of mass involved (e.g., a 150,000-dwt shuttle tanker) in close distance to an installation (FPSO) for a long period of time.

To analyze the nature of the incidents in maritime operations, it is necessary to define a complex relationship among design procedures, equipment, environmental conditions, and operational procedures. To gain a full understanding and comprehensive awareness of safety in each situation, it is necessary to use a systemic approach to consider all the aspects that may lead to hazardous events and to consider different uncertainty sources [6]. In complex system safety assessment, a systemic approach means to consider all functional entities that constitute the system, exploring patterns and inter-relationships within subsystems and seeing undesired events as the products of the working of the system.

In the 1980s and 1990s, the most risk analysts have been trained in the "classical" approach to risk analysis, where probability exists as a quantity characterizing the failure of the system being studied and independent of the analyst. This concept of probability is frequency based, and the results of the risk analyses provide estimates of these "true" probabilities. For operations involving complex nonlinearities and multicomponent system, especially, new techniques for risk analysis upon of abnormal event are needed. The quantification of risk cannot be handled with traditional statistical methods since it requires the quantification of the probability of accidental events that in most cases are rare [7].

The incidents in maritime operations often involve the analysis of low-probability events for which few data are available. Classical statistical methods are inefficient in these cases. Bayesian techniques are useful because of their ability to deal with sparse data and to incorporate a wide variety of information gained based on expert judgment. A further practical advantage of the subjective probability framework in risk assessment applications is that propagation of uncertainties through complex models is relatively simple.

In the last few decades, has been several studies examined trends about Bayesian techniques in risk assessment [7–13], such as those presented by Avan and Kvaloy [7] discussing some of the practical challenges of implementing Bayesian thinking and methods in risk analysis, emphasizing the introduction of probability models and parameters and associated uncertainty assessments. Siu and Kelly [8] present a tutorial on Bayesian parameter estimation especially relevant to probability risk assessment. Jun et al. [9] divide the system failure mode based on the criticality analysis using multistage event tree. They predict failure rates and the time to failures and consequently can predict the system reliability. Eleye-Datubo et al. [10] show in a marine evacuation scenario and that of authorized vessels to floating, production, storage, and offloading collision, based on a commercial computer tool. Meel and Seider [11] developed Bayesian model to predict the number of abnormal events in the next time interval utilizing information from previous intervals and determine fuzzy memberships to various critical zones to indicate the proximity of abnormal events to incipient faults, near misses, incidents, and accidents. Kalantarnia et al. [12], for example, use Bayesian theory to update the likelihood of the event occurrence and failure probability of the safety system and hence develop a dynamic failure assessment for a process. Yun et al. [13] use Bayesian estimation for insufficient LNG system failure data; the risk values estimated with these insufficient data may not show statistical stability or represent specific conditions of an LNG facility.

The quantification of risk requires the quantification of the likelihood of rare accidental events, which normally cannot be done without employing engineering judgment. In this paper the relationship between characteristics and causes of accidents and system components involved in hazardous offloading is analyzed about one type of consequence associated with the incident. This chapter presents a quantitative risk analysis based on Bayesian techniques; the relation between the probability of occurrence of each hazardous event and its consequence could be found; we have developed these concepts in [14]. The objective this approach is providing safety for offloading operations in deepwater oil fields. We consider both FPSO and shuttle as one integrated system. We present the application of risk-based analysis techniques to evaluate offloading operations between a FPSO and a shuttle tanker that could be used to develop actions and procedures to minimize the consequences of an accident for the operation. The methodology presented can provide a model in which reasoning is justified, while it enables a powerful marine decision-support solution that is simple to use, flexible, and appropriate for the risk assessment task. The methodology with Bayesian approach as for decision support is presented in Section 2; we presented the initials theoretically developed in [14], but we include it here again, for the sake of clarity. In Section 3, the application example is presented, and finally, in Section 4 the results and final comments are presented.

2. Dynamic risk assessment methodology

Risk can be represented by Eq. (1) which relates the undesired event's occurrence probability and the consequences:

$$Risk = (p_i, c_i)\mathrm{Risk} = (\mathrm{p_i}, \mathrm{c_i}) \tag{1}$$

where $\mathrm{p_i}$ is the ith event occurrence probability and $\mathrm{c_i}$ is the effect of the ith event occurrence [14].

For complex systems, the possibility that an unexpected scenario shows up is related to an initial event or failure which happens in a specific component. For each one of the system's or subsystems' components, it is necessary to know the probabilities that the unexpected condition (failure) shows up, and its consequences and states must be evaluated.

In this context, another important decision-making aspect in complex systems is the need for creating a model which can consider dynamic characteristics of system. In the case under analysis, these characteristics are given by the transition between states corresponding to safe operating zones [15].

Hence, let *ST* be a variable that represents a state of system, and let *K* be a scenario. The probability that *K* be true given the system is in the state *ST* can be represented by Eq. (2) [5]:

$$P(K|ST) = \frac{P(ST|K)\cdot P(K)}{P(ST)} \tag{2}$$

where *P(ST | K)* is the probability that the system was in the *ST* state given a scenario *K*, *P(K)* is the probability that a scenario *K* be true, and *P(ST)* is the probability that the system is in the state *ST*.

The method is based on probability risk assessment and Markovian process to aid decision-making (see **Figure 1**). To calculate the probability of accident scenario, the Bayesian approach is presented in detail in [5]. It is used to estimate the probabilities that the system is in each state stochastic model are applied. This methodology allows, quantitatively, to assess the consequences of the events of broad impact and to see relationship between the environment changes and those impacts. The methodology can be summarized in four steps: accident modeling, failure probability assessment with Bayesian techniques, evaluation of consequences, and Markovian process to aid decision-making.

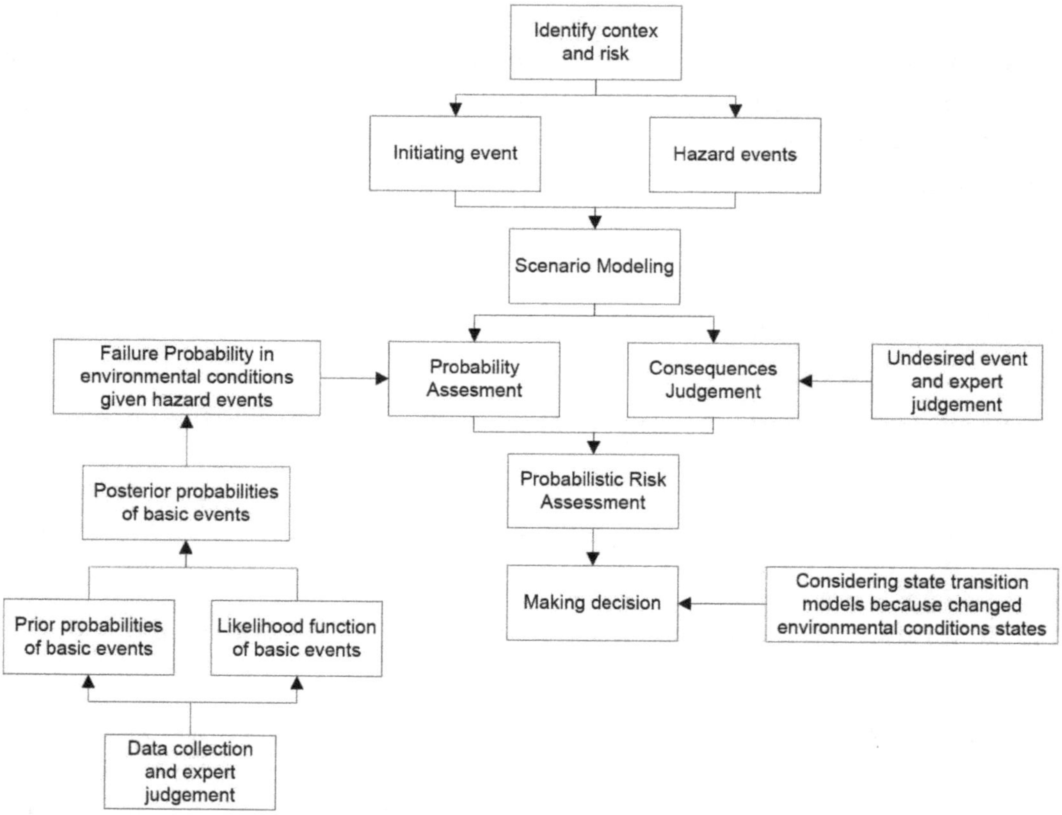

Figure 1. Probabilistic risk assessment methodology.

2.1. Accident modeling

The first step identifies the objective of the risk assessment and to identify and to select the undesirable consequences of interest. These consequences may include items like degrees of harm to environment or degrees of loss of operation. This step covers relevant design and operational information including operating emergency procedures.

In this same step, the hazard identification is based on techniques that allow, qualitatively, to assess the consequences of the events of broad impact and to see the effects on the environment, personnel, and facilities. It requires the identification of the hazard event that is one or more physical conditions with the potential to cause damaged. Aiming this stage is to depict the consequences and to determine their causes, because the procedure is based on the selection of hazard events [16].

To determine the hazard events, "brainstorming" technique is used involving experienced personnel as well as the procedures used for the practice of routine operations using a question-answer technique based on preliminary hazard analysis (PHA) concepts. Apart from human factors, failures of components installed in complex system are systematically considered by applying the methodology of failure modes and effect analysis, which usually starts from identifying failure modes of each item composing the whole system. Based on information about the system, interviews, and expert opinions, many hazards affecting the system are identified [15].

The accident modeling is finished with scenario modeling based on the use of the event tree. An event tree is used to identify the various paths that the system could take, starting with the initiating event and studying the failure progress as a series of successes or failures of intermediate events called hazard events, until an end state is reached. That sequence of events is named failure scenario for which the consequences are estimated.

2.2. Failure probability assessment

In this step the failure probability of occurrence of a failure scenario is calculated combining two conventional reliability analysis methods: fault tree analysis (FTA) and event tree.

The probability of each failure scenario is determined by summing the probability of each set of events which lead to this outcome. Each sequence probability is obtained by simply multiplying the probabilities of the events represented in each branch of the event tree in the case of independence case; if there is dependence between events, the Bayesian methods are used. The probabilities of the hazard event are obtained by solution of fault trees associated with each hazard event. Fault tree analysis is a systematic, deductive, and probabilistic risk assessment tool which clarifies the causal relations leading to a given undesired event. A fault tree is quantified considering that its basic events tend to follow a probability distribution. The failure probability of basic events is calculated using Bayesian methods.

2.2.1. Bayesian ideas and data analysis

The Bayesian techniques are appropriate for use in offshore offloading operation analysis because the Bayesian statistical analysis involves the explicit use of subjective information provided by the expert judgment, since initial uncertainty about unknown parameters of failure distribution of basic events must be modeled from a priori expert opinion or based on insufficient data and evidence collected. Bayes' theorem has been proven to be a powerful coherent method for probabilistically processing new data, as they become available over time, so that the current posterior distribution can then be used as the prior distribution when the next set of data becomes available.

The Bayesian method starts identifying the parameter to be estimated. This involves the consideration of the form of the likelihood function appropriate to the evidence that will be collected. The second step is development of prior probabilities to describe the system current state of knowledge. The next step incorporates information through the collection of evidence and construction of the likelihood function selected in the stage one. The final step results in new probabilities using Bayes' theorem, called posterior distribution, to describe your state of knowledge after combining the prior probabilities with the evidence [17].

The selection of an appropriate likelihood function requires engineering knowledge specific to the process being modeled, as well as the way the new data or evidences are generated. When modeling the number of failures associated with a given piece of equipment, the Poisson distribution is the proper likelihood function. While when modeling the number of failures on system demands, the binomial distribution is the proper likelihood function. For data in form of expert judgment, lognormal distribution is a proper likelihood function. For continuous data, for instance, time to failure, the exponential distribution is the proper likelihood [8]. However, situations can arise where more complicated likelihood functions need to be constructed. Given a process model, general approaches for developing functions of random variables can be used to develop likelihood functions [18].

Prior distributions can be specified in different forms depending on the type and source of information as well as the nature of the random variable of interest. The prior distributions can be informative prior distributions when it is one that reflects the analyst's beliefs concerning an unknown parameter or noninformative prior distributions when large amounts of data are available and when the analyst's prior beliefs are relatively vague. This paper deals with informative prior distributions deals. When it is assumed that the prior is a member of some parametric family of distributions, the form can be parametric and numerical. Among the parametric form are the gamma or lognormal for rates of events and beta for event probabilities per demand. Bayesian statistics combines knowledge about the parameter, which is reflected by the prior distribution, and information from the data, which is contained in the likelihood function. Using Bayes' theorem in its continuous form, the prior probability distribution of a continuous unknown quantity, $P_0(x)$, can be updated to incorporate new evidence E, as shown in Eq. (3):

$$P(x|E) = \frac{L(E|x) \cdot P_0(x)}{\int L(E|x) \cdot P_0(x) \cdot dx} \tag{3}$$

where $P(x|E)$ is the posterior probability distribution of the unknown quantity x given evidence E and $L(E|x)$ is the likelihood function.

For some combinations of likelihood functions and prior distributions, Eq. (3) must be evaluated numerically. For a given model, there is a family of distributions where if the prior distribution is a member of this family, then the posterior distribution will be a member of the same family. These families of distribution are called conjugate distribution [19]. The conjugate likelihood and prior are most commonly used in probability risk assessment as well as the form of the resulting posterior distributions. These combinations are shown in **Table 1**.

Prior $P_0(x)$	**Likelihood $L(E\mid x)$**	**Posterior $P(x\mid E)$**
Beta (α,β) $\frac{\Gamma(\alpha+\beta)}{\Gamma(\alpha)\cdot\Gamma(\beta)}\cdot x^{\alpha-1}\cdot(1-x)^{\beta-1}$	Binomial (r, n) $\frac{n!}{r!(n-r)!}x^r(1-x)^{n-r}$	Beta (α,β) $\frac{\Gamma(\alpha+\beta)}{\Gamma(\alpha)\cdot\Gamma(\alpha)}\cdot x^{\alpha-1}\cdot(1-x)^{\beta-1}$
Gamma (α,β) $\frac{x^{\alpha-1}}{\Gamma(\alpha)}e^{-\beta\cdot x}$	Poisson (x) $\frac{(x\cdot t)^r}{r!}e^{-x\cdot t}$	Gamma $(\alpha' = \alpha + r, \beta' = \beta + t)$ $\frac{x^{\alpha'-1}}{\Gamma(\alpha')}e^{-\beta'\cdot x}$
Lognormal (μ,σ) $\frac{1}{\sqrt{2\cdot\pi}\cdot\sigma\cdot x}e^{-\frac{1}{2}\left(\frac{\ln x-\mu}{\sigma}\right)}$	Poisson (x) $\frac{(x\cdot t)^r}{r!}e^{-x\cdot t}$	Numerical

Table 1. Typical prior and likelihood functions [19].

2.3. Evaluation of consequences and making decision

The effects on the system attributable to hazardous event are defined, and Markovian process is used to model the probability of changes during offloading operation that could cause changes in the risk profile developed in step 2. The decision-making theory is used to evaluate the possibility of emergency disconnection during the operation given the result of Markovian process.

Consequences of hazardous events or abnormal incidents on the shuttle tanker and offloading operation are described and explained. A severity numerical scale is defined for hazardous event classification. This scale was defined for three sets—safety of personal, facilities, and environment—the first is related to the damages or the lesions that can be caused to the employees and others, the second refers to damages in equipment or installations in shuttle tanker or FPSO, and the third is associated with the damages on fauna, flora, and ecosystem. That classification is presented in **Table 2**.

The risk is the combination between the failure probability and the severity magnitudes [20]. The decision-making part is related with accepting a certain risk scenario. The decision-making theory is used to evaluate the possibility of emergency disconnection during the operation. The risk is associated with an uncertain event or condition that, if it occurs, has a negative effect on system operational condition.

2.4. Markovian process

The state of a deterministic dynamical system is some variable which fixes the value of all present and future observables. Consequently, the present state determines the state at all future. However, strictly deterministic systems are rather thin on the ground, so a natural generalization is to say that the present state determines the distribution of future states.

Description		Set		
		Personal	Facilities	Environment
Insignificant	I	No significant harm to people, without removal of staff in the interior of the installation	No significant harm to installation	No significant harm to installation, contamination of environment in minimum concentration
Minor	II	Slight harm to people in installation, no significant harm to people outside installation	Minor damage or degradation of the installation, with repair at low cost	Contamination of environment below maximum concentration, though concentration between minimum and medium
Major	III	Serious harm to people in installation and/or slight harm to people outside installation	Major damage or degradation of the installation, with possible repair	Contamination of environment below maximum concentration, though concentration between medium and maximum
Catastrophic	IV	Single fatality or multiple severe harm to people inside and outside of installation	Damage or degradation without possible repair or repair take a long time to do	Contamination of environment above maximum concentration

Table 2. Relative severity criteria for hazardous event classification [15].

The probability of the system on "i state" is calculated as an approximate discrete model, based on that for small steps ($\Delta\theta$ toward zero) with recurrent algorithm. Assumed two states, the basic steps of the procedure are:

1. Declare initial variable counter $k = 0$, $\theta_k = 0$, and $\theta_{end.}$
2. Declare probability distribution of the initial state. In this case it is assumed that shuttle tank begins the offloading operation in operative zone: $P_1(\theta_k = 0) = 1$ and $P_2(\theta_k = 0) = 0$.
3. Select time steps ($\Delta\theta$).
4. Save t_k, $P_1(\theta_k)$, $P_2(\theta_k)$, and increment counter: $k = k + 1$.
5. Calculate $\theta_k = \theta_{k-1} + \Delta\theta$.
6. Calculate state transition rates ($p_{ijk}(\theta)$) for $\theta = (\theta_{k-1} + \theta_k)/2$.
7. Calculate transition matrix M_k for transition rates of step 4 using Eq. (5).
8. Calculate probability of the system state i at t_k as:
9. $P(\theta_k) = M_k \cdot P(\theta_{k-1})$

3. Return to step 4: The procedure continues until t = tend

The Markovian process shows the probability that the position of shuttle tanker will change from operational zone to alert zone in each environmental condition. That change affects the decision of continuing the offloading operation. The decision-making theory can be used to evaluate the need for disconnection in the case of occurrence of an environmental change coupled to a critical component failure in the shuttle tanker.

4. Application of the methodology

The method is applied on the analysis of the offloading operation, when the crude oil is transported to shore by shuttle tankers through an offloading arrangement with the use of a shuttle tanker with dynamic positioning systems (DP). From the point of view of the shuttle tanker, tandem offloading operation can in principle be summarized into the following five operational stages [15]: (1) approach, tanker approaches FPSO and stops at a predefined distance; (2) connection, messenger line, hawser, and loading hose are connected; (3) loading, oil is transferred from FPSO to tanker; (4) disconnection, manifold is flushed, and loading hose and hawser are disconnected; and (5) departure, tanker reverses away from FPSO while sending back hawser messenger line and finally sails away from oil field. In the first stage, the shuttle tanker approaches FPSO, at a maximum speed of 1.5

knots, and this stage finishes when shuttle tanker stood 50–100 m behind the FPSO; distance is considered appropriate to begin the connection stage. In the second stage, to physically connect shuttle tanker and FPSO, some activities are executed, for example, the messenger line crosses from one ship to the other allowing the mooring hawser and hose to be connected. The tanker may position itself by its own dynamic positioning system so that the hawser is not tensioned. As for safety reasons, a tug boat is also connected to the ship stern acting as a redundant component to control hawser tension. In the third stage, tests are realized, and the valves in vessels are open, and oil is transferred from FPSO to tanker. During this stage, transfer rates are slow initially as the integrity of both vessel systems are checked and gradually increased to a maximum transfer flow. When loading is completed and stopped, the hose is flushed, and the valves are closed. Finally, the hose is dropped and sends to FPSO the hose messenger line and the hawser. The shuttle tanker moves off away FPSO (MCGA [21]).

Patino Rodriguez et al. [15] found 56 hazardous events for shuttle tank. The connection stage is the phase with the highest number of hazardous event. In fact, this stage involves more activities associated with mooring hawser and hose connection, besides the smallest distance between shuttle tanker and FPSO. For all hazardous events, their causes were identified, as well as the activities executed aiming at minimizing the occurrence of these causes (mitigating scenarios). In a similar way, the consequences resulting from the hazardous event are identified. Some of these are characterized as catastrophic. Most of them are related to dynamic positioning system (DPS) failures. Considering that one of the most important aspects in the offloading operation is to keep the position between FPSO and shuttle tanker, the initiating event selected as for risk assessment is "DPS failure." The considered accident sequence is shown in **Figure 2** modeled as an accident progression of four hazard events: (1) auxiliary engine failure, (2) main engine failure, (3) tug failure, and (4) towing cable failure.

The fault tree for the four hazard events that appears in the event tree was developed. For all basic events of the four fault trees, the parameter to be estimated is failure rate, and the Poisson distribution is selected as likelihood function. Poisson distribution is considered as appropriate function given information available in database is the number of failures, r, in each time interval, t, [22, 23]. Analyzing the type and source of information (expert judgment and literature data) as well as the nature of the time to failure that is the random variable of interest, gamma distribution is selected as appropriate "prior distribution." The conjugate family with respect to the risk model is shown in **Table 1**. Using Bayes' theorem (Eq. 2) the posteriori distribution is obtained:

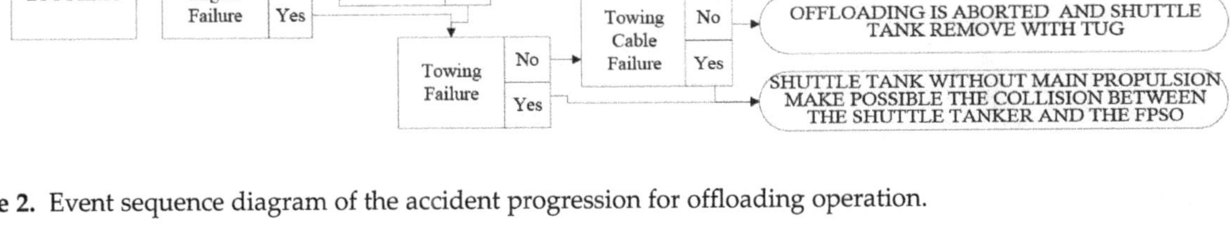

Figure 2. Event sequence diagram of the accident progression for offloading operation.

$$P(\lambda|E) = \frac{\left[\frac{(\lambda \cdot t)^r}{r!} \cdot e^{-\lambda \cdot t}\right] \cdot \left[\frac{\beta^\alpha \cdot \lambda^{\alpha-1}}{\Gamma(\alpha)} \cdot e^{-\beta \cdot \lambda}\right]}{\int_0^\infty \left[\frac{(\lambda \cdot t)^r}{r!} \cdot e^{-\lambda \cdot t}\right] \cdot \left[\frac{\beta^\alpha \cdot \lambda^{\alpha-1}}{\Gamma(\alpha)} \cdot e^{-\beta \cdot \lambda}\right] \cdot d\lambda} \Rightarrow$$

$$P(\lambda|E) = \left[\frac{(\beta + t)^{\alpha+r} \cdot \lambda^{\alpha+r-1}}{\Gamma(\alpha + r)}\right] \cdot e^{-(\beta+t)\cdot\lambda} \tag{4}$$

As an example, the posterior distribution is calculating for fuel system failure (see **Figure 3**) one component of main engine.

Aiming to obtain the probability that K be true given the system is in the state ST represented by Eq. (2), it is necessary to estimate the posterior mean value of failure rate. To calculate the failure probability of hazard events, we use fault tree analysis. Then for all basic events of the fault trees, the failure probability was determined using Bayesian inference. The posterior distribution is calculated, using the conjugate distribution. By analyzing the type of information availability, the Gamma distribution is selected as appropriate prior distribution, and Poisson distribution is selected as likelihood function. We calculate substituting in Eq. (4) the failure rates for fuel system failure (see **Table 3**). The prior distribution was estimated using databases that recorded the rate failure to equipment used in offshore industry.

The calculated probabilities for the basic events are used as input to a fault tree to determine the probability of the event hazard: "no fuel flow." Using probability theory and assuming that the fuel system is operated for t = 43,800 h (time between maintenance), the probability of "no

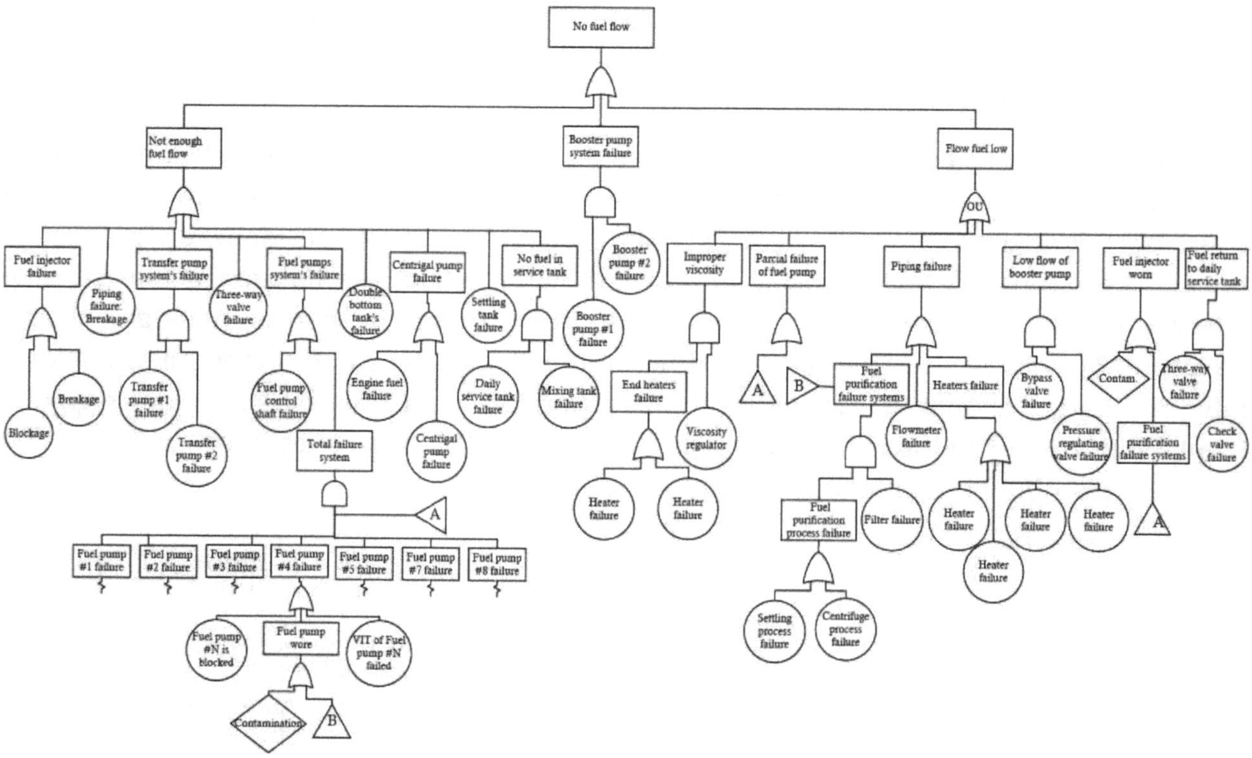

Figure 3. Fault tree for fuel system failure.

Equipment	$E[P_0(\lambda)]$ [failure/h]	$ST[P_0(\lambda)]$ [failure/h]	$P(\lambda \mid E)$ [failure/h]	Equipment	$E[P_0(\lambda)]$ [failure/h]	$ST[P_0(\lambda)]$ [failure/h]	$P[\lambda \mid E]$ [failure/h]
Booster pump	1.10E-03	1.10E-03	2.24E-05	Fuel pumps	1.43E-03	1.13E-03	3.55E-05
Bypass valve	2.28E-05	1.50E-05	1.59E-05	Heater	4.54E-05	3.74E-05	1.93E-05
Centrifugal pump	7.36E-04	1.20E-04	3.95E-04	Main tank	2.13E-04	2.13E-04	2.06E-05
Centrifuge	1.69E-05	5.94E-06	1.55E-05	Mixing tank	9.50E-06	9.11E-06	6.87E-06
Check valve	3.60E-07	5.10E-07	3.49E-07	Piping: blockage	3.70E-07	6.18E-07	3.54E-07
Daily service tank	9.50E-06	9.11E-06	6.87E-06	Piping: breakage	4.40E-07	9.57E-07	4.03E-07
Fuel pump control shaft	3.00E-05	3.00E-05	1.30E-05	Pressure regul. Valve	8.81E-06	1.25E-05	4.98E-06
Engine centrif. Pump	1.13E-04	2.81E-05	8.62E-05	Settling	4.37E-04	6.26E-04	1.08E-05
Filter heated	2.00E-06	2.00E-06	1.84E-06	Settling tank	6.26E-05	1.12E-04	6.43E-06
Flow meter	1.32E-05	3.26E-06	1.27E-05	Three-way valve	2.28E-05	1.50E-05	1.59E-05
Fuel injector: blockage	7.24E-06	1.02E-05	4.43E-06	Transfer pump	7.36E-04	1.20E-04	3.95E-04
Fuel injector: breakage	2.00E-07	2.00E-07	1.98E-07	Viscosity regulator	6.39E-06	8.96E-06	4.12E-06
Fuel Pumps	1.43E-03	1.13E-03	3.55E-05	VIT system	2.06E-07	2.06E-07	2.04E-07

Table 3. Failure rates and standard deviations of the basic events of fault tree for fuel system failure.

fuel flow" is 8.390E-04. The prior and posterior density of basic event that has more influence on system failure is shown in **Figure 4**, associated with the failure of the centrifugal pump. A 90% interval estimate for failure rate is found by computing the 5th and 95th percentiles of gamma distribution, and the interval is between 2,96E-04 and 5,08E-04.

The same procedure is used for other subsystems, and the probability of hazard event "main engine failure" is found by solving the fault tree associated with that failure. In the same way, that procedure is applied to find the probability of all hazard events as shown in **Table 4**.

Connected to the hazard event, the operation involves risks related to collisions during the offshore operation as presented in **Figure 2**. The event tree in **Figure 5** is the failure scenario development associated with the failure in DPS, considering the probabilities presented in **Table 4**.

The proposed method for risk assessment seems to be suitable for complex systems analysis, since it not only allows for the identification of critical consequences, but it is also a tool to make decisions, because it enables a quantitative evaluation of accident progression in systems that change their operational condition throughout time.

The sequence of abnormal events is determined, and the consequences are estimated using the event tree. The initiating event selected is the shuttle tanker change from operational zone to alert zone. The accident sequence considered is modeled as an accident progression of five hazard events, and we have four consequence categories. The fault tree for the five hazard events was developed as shown in **Figure 5**. The shuttle tanker is loss of position in powered condition, and its subsequent collision with the FPSO is the most significant risk.

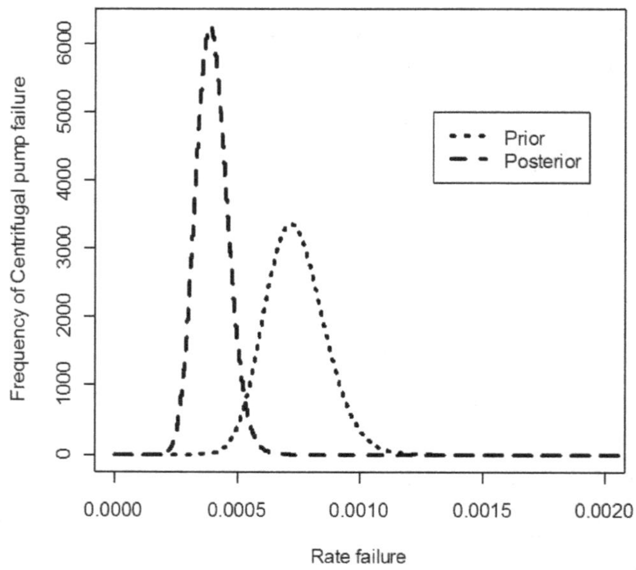

Figure 4. The prior density and posterior density for centrifugal pump failure rate.

Hazard event	P(λ I E) [failure/h]	90% interval estimate for rate failure	
		5%	95%
Dynamic positioning system (DPS) failure	1.58E-05	3.18E-07	5.29E-05
Auxiliary engine failure	1.97E-04	1.01E-04	3.18E-04
Main engine	4.95E-05	9.70E-06	1.14E-04
Tug failure	2.28E-05	1.17E-06	6.82E-05
Towing cable failure	2.18E-03	0.001837	0.002555

Table 4. Posterior probabilities for hazard events involved in the offloading operation and a 90% interval estimate for failure rate.

The failure scenario presented in **Figure 5** can occur at any time during offloading operation. The position of the tanker in relation to FPSO during offloading is controlled. In case it reaches the alert zone, as shown in **Figure 6**, the tanker can be disconnected and the offloading is aborted. So the consequence of the failures considered in the study can be more severe depending of the relative position of the tanker.

It is essential to consider the probability of the change of the shuttle tanker position from operational zone to alert zone, as shown in **Figure 6**, during offloading. The distribution parameters are estimated using a simulator that reproduces ship motions in a specific operation condition and environmental condition. We used these conditions of waves, wind, and currents.

After finding the failure probability of all hazard events, the failure probability for scenarios is calculated by multiplying hazard events. The probability of each consequence

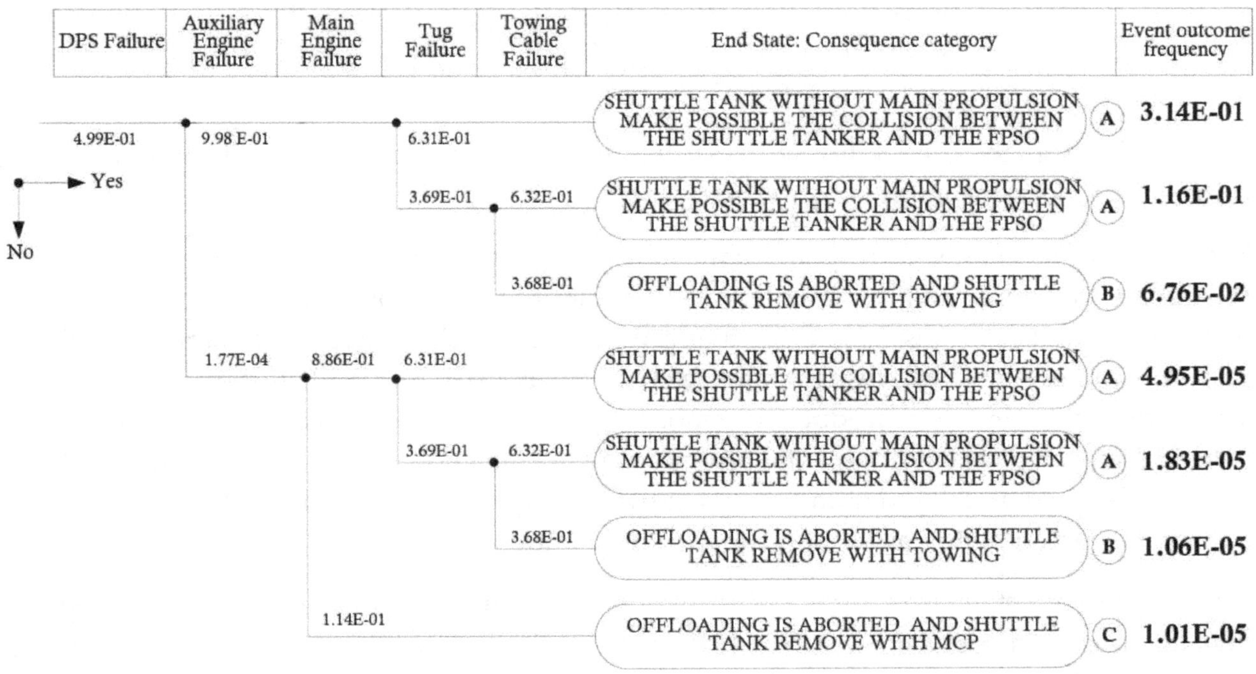

Figure 5. Event tree for the offloading operation.

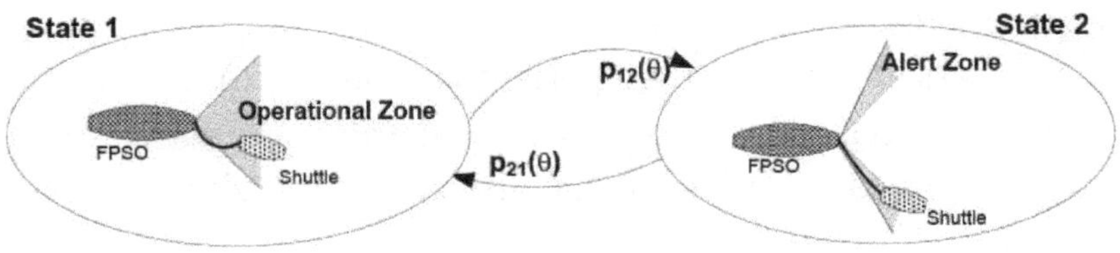

Figure 6. Markov state transition diagram.

category is calculated by adding the probabilities of the scenarios with the same consequence category. The random variable that corresponds to the angle between the FPSO and shuttle tanker during offloading operation is modeled as Weibull distribution. According to the standards of the offloading operation in Brazil, the angle in the operational zone should not be greater than 45 degrees; as a result of these conditions, the parameter of four consequence categories was estimated, and the equation for transition rate is determined. Let us consider the two states established before: operational zone and alarm zone.

The transition rates between states are not constant; then the stochastic process can be modeled as semi-Markov process which shows the probability of the position of the shuttle tanker changing from operational zone to alert zone in a given environmental condition.

By applying the results obtained from the simulation, Markovian analysis, and event tree, the probability that a K scenario is true is obtained, given the system is in the ST state.

In Eq. (5) we define a $K \times K$ state transition probability matrix M_k.

$$M_k = \begin{bmatrix} 1 - p_{12k} \cdot \Delta\theta & p_{21k} \cdot \Delta\theta \\ p_{12k} \cdot \Delta\theta & 1 - p_{21k} \cdot \Delta\theta \end{bmatrix} \quad (5)$$

where $p_{ij}(\theta)\Delta\theta$ is the probability of the system, which is operational zone at position θ, will come alert zone in the interval $(\theta, \theta+\Delta\theta)$.

The state transition rates correspond to the following event rates: the shuttle tanker gets out of the operational zone, and the shuttle tank gets into the operational zone. In each state (*ST*) there are a number of possible events that can cause a transition. A ship dynamics simulator that determines ship maneuvering characteristics was used to calculate the transition. The simulator can accurately reproduce ship motion in the presence of waves, wind, and currents. **Table 5** shows typical environmental conditions in the fall and in the spring for Campos Basin (Brazil). Hence, with the program outputs, it was possible to calculate the angle between FPSO and shuttle tanker at any moment during the offloading operation.

According to the standards of the offloading operation in Brazil, this angle within the operational zone should not be greater than 45 degrees. Weibull probability functions were found as proper distributions to represent the angle between FPSO and shuttle tanker during the offloading operation both inside and outside the operational zone. The parameters and transition rate equation are shown in **Table 6**.

Then, using the recurrent algorithm shown in the section of Markovian process, the probability (*P(ST)*) that the shuttle tanker is inside the operational zone, without any failure, is 0.7918. In the same way, inducing the hazard events in ship dynamics simulator is possible to simulate the consequence categories and to determine the probability that the system was in the *ST* state given a scenario *K* as shown in **Table 7**.

Applying Eq. (2) the probability that a scenario *K* is true given the system is in the state *ST* is obtained. For instance, the probability that shuttle tanker is without main propulsion, making

Current [m/s]	Wind [m/s]	Wave [m]
0.71 S	11.16 SE	2.9 SE

Table 5. Environmental conditions.

State	Parameter Weibull distribution				Transition rate equation
	Consequence category				
	0	C	B	A	
Inside the operational zone	$\beta = 1.641$; $\eta = 12.97$	$\beta = 1.596$; $\eta = 13.05$	$\beta = 1.473$; $\eta = 12.01$	$\beta = 1.691$; $\eta = 14.34$	$\frac{\beta}{\eta} \cdot \left(\frac{\theta_k}{\eta}\right)^{\beta-1}$
Outside the operational zone	$\beta = 10.99$; $\eta = 30.07$	$\beta = 8.604$; $\eta = 60.51$	$\beta = 8.499$; $\eta = 60.40$	$\beta = 7.259$; $\eta = 63.21$	

Table 6. Parameters and transition rate for offloading operation.

State	Consequence category			
	P(ST)	P(K = C)	P(K = B)	P(K = A)
Inside the operational zone	0.7918	0.19546	0.039312	0.03528
Outside the operational zone	0.2082	0.80454	0.96069	0.96472

Table 7. Probabilities that the tanker is inside a given location each for each consequence category.

possible the collision between the shuttle tanker and the FPSO, given that shuttle tanker is in the inside the operational zone is

$$P(K = C|ST = 1) = \frac{(0.1954)\cdot(0.43)}{0.7918} = 0.1059$$

5. Conclusion

The tandem offloading operation is a complex and difficult marine operation. It may range from once every 3 to 5 days, depending on the production rate, storage capacity of FPSO, and shuttle tanker size. The duration of the operation takes about 24 hours based on FPSO storage capacity and oil transfer rate. Meanwhile, a suitable environmental condition is required. Shuttle tanker loss of position in powered condition and subsequently collision with FPSO is the most significant risk.

The proposed method for risk assessment seems to be suitable for complex systems since it allows not only the identification of critical consequences to analyze this kind system but also is a tool to make decision because it allows a quantitative evaluation of accident progression in system that change its operational condition during the time.

The development of the fault tree and event tree is important for the understanding of the functional relation between system components and the relationship with accident progression. Based on the modeling of each accident scenario, the Bayesian analysis is performed considering the evidence of database and knowledge of offloading operation. The objective of Bayesian estimation was to develop a posterior distribution for a set of uncertain parameters allowing estimating a probability for several consequence categories as an integral part of current theories on decision-making under uncertainty.

Based on results of a ship dynamics simulator, the method allows to carry out the probability that the shuttle tanker was in a given position, indicating the variation of the position of the tanker in relation to the FPSO due to environmental conditions.

For the case under analysis, which considered the position between FPSO and shuttle tanker during offloading operation, defined by two operational states, the probability that a failure scenario is true given the system is in a specific operational state is obtained. Both states have the distribution of positions represented by a Weibull probability function.

The method is a proactive methodology to prevent accidents through risk assessment aiming at identifying and depicting a system, to reduce failures and to minimize consequences of the hazardous events. The results of the analysis support the development of mitigating scenarios for the causes of hazardous events and contingency scenarios for the consequences of hazardous events.

Author details

C. E. Patiño Rodriguez

Address all correspondence to: elena.patino@udea.edu.co

Department of Industrial Engineering, Engineering College, University of Antioquia, Medellín, Colombia

References

[1] Hujer K. Trends in Oil Spills from Tanker Ships 1995–2004. London: International Tanker Owners Pollution Federation; 2005

[2] Tanker Operator. Shuttles forged in the crucible. Tanker Operator, April 2003

[3] ONIP. Programa Nacional de Mobilização da Indústria Nacional do Petróleo e Gás Natural, 01 11 2002. [En línea]. Available from: www.onip.org.br. [Último acceso: 25 05 2008]

[4] Reis SP. Transporte marítimo de petróleo e derivados na costa brasileira: Estrutura e implicações ambientais. Rio de Janeiro: Universidade Federal do Rio de Janeiro; 2004

[5] Patino-Rodriguez C. Análise de risco em operações de "offloading" – um modelo de avaliação probabilística dinâmica para a tomada de decisão. Sao Paulo: Universidade de Sao Paulo; 2012

[6] Nilson F. Risk-based approach to plant life management. Nuclear Engineering and Design. 2003:293-300

[7] Aven y T, Kvaloy JT. Implementing the Bayesian paradigm in risk analysis. Reliability Engineering and System Safety. 2002;**78**:195-201

[8] Siu NO, Kelly DL. Bayesian parameter estimation in probabilistic risk assessment. Reliability Engineering and System Safety. 1998;**62**:89-115

[9] Jun C-H, Chang SY, Hong Y, Yang H. A Bayesian approach to prediction of system failure rates by criticalities under event trees. International Journal Production Economics. 1999; **60**:623-628

[10] Eleye-Datubo AG, Wall A, Saajedi A, Wang J. Enabling a powerful marine and offshore decision-support solution through Bayesian network technique. Risk Analysis. 2006;**26**(3): 695-721

[11] Meel y A, Seider WD. Plant-specific dynamic failure assessment using Bayesian theory. Chemical Engineering Science. 2006;**61**:7036-7056

[12] Kalantarnia M, Khan F, Hawboldt K. Dynamic risk assessment using failure asssessment and Bayesian theory. Journal of Loss Prevention in the Process Industries. 2009;**22**:600-606

[13] Yun GW, Rogers WJ, Mannan MS. Risk assessment of LNG importation terminals using the Bayesian-LOPA methodoloy. Journal of Loss Prevention in the Process Industries. 2009;**22**:91-96

[14] Patino-Rodriguez C, Souza G. Decision-making model for offshore offloading operations based on probabilistic risk assessment. Vulnerability Uncertainty and Risk: Analysis, Modeling, and Management. 2011:382-393

[15] Patino-Rodriguez CE, Souza GFM, Martins MR. Risk-based analysis of offloading operations with FPSO production units. In: Proceedings of 20th International Congress of Mechanical Engineering. Gramado; 2009

[16] Millan J, O'Young S. Hybrid system modeling of tandem dynamically-positoned vessels. In: Proceedings of the 39th IEEE Conference on Decision and Control. Sydney, Australia; 2000

[17] Singpurwalla ND. Reliability and Risk: A Bayesian Perspective. London: John Wiley & Son Ltd; 2006

[18] Papoulis y A, Unnikrishna S. Probability, Random Variables, and Stochastic Processes, Boston: McGraw-Hill, 2002

[19] Lindley DV. Introduction to Probability and Statistics : From a Bayesian Viewpoint. Cambridge: Cambridge University Press; 1965

[20] Kumamoto y H, Henley E. Probabilistic Risk Assessment and Management for Engineers and Scientists, New York: IEEE Press, 1996

[21] MCGA. Ship and Cargoes. [En línea]. Available from: http://www.mcga.gov.uk/c4mca/stscontingencyplan291105.pdf. [Último acceso: 2008] 2005

[22] Lee FP. Loss Prevention in the Process Industries. Vol. 3. Oxford: Butterworth Heinemann; 1996

[23] OREDA. Offshore Reliability Data Handbook, Norway: OREDA Participants; 2002

Permissions

All chapters in this book were first published in PMSE, by InTech Open; hereby published with permission under the Creative Commons Attribution License or equivalent. Every chapter published in this book has been scrutinized by our experts. Their significance has been extensively debated. The topics covered herein carry significant findings which will fuel the growth of the discipline. They may even be implemented as practical applications or may be referred to as a beginning point for another development.

The contributors of this book come from diverse backgrounds, making this book a truly international effort. This book will bring forth new frontiers with its revolutionizing research information and detailed analysis of the nascent developments around the world.

We would like to thank all the contributing authors for lending their expertise to make the book truly unique. They have played a crucial role in the development of this book. Without their invaluable contributions this book wouldn't have been possible. They have made vital efforts to compile up to date information on the varied aspects of this subject to make this book a valuable addition to the collection of many professionals and students.

This book was conceptualized with the vision of imparting up-to-date information and advanced data in this field. To ensure the same, a matchless editorial board was set up. Every individual on the board went through rigorous rounds of assessment to prove their worth. After which they invested a large part of their time researching and compiling the most relevant data for our readers.

The editorial board has been involved in producing this book since its inception. They have spent rigorous hours researching and exploring the diverse topics which have resulted in the successful publishing of this book. They have passed on their knowledge of decades through this book. To expedite this challenging task, the publisher supported the team at every step. A small team of assistant editors was also appointed to further simplify the editing procedure and attain best results for the readers.

Apart from the editorial board, the designing team has also invested a significant amount of their time in understanding the subject and creating the most relevant covers. They scrutinized every image to scout for the most suitable representation of the subject and create an appropriate cover for the book.

The publishing team has been an ardent support to the editorial, designing and production team. Their endless efforts to recruit the best for this project, has resulted in the accomplishment of this book. They are a veteran in the field of academics and their pool of knowledge is as vast as their experience in printing. Their expertise and guidance has proved useful at every step. Their uncompromising quality standards have made this book an exceptional effort. Their encouragement from time to time has been an inspiration for everyone.

The publisher and the editorial board hope that this book will prove to be a valuable piece of knowledge for researchers, students, practitioners and scholars across the globe.

List of Contributors

Berg Heinz-Peter
Bundesamt für kerntechnische Entsorgungssicherheit (BfE), Salzgitter, Germany

Roewekamp Marina
Gesellschaft für Anlagen-und Reaktorsicherheit (GRS) gGmbH, Köln, Germany

Nikolay Paramonov
Moscow Technological University Mirea, Russia

Alexey Markov
Bauman Moscow State Technical University, Moscow, Russia

Alexander Barabanov and Valentin Tsirlov
NPO Echelon, Moscow, Russia

Anatoly Lepikhin and Vladimir Moskvichev
Institute of Computational Technologies SB RAS, Krasnoyarsk, Russia

Vladimir Chebotarev and Kseniya Kablukova
CC FEB RAS, Khabarovsk, Russia

Boris Davydov
FESTU, Khabarovsk, Russia

Igor Goncharov, Nikita Goncharov and Pavel Parinov
JSC "NGO" Infosecurity, Voronezh, Russia

Sergey Kochedykov and Alexander Dushkin
Voronezh Institute of the Federal Penitentiary Service of Russia, Voronezh, Russia

Dmitry O. Reznikov, Nikolay A. Makhutov and Rasim S. Akhmetkhanov
Mechanical Engineering Research Institute, Moscow, Russia

Yuriy V. Lisin, Nikolay A. Makhutov and Dmitriy A. Neganov
Pipeline Transport Institute (PTI, LLC), Moscow, Russia

Vladimir A. Nadein
LLC "Oil and Gas Safety – Energodiagnostika", Moscow, Russia

Vsevolod Kershenbaum
National Institute of Oil and Gas, Moscow, Russia
Gubkin Russian State University of Oil and Gas, Moscow, Russia

Leonid Grigoriev
Gubkin Russian State University of Oil and Gas, Moscow, Russia

Petr Kanygin
Chamber of Commerce and Industry of Russian Federation, Moscow, Russia

Andrey Nistratov
Chamber of Commerce and Industry of Russian Federation, Moscow, Russia
The Federal State Organization, Russian Energy Agency of the Ministry of Energy of the Russian Federation, Moscow, Russia

Vladimir Artemyev and Jury Rudenko
JSC "SUEK", Moscow, Russia

George Nistratov
Scientific Research Institute of Applied Mathematics and Certification, Moscow, Russia

C. E. Patiño Rodriguez
Department of Industrial Engineering, Engineering College, University of Antioquia, Medellín, Colombia

Index

www.ingramcontent.com/pod-product-compliance
Lightning Source LLC
LaVergne TN
LVHW060513210926
848485LV00005B/1074